iOS 6 Programming
Pushing the Limits

iOS 6 Programming
Pushing the Limits

ADVANCED APPLICATION DEVELOPMENT FOR APPLE IPHONE®, IPAD®, AND IPOD TOUCH®

Rob Napier and Mugunth Kumar

A John Wiley and Sons, Ltd, Publication

This edition first published 2013

© 2013 Rob Napier and Mugunth Kumar

Registered office

John Wiley & Sons Ltd, The Atrium, Southern Gate, Chichester, West Sussex, PO19 8SQ, United Kingdom

For details of our global editorial offices, for customer services and for information about how to apply for permission to reuse the copyright material in this book please see our website at www.wiley.com.

ISBN 978-1-118-44995-0 (pbk); ISBN 978-1-118-44996-7 (ebk); ISBN 978-1-118-44997-4 (ebk); ISBN 978-1-118-44998-1 (ebk)

A catalogue record for this book is available from the British Library.

Set in 9.5pt Myriad Pro Regular by Wiley Publishing

Printed in United States by Bind-Rite

Publisher's Acknowledgements

Some of the people who helped bring this book to market include the following:

Editorial and Production

VP Consumer and Technology Publishing Director: Michelle Leete
Associate Director–Book Content Management: Martin Tribe
Associate Publisher: Chris Webb
Acquisitions Editor: Craig Smith
Assistant Editor: Ellie Scott
Development Editor: Tom Dinse
Copy Editor: Melba Hopper
Technical Editor: Jay Thrash
Editorial Manager: Jodi Jensen
Senior Project Editor: Sara Shlaer
Editorial Assistant: Leslie Saxman

Marketing

Associate Marketing Director: Louise Breinholt
Marketing Manager: Lorna Mein
Senior Marketing Executive: Kate Parrett

Composition Services

Compositor: Indianapolis Composition Services
Proofreader: Cynthia Fields
Indexer: Potomac Indexing, LLC

About the Authors

Rob Napier is a builder of tree houses, hiker, and proud father. He began developing for the Mac in 2005, and picked up iPhone development when the first SDK was released, working on products such as The Daily, PandoraBoy, and Cisco Mobile. He is a major contributor to Stack Overflow and maintains the *Cocoaphony* blog (cocoaphony.com).

Mugunth Kumar is an independent iOS developer based in Singapore. He graduated in 2009 and holds a Masters degree from Nanyang Technological University, Singapore, majoring in Information Systems. He writes about mobile development, software usability, and iOS-related tutorials on his blog (blog.mugunthkumar.com). Prior to iOS development he worked for Fortune 500 companies GE and Honeywell as a software consultant on Windows and .NET platforms. His core areas of interest include programming methodologies (Object Oriented and Functional), mobile development and usability engineering. If he were not coding, he would probably be found at some exotic place capturing scenic photos of Mother Nature.

Dedication

To Neverwood. Thanks for your patience.
Rob

To my mother who shaped the first twenty years of my life
Mugunth

About the
Technical Editor

Jay Thrash is a veteran software developer who has spent the past three years designing and developing iOS applications. During his career, he developed a keen interest in the areas of user interaction and interface design.

Prior to settling down as an iOS developer, Jay has worked on a variety of platforms and applications, including flight simulators and web application development. He has also spent over six years in the PC and console gaming industry.

Authors' Acknowledgments

After spending much of last year writing for iOS 5, Rob thanks his family for their patience as he did it all again for iOS 6. Mugunth thanks his parents and friends for their support while writing this book. Thanks to Wiley, especially Craig Smith and Ellie Scott, for the continued support, encouragement, and nudging that it takes to get a book out the door. Thanks to Jay Thrash for jumping into the fray and tech editing for us, and to Tom Dinse who made sure that it was intelligible. Thanks to the Apple engineers at WWDC who put up with a deluge of questions, probing, and occasional complaining. And special thanks to Steve Jobs for a lifetime of elegant boxes. He will be missed.

Contents

Introduction

Apple has a history of alternating its releases between being user-focused and developer-focused. iOS 6 is mostly about users. There's a totally new map interface with turn-by-turn driving directions. Social networks permeate the interface. Pass Kit opens up whole new ways to integrate the iPhone with daily life. The Siri enhancements make the iPhone 4S even more compelling. Expect users to upgrade quickly.

That's great news, because iOS 6 is an exciting release for developers, too. So many things have gotten so much easier. It's easier to lay out your UI with collection views and auto layout. UIKit has added rich text support. When your apps go into the background, state preservation makes it easy to keep track of where you were. Many things that were hard or tedious just became a lot simpler.

If you're ready to take on the newest Apple release and push your application to the limits, this book will get you there.

Who This Book Is For

This is not an introductory book. Many other books out there cover Objective-C and will walk you step by step through Interface Builder. However, this book assumes that you have a little experience with iOS. Maybe you're self-taught, or maybe you've taken a class. You've hopefully written at least most of an application, even if you haven't submitted it yet. If you're ready to move beyond the basics, to learn the best practices and the secrets that the authors have gleaned from practical experience writing real applications, then this is the book for you.

This book also is not just a list of recipes. There's plenty of sample code here, but the focus is on discovering how to design, code, and maintain great iOS apps. Much of this book is about *why* rather than just *how*. You'll find out as much about design patterns and writing reusable code as about syntax and new frameworks.

What This Book Covers

The iOS platforms always move forward, and so does this book. Most of the examples here require at least iOS 5, and many require iOS 6. All examples use Automatic Reference Counting, automatic property synthesis, and object literals. Except in a very few places, this book will not cover backward compatibility. If you've been shipping code long enough to need backward compatibility, you probably know how to deal with it. This book is about writing the best possible apps using the best features available.

This book focuses on iPhone 4, iPad 2, and higher models. Most topics here are applicable to the original iPad, iPod touch, iPhone 3GS, and Apple TV. Chapter 16 is devoted to dealing with the differences between the platforms.

What Is New in This Edition

This edition covers most of the newest additions to iOS 6, including collection views and Auto Layout (Chapter 7), and a complete chapter is dedicated to Automatic Reference Counting. We have added a whole new chapter explaining how to move your data to the cloud, covering some third-party backend as a service (BaaS) offerings. More relevant to iOS, this year saw the greatest improvement in the LLVM compiler and debugger. We expanded the LLVM section in Chapter 2 to highlight the important features of LLVM that you could use to write the next-generation apps. We also added a whole new chapter dedicated to debugging with LLDB (Chapter 19), covering advanced topics, including Python scripting, knowing the layout of registers, and reading the registers to solve the most difficult debugging problems. That chapter also provides a brief look at some command-line tools such as `otool` and `atos` that can help you with debugging. Lastly, Chapter 20 is dedicated to showing how to optimize your CPU performance, memory usage, network bandwidth, and battery-using Instruments.

Even if you have the previous edition of this book, the new coverage on debugging (Chapter 19), performance tuning (Chapter 20), and collection views and Auto Layout (Chapter 7) will probably be enough to convince you that you need a copy of this edition.

How This Book Is Structured

iOS has a very rich set of tools, from high-level frameworks like UIKit to very low-level tools like Core Text. Often, there are several ways to achieve a goal. As a developer, how do you pick the right tool for the job?

This book separates the everyday from the special purpose, helping you pick the right solution to each problem. You'll discover why each framework exists, how the frameworks relate to each other, and when to choose one over another. Then you'll learn how to make the most of each framework for solving its type of problem.

There are four parts to this book, moving from the most common tools to the most powerful. Chapters that are new in this edition or have been extensively updated are indicated.

Part I: What's New?

If you're familiar with iOS 5, then this section quickly introduces you to the new features of iOS 6.

- **(Updated) Chapter 1: "The Brand New Stuff"**—iOS 6 adds a lot of new features, and here you get a quick overview of what's available.

- **(Updated) Chapter 2: "Getting Comfortable with Xcode 4 and the LLVM compiler"**—Xcode 4 is a huge IDE, and many of its features aren't obvious. Whether you're new to Xcode or a veteran from earlier versions, this chapter will help you make the most of this powerful (and sometimes frustrating) tool.

Part II: Getting the Most Out of Everyday Tools

As an iOS developer, you've encountered a wide variety of common tools, from notifications to table views to animation layers. But are you using these tools to their full potential? In this part, you find the best practices from seasoned developers in Cocoa development.

- **Chapter 3: "Everyday Objective-C"**—If you're ready to move to the next level in Objective-C, this chapter introduces you to the tools experienced developers use every day to improve application design, maintainability, and reusability.

- **Chapter 4: "Hold On Loosely: Cocoa Design Patterns"**—Cocoa relies on a number of common and consistent design patterns. You discover what they are so you can solve problems the same way Apple does.

- **(New) Chapter 5: "Memory Management with Objective-C ARC"**—Automatic Reference Counting has radically changed how iOS developers work. While most of the conversion is easy, migrating your code and your practices can create some subtle problems. You find out how to avoid the problems and get the most out of this powerful technology.

- **Chapter 6: "Getting Table Views Right"**—Table views are perhaps the most complex and commonly used UI element in iOS. They are simple and elegant in design, but confusing to developers who don't understand how they work. You learn how to use them correctly and how to solve some special problems like infinite scrolling.

- **(New) Chapter 7: "Great at Any Angle: Collection Views and Auto Layout"**—Nothing created as much buzz at WWDC 2012 as `UICollectionView`. Up until now, `UITableView` has dominated iOS interfaces, but that will likely change in iOS 6 as developers master this new control. With the addition of constraint-based layout from OS X, incredibly complex and beautiful interfaces are now much simpler to implement. This chapter shows you the ropes.

- **Chapter 8: "Better Drawing"**—Custom drawing is intimidating to many new developers, but it's a key part of building beautiful and fast user interfaces. You'll discover the available drawing options from UIKit to Core Graphics, and how to optimize them to look their best while keeping them fast.

- **Chapter 9: "Layers Like an Onion: Core Animation"**—iOS devices have incredible facilities for animation. With a powerful GPU and the highly optimized Core Animation, you can build engaging, exciting, and intuitive interfaces. In this chapter, you go beyond the basics and discover the secrets of animation.

- **Chapter 10: "Tackling Those Pesky Errors"**—You try to write perfect code, but sometimes things go wrong. How your application reacts to the unexpected is what separates decent apps from extraordinary apps. You'll find the common patterns for error handling, how to log, and how to make your code more resilient against the unexpected.

- **(New) Chapter 11: "Location Services: Know Where You Are"**—Many applications need to know where the user is. While iOS makes this easy, it is important not to drain the battery in the process. You find out how to balance functionality with efficiency and give the user the best possible experience.

Part III: The Right Tool for the Job

There are tools that are part of nearly every application, and there are tools that you only need from time to time. In this section, you learn about the tools and techniques that are a little more specialized.

- **(New) Chapter 12: "Common UI Paradigms Using Table Views"**—Exciting as collection views are, the table view is still the workhorse of iOS. You find out how to take table views beyond the basics with infinite scrolling views and other advanced topics.

- **(Updated) Chapter 13: "Controlling Multitasking"**—Multitasking is an important part of many applications, and you discover how to do multiple things at once while your application is running and when your application is in the background. The new state preservation system of iOS 6 makes this much easier, and you learn how to use it for common situations.

- **(Updated) Chapter 14: "REST for the Weary"**—REST-based services are a mainstay of modern applications, and you learn how to best implement them in iOS.

▦ **(Updated) Chapter 15: "Batten the Hatches with Security Services"**—User security and privacy are paramount today, and you find out how to protect your application and user data from attackers with the keychain, certificates, and encryption.

▦ **Chapter 16: "Running on Multiple iPlatforms and iDevices"**—The iOS landscape gets more complex every year with iPod touch, iPhone, iPad, Apple TV, and a steady stream of new editions. It's not enough just to write once, run everywhere. You need your applications to be their *best* everywhere. You learn how to adapt your apps to the hardware and get the most out of every platform.

▦ **Chapter 17: "Internationalization and Localization"**—Although you may want to focus on a single market today, you can do small things to ease the transition to a global market tomorrow. Save money and headaches later, without interrupting today's development.

▦ **Chapter 18: "Selling Past the Sale with In App Purchases"**—In App Purchases are still an untapped market for many developers. Users like the add-on content, and developers love the extra revenue. You discover the best ways to make this important feature a reality in your application.

▦ **(New) Chapter 19: "Debugging"**—If only every application were perfect the first time. Luckily, Xcode and LLDB provide many tools to help you track down even the trickiest of bugs. You go beyond the basics and find out how to deal with errors in development and in the field.

▦ **(New) Chapter 20: "Performance Tuning Until It Flies"**—Performance separates the "okay" app from the exceptional app. It's critical to optimizing CPU and memory performance, but you also need to optimize battery and network usage. Apple provides an incredible tool for this in Instruments. You discover how to use Instruments to find the bottlenecks, and then how to improve performance once you find the problems.

Part IV: Pushing the Limits

This section is what this book is all about. You've learned the basics. You've learned the everyday. Now push the limits with the most advanced tools available. You discover the ins and outs of deep iOS.

▦ **(New) Chapter 21: "Storyboards and Custom Transitions"**—Storyboards can still be confusing and a bit intimidating for developers familiar with nib files. You learn how to use storyboards to your advantage and how to push them beyond the basics.

▦ **Chapter 22: "Cocoa's Biggest Trick: Key-Value Observing"**—Many of Apple's most powerful frameworks rely on KVO for their performance and flexibility. You find out how to leverage the flexibility and speed of KVO, as well as the trick that makes it so transparent.

▦ **Chapter 23: "Think Different: Blocks and Functional Programming"**—Many developers are still absorbing the addition of blocks to Objective-C. They're valuable for interacting with Apple frameworks, but they also open new ways of thinking about your program. Embrace a new style, and maximize its benefits in your next project.

▦ **Chapter 24: "Going Offline"**—Network programming is difficult, but even harder is providing a seamless offline experience. Discover how to best cache your data and integrate it into your network engine.

▦ **(New) Chapter 25: "Data in the Cloud"**—Developers are still adapting the new iCloud service. Supporting it quickly can move your app ahead of the pack. You learn your way around this powerful system.

▦ **(Updated) Chapter 26: "Fancy Text Layout"**—From UIKit to Core Text, iOS is full of ways to display text. iOS adds rich text support throughout UIKit. You find out how to use attributed strings, web views, and Core Text to handle the toughest layout problems.

- **Chapter 27: "Building a (Core) Foundation"**—When you want the most powerful frameworks available on iOS, you're going to want the Core frameworks like Core Graphics, Core Animation, and Core Text. All of these rely on Core Foundation. In this chapter, you discover how to work Core Foundation data types so you can leverage everything iOS has to offer.

- **Chapter 28: "Deep Objective-C"**—When you're ready to pull back the curtain on how Objective-C really works, this is the chapter for you. You find out how to use the Objective-C runtime directly to dynamically modify classes and methods. You also learn how Objective-C method calls are dispatched to C function calls, and how you can take control of the system to extend your programs in incredible ways.

You can skip around in this book to focus on the topics you need most. Each chapter stands alone, except for those that require Core Foundation data objects (particularly Core Graphics, Core Animation, and Core Text). Those chapters direct you to Chapter 27, "Building a (Core) Foundation," when you need that information.

What You Need to Use This Book

All examples in this book were developed with Xcode 4.5 on Mac OS X 10.8 and iOS 6. You need an Apple developer account to access most of the tools and documentation, and you need a developer license to run applications on your iOS device. Visit `developer.apple.com/programs/ios` to sign up.

Most of the examples in this book will run in the iOS Simulator that comes with Xcode 4.5. You can use the iOS Simulator without an Apple developer license.

Finding Apple Documentation

Apple provides extensive documentation at its website and within Xcode. The URLs change frequently and are often very long. This book refers to Apple documents by title rather than by URL. To find documents in Xcode, press Cmd-Option-? or click Help⇨Documentation and API Reference. In the Documentation Organizer, click the Search icon, type the name of the document, and then select the document from the search results. See Figure 1 for an example of how to search for the *Coding Guidelines for Cocoa.*

To find documents at the Apple developer site, visit `developer.apple.com`, click Member Center, and log in. Select the iOS Dev Center, and enter the document title in the Search Developer search box.

The online documentation is generally identical to the Xcode documentation. You may receive results for both iOS and Mac. Be sure to choose the iOS version. Many iOS documents are copies of their Mac counterparts and occasionally include function calls or constants that are not available on iOS. This book tells you which features are available on iOS.

Source Code

As you work through the examples in this book, you may choose either to type in all the code manually or to use the source code files that accompany the book. All of the source code used in this book is available for download at either `http://iosptl.com/code` or `www.wiley.com/go/ptl/ios6programming`. For example, you will find the following sample code online in the Chapter 26 folder, in the `SimpleLayout` project, and the `CoreTextLabel.m` file.

CoreTextLabel.m (SimpleLayout)

```
- (id)initWithFrame:(CGRect)frame {
  if ((self = [super initWithFrame:frame])) {
    CGAffineTransform
    transform = CGAffineTransformMakeScale(1, -1);
    CGAffineTransformTranslate(transform,
                               0, -self.bounds.size.height);
    self.transform = transform;
    self.backgroundColor = [UIColor whiteColor];
  }
  return self;
}
```

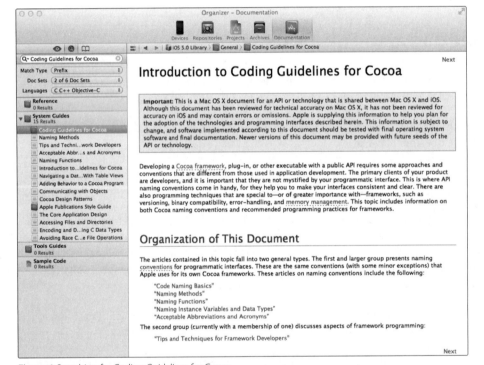

Figure 1 Searching for Coding Guidelines for Cocoa

Some source code snippets shown in the book are not comprehensive and are meant to help you understand the chapter. For those instances, you can refer to the files available on the website for the complete source code.

Errata

We try to get everything right, but sometimes things change, sometimes we mistype, and sometimes we're just mistaken. See http://iosptl.com/code for the latest updates, as well as blog posts on topics that haven't made the book yet. Send any mistakes you find to robnapier@gmail.com and contact@mk.sg.

Part I

What's New?

Chapter 1

The Brand New Stuff

In 2007, Steve Jobs went on stage at Macworld and proclaimed that software running on the iPhone was at least five years ahead of the competition. Since its initial release up to the time of this writing, Apple has been iterating the operating system year after year and has even added two new devices, the iPad and the Apple TV, to the list of products capable of running iOS. Over time, as the operating system was customized to run on more devices than the iPhone, it was rebranded as iOS. Now, after five years and true to the words of Steve Jobs, iOS is still the most advanced operating system around. iOS 6 takes the most advanced operating system to the next level, adding great features, including some developer-centric features like collection views, Auto Layout, and improved data privacy.

This book is about programming with iOS 6, and in this chapter, I go through the main features of iOS 6 and refer you to the chapters covering them in detail.

What's New

Every version of iOS has introduced several key features and other minor features. The second version, iPhone OS 2, was the first to have a public SDK.

iPhone OS 3 brought Core Data from the Mac to the iPhone; other additions included Push Notifications, `ExternalAccessory` Kit, In App Purchases through `StoreKit.framework`, in app e-mail sheets, `MapKit.framework` that allows developers to embed Google Maps into their apps, read-only access to the iPod library, and keychain data sharing.

A minor update, 3.1, added video editor support, and another minor (yet so major) update, 3.2, added Core Text and gesture recognizers, file sharing, and PDF generation support. The 3.2 update also added a whole new product, iPad, and support for developing apps that run on iPad, as well as Universal apps that run on iPad (3.2) and iPhone (3.1.3). However, 3.2 was iPad-only and didn't run on iPhone or iPod touch devices.

iPhone OS 4 (rebranded as iOS 4) introduced the much-awaited multitasking support, local push notifications, read-only access to calendar (`EventKit.framework`), blocks and Grand Central Dispatch, In app SMS sheets, and Retina display support. This version was iPhone-only and didn't support developing apps for iPad. A minor update to this, iOS 4.2, unified iPhone and iPad operating systems.

iOS 5 introduced several important features like iCloud, Automatic Reference Counting, built-in Twitter framework, and several other minor features.

iOS 6 introduces Pass Kit, a framework which providers can use to push tickets, passes, or coupons directly to the iPhone. For example, a hotel reservation iOS app can send a confirmation of registration as a pass to the user's iPhone.

iOS 6 also introduces a brand-new maps application for end users. To developers, there is a new SDK that allows you to show directions. You open maps by calling the method `openMapsWithItems:launchOptions:` on `MKMapItem`:

```
+ (BOOL)openMapsWithItems:(NSArray *)mapItems launchOptions:
    (NSDictionary *)launchOptions
```

However, iOS 6 doesn't support opening maps through a URL scheme. So, you need to fall back to Google for iOS 5 devices, as illustrated in the following code fragment:

```
if ([[UIApplication sharedApplication] canOpenURL:
    [NSURL URLWithString:@"maps"]]) {
    // ios 6 specific
    [MKMapItem openMapsWithItems:<items> launchOptions:<options>]; } else
{
    // ios 5 specific, fall back to Google
    [[UIApplication sharedApplication] openURL:
    [NSURL URLWithString:@"http://maps.google.com/<your url>"]]
}
```

This book focuses primarily on the new features added in iOS 6. In the next few sections, I introduce you to the key features added to iOS 6 and provide guidance for using them to push your apps to the next level.

Collection Views

Arguably the most important addition to iOS 6 is a brand-new controller for displaying and managing collection views. Collection views finally provide a way to lay out your UI elements as a grid view like the built-in Photos app or the iBooks app. You no longer have to depend on third-party grid view frameworks for creating grids or mess around with scroll views. Collection views also provide an easy way to animate content (similar to `UITableView`'s row animation) and to insert, delete, and reorder them. (See Chapter 7 for more on collection views.)

Auto Layout

Auto Layout is a new way to define rules that determine how elements in your user interface should be laid out. This capability was introduced in the Mac OS X SDK last year and brought to iOS with iOS 6. Auto Layout is more intuitive than the previously used "springs and struts" model and provides built-in support for mirroring UI elements for right-to-left languages. Arguably, the most important feature that Auto Layout brings in is the capability to swap strings when you localize your nib files.

Automatic Reference Counting

Though Automatic Reference Counting (ARC) was introduced with iOS 5, information about ARC and internal workings of the LLVM compiler has changed considerably, including the default storage type of properties. This edition of the book adds a whole new chapter (Chapter 5) that explains ARC, the internals of how ARC works and how to convert your code base to Objective-C ARC using the Convert to Objective-C ARC tool.

> ARC is *not* like garbage collection offered on Mac OS X from version 10.5 (Leopard). Garbage collection is automatic memory management, whereas ARC is Automatic Reference Counting. This means that you, as a developer, don't have to write a matching release for every retain statement. The compiler automatically inserts them for you.

ARC adds two new lifetime qualifiers, `strong` and `weak`, and also imposes certain new rules—for example, you can no longer invoke release, retain on any object. This applies to custom `dealloc` methods as well. When using ARC, your custom `dealloc` methods should release only resources (files or ports) and not instance variables. (For more on ARC, see Chapter 5.)

In App Purchases Hosted Content

iOS 6 allows you to host downloadable content for your In App Purchases with Apple. For example, if you're making a racing game and you're selling "cars" and "tracks" as In App Purchases, you can now host the map tiles for the tracks and the model file for the car with Apple's servers. When the purchase is complete, you can initiate a download, all without paying a dime for hosting those files. (See Chapter 18 for information on In App Purchases and how to download hosted content.)

Social Framework

iOS 5 integrated Twitter experience right into the OS. This made sending a tweet from your app as easy as sending an e-mail using in app e-mail sheets. The framework also handled authentication for you. The Twitter framework on iOS 5 integrates with the Accounts framework to provide account authentication. In iOS 6, Apple supplanted `Twitter.framework` with `Social.framework`, which is more generic and handles other social networks like Facebook and Weibo in addition to Twitter. What this means to you as a developer is that Facebook is now tightly integrated right into the OS. So, posting to Facebook and authenticating users with Facebook is easier and doesn't require Facebook's official Facebook Graph SDK. However, you will still need the Graph SDK if you're supporting iOS 5 devices. The `UIKit.Framework` introduces a new view controller called `UIActivityViewController` that is used in conjunction with the `Social.framework` to post content to a social network.

UI State Preservation

In iOS 3 and older operating systems, restoring the state of your UI was hard. When Apple introduced multitasking in iOS 4, apps automatically preserved and restored their state as long as they were running in the background. But low memory situations can kill your app while it's running in the background and your app gets no notification about it. When an app was killed in the background, it wasn't easy to restore the state of the UI, and the only way to restore the UI state was to implement a variety of hacks that developers implemented prior to iOS 4. In iOS 6, Apple made state preservation easier by providing state restoration support built into `UIKit`. (See Chapter 13 for more on UI state preservation.)

Other New Features

iOS 6 also introduces several other minor features including Pass Kit, increased privacy, support for interacting with Reminders, and more importantly, rich text support for `UITextField`.

Pass Kit

Pass Kit is a new framework that allows users to store items like coupons, boarding passes, and event tickets on their iOS devices, instead of a physical equivalent. Passes are created on the server and delivered to the user's device, and as an iOS developer, you will not be writing code to handle them. Pass Kit, while a major feature in iOS 6, is more like APNS, heavy on the server side but little or no coding effort on the client side. As such, Pass Kit isn't covered in this book.

Reminders

iOS 4 introduced `EventKit.framework` that allowed third-party apps to access a user's calendars. In iOS 6, `EventKit.framework` has methods to interact with a user's Reminders as well.

Privacy

Privacy has always been a monumental concern with apps. Prior to iOS 6, some apps, including those from high-profile social networking companies like Path, uploaded contact information to their servers without explicit consent from users. In iOS 6, access to Contacts, Photo Library (either through `AssetsLibrary.framework` or through `UIImagePickerController`), reminders, and calendars requires explicit permission from the user. This means that, if you've previously accessed this information, be prepared to receive empty data from the user. You can also customize why you need access to such information by setting the values for the keys `NSLocationUsageDescription`, `NSPhotoLibraryUsageDescription`, `NSCalendarsUsageDescription`, `NSContactsUsageDescription`, and `NSRemindersUsageDescription` in your application's `Info.plist` file. You can also set the values from the UI using one of the options shown in Figure 1-1.

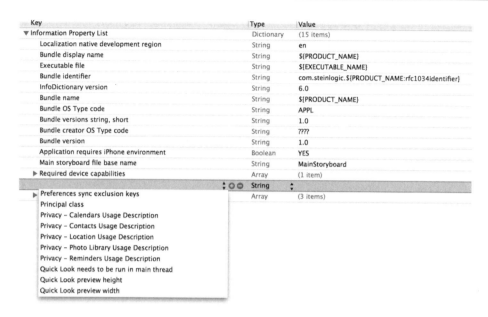

Key	Type	Value
▼ Information Property List	Dictionary	(15 items)
Localization native development region	String	en
Bundle display name	String	${PRODUCT_NAME}
Executable file	String	${EXECUTABLE_NAME}
Bundle identifier	String	com.steinlogic.${PRODUCT_NAME:rfc1034identifier}
InfoDictionary version	String	6.0
Bundle name	String	${PRODUCT_NAME}
Bundle OS Type code	String	APPL
Bundle versions string, short	String	1.0
Bundle creator OS Type code	String	????
Bundle version	String	1.0
Application requires iPhone environment	Boolean	YES
Main storyboard file base name	String	MainStoryboard
▶ Required device capabilities	Array	(1 item)
	String	
▶ Preferences sync exclusion keys	Array	(3 items)
Principal class		
Privacy – Calendars Usage Description		
Privacy – Contacts Usage Description		
Privacy – Location Usage Description		
Privacy – Photo Library Usage Description		
Privacy – Reminders Usage Description		
Quick Look needs to be run in main thread		
Quick Look preview height		
Quick Look preview width		

Figure 1-1 Info.plist showing the newly added Usage Description keys

Displaying Rich Text

`UITextView` and `UITextField` support `NSAttributedString`, and you can display rich text by creating an `NSAttributedString`. You create an `NSAttribtedString` using `CoreText.framework`. (See Chapter 26 for a discussion of Core Text.)

Summary

Adoption rates of iOS have always been way ahead of the competition. A couple of years ago, when iPhone OS 3.0 was launched, adoption rates were partly hindered on iPod touches because it was a $10 paid upgrade. However, Apple soon made it free, and adoption rates increased. Similarly, when Apple released iOS 4, the adoption rate was initially slow because of performance issues on older phones, namely iPhone 3G and the original iPhone (and equivalent iPod touches). Some features, mainly multitasking, were also not available for older devices. Nevertheless, the latest iOS usually is adopted on more than 90 percent of devices within the first two months of their launch. The iOS 5 adoption rate wasn't hindered, and the adoption rates were faster than those for iOS 4. This was probably because there weren't any missing features for older phones (like iPhone 4). Also, older devices (like iPhone 3GS) didn't run slower on iOS 5, unlike with iOS 4. Finally, cleaner notifications, iTunes, WiFi-sync, and iMessage were killer features that accelerated end users' adoption ofiOS 5.

iOS 5.1 was the first over-the-air update, and adoption rates were even higher than for iOS 5. For iOS 6, I foresee similar high adoption rates. All this means that you want to start using every possible iOS 6 feature as soon as possible to get your app shining in all its glory. Features like collection views, and using Auto Layout to support the new taller iPhone 5 should be reason enough to update your apps to iOS 6. Having said that, it's time to start your iOS 6 journey.

Further Reading

Apple Documentation

The following document is available in the iOS Developer Library at `developer.apple.com` or through the Xcode Documentation and API Reference.

What's New in iOS 6

Chapter 2

Getting Comfortable with Xcode 4 and the LLVM Compiler

Apple officially announced Xcode 4 at WWDC 2010 (June 2010), and the beta version was available to attendees. It was in beta for quite a while (around nine months), and a Gold Master was made available through the iOS/Mac developer center in February 2011. Weeks later, in March, Xcode 4 was officially released, and developers who subscribe to the iOS or Mac developer programs were able to get it for free. Others were able to buy it from the Mac App Store.

Xcode 4 is a completely rewritten IDE (integrated development environment) replacing Xcode 3. The major features include, but are not limited to, single window editing, navigators, integrated Interface Builder, an integrated Git version control system, and *schemes* (a new way to configure and share build settings in your product). You learn in detail about every major feature in this chapter.

Xcode 4 features are not just skin deep—they come with some huge compiler-level changes as well. The LLVM compiler is the new brain behind Xcode. Apple made LLVM-GCC the default compiler in the original version of Xcode. Beginning with Xcode 4.0 and in the version that is released with iOS 5 (Xcode 4.2), LLVM 3.0 was the default compiler; it uses Clang as its front end. Using Clang as the front end over GCC has several advantages, and several new features of Xcode 4 were added because of this change. With Xcode 4.5 (released with iOS 6), Apple made LLVM 4 the default compiler. LLVM 4 brings in some new literals giving some syntactic sugar to Objective-C. Later in this chapter, I'll introduce you to the new Objective-C literals added in LLVM 4.

Because Apple has switched both the default compiler and the debugger to LLVM, you should know how to harness the power of the new compiler to increase your coding and debugging speed and how to use the IDE to be more productive. The most important feature of the LLVM compiler is better and faster compilation with the Clang front end, which provides better code completion support.

iOS 5 introduced a new memory management called *Automatic Reference Counting* or ARC, and moving forward, ARC will probably be the only memory management technique supported by Apple. Automatic Reference Counting is a memory management model where the compiler automatically inserts retains and releases appropriately for you. This means that you don't have to worry about managing memory and focus or writing the next great app. You're relieved from worrying about leaks, for the most part. While that is true to some extent, migrating an existing application written using the older, reference-counting memory management model isn't easy. Fortunately, Xcode 4.2 and above has a built-in migration tool (Convert to Objective-C ARC) that will help you to convert your existing code base to support the ARC compiler.

This chapter covers the important features of the IDE, the new features offered by the LLVM 4 Compiler, the built-in integrated version control system, schemes (new to Xcode 4), writing readable and commentable project configuration files, the features of the Xcode 4 Organizer, and finally, the Convert to Objective-C ARC migration tool.

Getting to Know the User Interface

Xcode 4 features a whole new iTunes-like user interface (UI). The toolbar is gone in favor of iTunes-like Play/Stop buttons. The build setting chooser is gone in favor of the new schemes selector. There's a new LCD-like status display similar to iTunes. In this section, I briefly show you the five most important aspects of the Xcode 4 UI and how to get the most out of it.

You can access seven navigators by clicking the buttons highlighted in Figure 2-1. You can also access them via the shortcut keys Cmd-1 through Cmd-7.

Figure 2-1 The Xcode navigator items

You can access the same navigators from Xcode's View menu item, as shown in Figure 2-2.

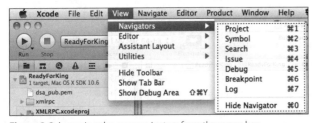

Figure 2-2 Accessing the new navigators from the menu bar

Figure 2-3 highlights the three important parts of Xcode 4's user interface. The first view is the navigator pane showing the first navigator, the project navigator. You access the list of files and frameworks in your project from the project navigator. The panel on the right side is the Xcode 4 workspace pane. The Xcode 4 workspace can edit more types of files than just Objective-C or property list files, and the project and build settings editor just happens to be one such editor. You can access the project and build settings editor by selecting the project file from the project navigator. Additional navigators are symbol, search, issue, debug, breakpoint, and log. You look at them in detail later in this chapter.

Workspace pane

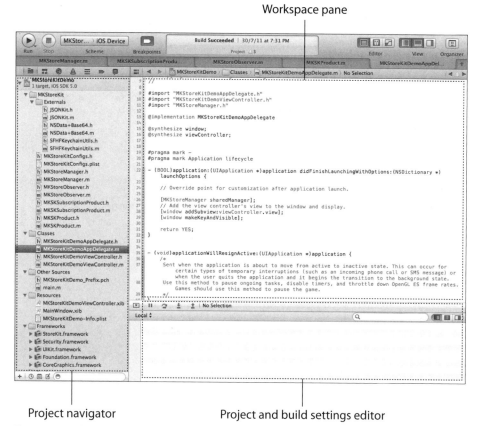

Project navigator   Project and build settings editor

Figure 2-3 Xcode 4 window showing several opened tabs and important parts of the user interface

Tabbed Editor

Xcode 4 supports opening multiple tabs within the same window. This means that you will spend less time searching for the window that displays your source file. Figure 2-3 shows Xcode 4's tabbed editor in action.

The tabbed editor in Xcode 4 behaves differently from tabbed editors in, for example, Eclipse, Visual Studio, or TextMate. Think of Xcode 4's tabs as virtual workspaces instead of just file editors. The navigator pane's file selection and search criteria are preserved when you switch back and forth between tabs. I recommend opening three to four tabs, each showing a related group of files. For example, you can use one tab to show your model classes, another tab for your view controllers, a third tab for Interface Builder files, and maybe, if you use Core Data, a fourth tab for showing Core Data-related files. Use tabs to make the workspace to suit your thought process or workflow. Going to work on your Core Data files? Switch to the Core Data tab. Going to work on Interface Builder? Switch to the Interface Builder tab.

Changes to Key Bindings

Xcode 4 has changed most of its keyboard shortcuts from Xcode 3 and they are different from other IDEs as well. Three commonly used shortcuts for debugging are changed to F6 (Step over), F7 (Step into), and F8

(Step out). The shortcut for *Build and Run* is Cmd-R. Initially when you get started with Xcode 4, these are the most important shortcuts that you should remember.

For a good, comprehensive list of keyboard shortcuts, I recommend Cocoa Samurai's list, available for download at `http://cocoasamurai.blogspot.com/2011/03/xcode-4-keyboard-shortcuts-now.html`, and another by The Pragmatic Studio, available at `http://pragmaticstudio.com/media/Xcode4Shortcuts.pdf`.

Project Settings Editor

The project settings editor also allows you to edit your project and target's settings. Furthermore, adding additional frameworks to your product and passing command-line arguments are all now a part of this build settings panel of the project settings editor.

Integrated Version Control

Xcode 4 integrates the Git version control system. Git is arguably one of the most powerful distributed version control systems as of this writing. Later in this chapter, you find out how to get the best out of your version control system.

Workspaces

Xcode 4 supports the concept of workspaces and allows you to create a workspace and add multiple projects within it. For example, if you're writing a Mac + iOS app, you can share a wealth of code. Instead of manually copying and pasting code, you can extract the common code into a separate static library project and add it to the workspace. The primary advantage of a workspace is *implicit dependencies*. This means that when you build your Mac app (or iOS app), Xcode 4 automatically detects that it's dependent on your static library project and builds it first, without you explicitly requesting it to do so.

All in One Window

Xcode 4 is a single window IDE like Eclipse or Visual Studio. Every file you use in your project can be opened without "switching" to it using Mac's Exposé. With full-screen editing in Lion, you will appreciate the single window IDE and find yourself spending less time switching or searching for a window.

The IDE, as you saw earlier in this chapter, consists of a set of navigators, a workspace area, and the utility area. There are seven navigators that replace the functionality of the Groups and Files view.

The workspace area is where you edit your files, which can be either source code or `.plist` files or even Interface Builder (IB) files. There are three different kinds of editors: the standard editor that you use for editing files, the assistant editor for editing files related to the current file in standard editor, and a versions editor that shows version history of a file.

The utility area is akin to the Inspector pane found on most other applications. Below the utility area is the library from which you can drag user interface elements into your IB or code snippets into your source code. It's time to take a closer look at the different navigators available in Xcode 4.

Navigating the Navigators

Seven navigator panes are built into Xcode 4. These features were also present in Xcode 3 (in the Groups and Files view), but are presented in a more meaningful way in Xcode 4.

In Xcode 3, there was a single view—the Groups and Files view—where you did pretty much everything. You chose the file to edit, edited project settings, added frameworks, accessed breakpoints and debug logs, and did a lot more all from the same view. Xcode 4 groups these actions into seven different navigators.

All navigators have a filter and scope box (shown in Figure 2-4) that can be accessed using Cmd-Option-J. This shortcut puts the focus on the filter and scope box below the navigator from where you can search for a project file. The keyboard shortcut works for all navigators, so if you're in, say, the symbol navigator, you can use this to quickly filter symbols. The filter box might have additional buttons to restrict the scope, and sometimes (as in the debug navigator) it might be replaced with a UI that looks different but offers the same functionality.

Figure 2-4 Xcode navigator showing the filter and scope bar

You can press the shortcut key Cmd-Option-J to quickly jump to the filter and scope box.

Navigators provide a clear separation of duties on the UI. For example, there is a dedicated navigator for breakpoints, a dedicated navigator for issues, and another for logs. Although it might be difficult to switch to the new separate navigators, you will appreciate them once you become accustomed to them.

Some features, like the capability to create smart groups, were dropped probably because few people use them. However, if you have a project in Xcode 3 that uses smart groups, opening it in Xcode 4 and saving it doesn't remove them from the project file. So when you open the project again on Xcode 3, you will still see your smart groups.

Project Navigator

The project navigator, as the name suggests, helps you locate your source code files, frameworks, and targets. Similarly to Xcode 3 and most other IDEs, the project navigator also serves as the source code control UI. This means that when you add files to the project navigator, they are automatically added to your source control (if you use one), and the project navigator also updates the UI with the files' source control status. You can jump to the project navigator with the Cmd-1 shortcut.

Symbol Navigator

You can jump to the symbol navigator with the Cmd-2 shortcut. The symbol navigator makes it easy to locate a specific symbol or class in your project. The Clang front end of the LLVM compiler integrates well with Xcode 4 and has made browsing through symbols in the project faster.

Search Navigator

The search navigator is functionally exactly the same as the Xcode 3's Find and Replace feature. You can access this navigator with the traditional Cmd-Shift-F shortcut or the navigator shortcut Cmd-3.

Issue Navigator

When you build your project, compiler warnings, error messages, or analyzer warnings appear on the issue navigator. The issue navigator on Xcode 4 is clear of build log messages and they are moved to a separate navigator called the log navigator. The log navigator maintains every build log in chronological order. You can access the issue navigator using the navigator shortcut Cmd-4.

Debug Navigator

The debug panel in Xcode 3, which you access by pressing Cmd-Shift-Y, is equivalent to Xcode 4's debug navigator. You can access the debug navigator in Xcode 4 using the navigator shortcut Cmd-5. The most important addition is the scope slider. Instead of the filter and scope search box present in other navigators, the debug navigator uses a scope slider. Drag the scope slider to customize your scope preference.

Breakpoint Navigator

The sixth navigator is the breakpoint navigator. On Xcode 3, this was managed in a separate window. The nifty addition here is the capability to quickly add a symbolic breakpoint or an exception breakpoint. You can access the breakpoint navigator using the shortcut Cmd-6.

> **A noteworthy feature of the breakpoint navigator is the capability to share your breakpoints with coworkers. From the breakpoint navigator, Cmd-click the project file and click Share Breakpoints.**

Log Navigator

The log navigator maintains a log of every build, and you can even search for entries in a log that was created several builds ago. You can access the log navigator using the shortcut Cmd-7.

Help from Your Assistant

Xcode 4 has three main editors, and they are akin to multiwindow document editing present in other competing IDEs. The two editors that augment Xcode 4's standard editor are the assistant editor and the versions editor. The best thing about the assistant editor is that, when you turn in on, it intelligently knows the file most relevant to the file you're currently working with.

For example, when you're editing a Core Data model, turning on the assistant editor opens the corresponding Core Data's model file. Similarly, when you're editing an Interface Builder file, it opens the corresponding header file.

A common action like adding an `IBAction` declaration in your header and coming back to the Interface Builder to connect it can be done easily within the same window using the assistant editor. The assistant editor also provides quick access to other related files like the superclass, subclass, and siblings.

Integrated Interface Builder

Prior to Xcode 4, Interface Builder was a standalone application, and the most common mistake a programmer made was failing to sync Interface Builder and Xcode properly. For example, forgetting to save an Interface Builder connection could crash your app at runtime. Additionally, the very fact that there are two applications for writing iOS apps confuses developers coming from an Eclipse or Visual Studio background. Those difficulties are in the past because Xcode 4 integrates Interface Builder right into the main IDE, and it's now very easy to sync your user interface with the controller code.

Interface Builder Panels

Interface Builder on Xcode 3 usually has multiple windows floating around. At a bare minimum, you have the main document window, the library panel, the inspector, and the actual user interface view. In Xcode 4, the library and inspector are brought into the utility area. The document window is docked to the left. Figure 2-5 shows a classical Interface Builder file open in Xcode 4. The navigator panel, which is normally open, is closed in Figure 2.5 to illustrate the other panels of Interface Builder.

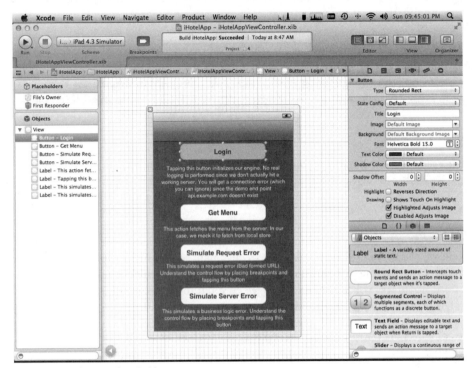

Figure 2-5 Xcode 4 Interface Builder

The left pane now contains the objects in the Interface Builder file. The utility area shows properties for the selected object.

When you turn on the assistant editor, Xcode 4 will automatically open the correct header file for you.

Generating Code Using Assistant Editor and Integrated Interface Builder

The most important feature that has been added to Xcode 4's integrated Interface Builder is the capability to generate properties (IBOutlet) or event handling (IBAction) code directly from IB. When the assistant editor is open, all you have to do is command-click and drag objects to the header file to generate properties, or command-click and drag events from the utility area to the header to generate IBActions. If your drag destination is valid, Xcode shows an insertion marker and adds the code right in—no connections needed. It's all done for you.

LLVM Compiler: A Tryst with the Brain

Xcode 3 and prior versions were not as "intelligent" compared to competing IDEs such as Eclipse or Visual Studio. The main reason for this is that Xcode 3 used GCC as the compiler. Although GCC is a good compiler, it doesn't offer much interoperability with the IDE. A compiler normally has a front end that converts source code into an intermediate representation and expands preprocessor macro definitions. It also has a back end that generates code and optimizes it. The GCC compiler is essentially a back-end compiler, which means that when you provide it with source code, it generates compiled binary for the given source. The GCC compiler was primarily developed for compiling code and not for parsing it. Essentially, this means that Apple has to write its own version of parsers to assist you with debugging. GCC is GPL licensed and what that means is that Apple's version of parsers cannot use the same GCC code without changing the Xcode license to GPL. Because Apple's parsers and GCC are from different code bases, there were always some discrepancies between what the GCC compiler "thinks" and what Xcode "thinks."

The Clang Front End

To alleviate this problem, Apple switched to the LLVM compiler last year. When Xcode 4.2 debuted with iOS 5, the default compiler was LLVM compiler 3.0. Although LLVM is not as "efficient" as GCC in code generation, it's more modular and extensible. LLVM is also more than twice as fast in terms of compile time (thereby increasing your productivity). A number of front ends have been developed for LLVM, and one of these is Clang, which is heavily funded by Apple. (Clang stands for *C language*.) Clang supports incremental compilation, which means that the IDE can actually compile the code as you type and show you near-instantaneous compilation errors. You will find this very useful when you start using Xcode 4. Moreover, the modular nature of Clang makes it easy to support code refactoring and features like Edit All in Scope.

I'm a Bug! Fix Me

LLVM's tighter integration with the IDE also helps Xcode 4 to offer suggestions about what the developer must do when the compiler encounters an error. This feature is called Fix-it.

Figure 2-6 shows a suggestion to remove the closing square brackets to match the number of opening brackets.

Figure 2-6 Xcode 4 Fix-it in action

What's New in LLVM 4

One other reason for Apple to move from GCC to LLVM is to have more control over the Objective-C programming language. With Xcode 4.4, Apple is aiming to make Objective-C more concise by adding literals. Apple can make this kind of language change only if it has control over the compiler, and LLVM gives Apple the power to change Objective-C.

Literals

Literals aren't something new, and Objective-C has had literals for a long time. Take a look at the declaration syntax of a simple `NSString`.

NSString

String Declaration

```
NSString *greeting = [[NSString alloc] initWithCString:"Hello World!"
    encoding:NSUTF8StringEncoding]; // without using literals

// second method using literals
NSString *greeting = @"Hello World!";
```

Now, note that, instead of using a method to create a string, we prefix it with an @ symbol. This kind of declaration is called a *literal*. In LLVM 4, Apple has added similar syntax for creating an `NSNumber` directly from scalar values, collection literals for `NSArray` and `NSDictionary` that help in concise creation of objects, and finally object subscripting that adds a concise syntax for accessing arrays and dictionaries. I show you these syntactic changes in the next couple of sections.

NSNumber from Scalars

`NSNumber` is a Foundation object while the numbers 1, 2, and 3 are scalars. To convert a scalar value to a Foundation object, you have to box the scalar value (in any language, not just Objective-C). You used to declare a `NSNumber` using the following syntax:

```
NSNumber *myNumber = [NSNumber numberWithInt:3];
NSNumber *yesValue = [NSNumber numberWithBOOL:YES];
```

With LLVM 4, you can simplify this couple of lines using the following syntax:

```
NSNumber *myNumber = @3;
NSNumber *yesValue = @YES;
```

The compiler automatically takes care of boxing the scalar values to Foundation objects. Note that you *shouldn't* wrap the scalars within double quotes. Doing so, like @"3", will make it an `NSString` whereas @3 is an `NSNumber`. By default, any scalar number is treated as a signed integer and any scalar decimal is treated as a double. As shown in the following code, you can instruct the compiler to convert the scalar to a float (instead of double) by suffixing the decimal with F and to a unsigned integer (instead of signed) by suffixing it with a U. On similar lines, L and LL, respectively, mean long and long long.

```
NSNumber *valueOfPi = @3.14F // declaring a float
NSNumber *radius = @3U // declaring a unsigned integer
```

If you're using Objective-C++, you can use `@true` and `@false` for declaring Boolean true and false.

The next literal that you learn will help you declare collections in a concise way.

Collection Literals (NSArray and NSDictionary)

Collection literals added to LLVM 4 allow you to declare an NSArray using the following syntax.

Creating an Array of Strings Using Literals

```
NSString *str1 = @"Hello";
NSString *str2 = @"World";
NSString *str3 = @"!";

NSArray *myArray1 = [NSArray arrayWithObjects:str1, str2, str3, nil];
// create a new array with old method
NSArray *myArray2 = @[str1, str2, str3]; // create a new array
```

The preceding code snippet shows how to create a new array using literals. myArray1 is created using the old method, and myArray2 is created using literals. With literals, you no longer have to nil terminate your arrays. The compiler does that automatically for you. If str2 or str3 were nil, and if the compiler could deduce that, a warning would be issued. Otherwise, a runtime error would occur.

You can combine collection literals with the NSNumber scalar conversion literals that you read about in the previous section to easily create an array of numbers, as illustrated in the following code snippet:

```
NSArray *arrayOfIntegers = @[@1, @2, @3, @4, @5];
```

Creating a dictionary also follows a similar syntax.

Creating a Dictionary Using Literals

```
NSDictionary *dictionary = @[@"key1":@"value1",
                             @"key2":@"value2",
                             @"key3":@"value3"];
```

Again, no element within this can be nil.

Object Subscripting (NSArray and NSDictionary)

In traditional C or C++, you access elements within an array using the subscript operator.

```
int a[5] = {1, 2, 3, 4, 5};
int firstElement = a[0];
```

Although this works for scalar arrays, until LLVM 4, NSArray and NSDictionary did not support subscript access to elements. For accessing an element within an NSArray, you use the objectAtIndex: method, and for accessing the value of an object associated with a key, you use the valueForKey: or objectForKey: method. With LLVM 4, Apple has introduced literals that allow subscripted object access. This is illustrated in the following code.

Accessing an Array Element Using a Subscript

```
NSArray *array = @[@1, @2, @3, @4, @5];
int elementAt3 = array[3];
```

LLVM 4 supports array-style subscripts that take integers as subscripts and dictionary-style subscripts that take a Foundation object pointer as a subscript. You use the dictionary style subscript to access values in a dictionary.

Accessing a Dictionary Element Using a Subscript

```
NSDictionary *dictionary = @[@"key1":@"value1",
                             @"key2":@"value2",
                             @"key3":@"value3"];
int elementAt3 = dictionary [@"key3"];
```

Boxed Literals

Clang 3.2 also supports something called boxed literals that allow enums to be accessed using a literal. As of this writing, Apple's LLVM compiler (including LLVM 4) doesn't support this feature. It might in a future release.

> Any array or dictionary created using literals is immutable by default. To create a mutable array, you have to resort back to `[NSMutableArray arrayWithObjects:…, nil]` or make a mutable copy using the `mutableCopy` method.

Literals and Availability

The new literals just explained are available from LLVM 4 onward. However, if you're a purist and want to write code that works with the older LLVM 3.1 and LLVM 4 compilers, you can use the __has_feature macro.

__has_feature Macro

```
#if __has_feature(objc_array_literals)
//code that uses LLVM 4 literals
#else
//fallback code that uses the older technique
#endif
```

Two other similar macros can be used to check the availability of dictionary literals and object subscripting, as illustrated in the following code.

Macro to Check Availability of Dictionary Literals and Object Subscripting

```
#if __has_feature(objc_dictionary_literals)
#endif
#if __has_feature(objc_subscripting)
#endif
```

I recommend avoiding conditional compilation. Since the compiler is available free of cost, most developers will be upgrading soon. If you still think you need to support the older LLVM 3.1 compiler, avoid using the new literal syntax altogether.

Auto Synthesis of Instance Variables

The LLVM 4 compiler adds one other nifty feature that automatically synthesizes ivars (instance variables) if you don't declare them explicitly.

For example, for the list of properties shown here

```
@property (strong, nonatomic) MPMoviePlayerController *moviePlayer;
@property (assign, nonatomic) int currentSegment;
@property (nonatomic, copy) void (^completionHandler)();
```

you would normally have to synthesize ivars in the implementation file, as illustrated here.

```
@synthesize moviePlayer = _moviePlayer;
@synthesize currentSegment = _currentSegment;
@synthesize completionHandler = _completionHandler;
```

With the new LLVM compiler, you can now type less code and can omit this ivar synthesis. The LLVM 4 compiler automatically synthesizes them for you. If you explicitly write a synthesizer, the LLVM 4 compiler is intelligent enough not to synthesize the ivar again. Do note that the default synthesizer generated by the LLVM 4 compiler has a leading underscore prefix.

The runtime method, `setValue:forKey:` used by KVC, writes to an ivar specified by key prefixed with a leading underscore if an ivar specified in the `forKey` argument is not declared. As such, `[myObject setValue:@1 forKey:@"number"];` will set the value of 1 to the ivar `_number` if an ivar named `number` (without the leading underscore) is not declared. The auto-synthesized ivars in LLVM 4.0 might be troublesome in some cases like lazy read/write ivars. In those cases, synthesize your ivars explicitly by prefixing them with a double underscore. This edge case applies to read-only properties as well. To prevent your read-only properties from being written to using runtime KVC methods, prefix their synthesizers with a double underscore. Apple uses the double underscore prefix for lazily loaded properties (`NSManagedObjectContext` and `NSPersistentStoreCoordinator`) in your application delegate if you use Core Data.

Git Your Versions Here

Another interesting feature addition to Xcode 4 is the integrated Git version control system. Git is a distributed version control system written by Linus Torvalds (yes, the same guy behind the Linux kernel) primarily to maintain the Linux kernel repository. The distributed nature, speed, reliability, cheap branching, and ability to easily do nonlinear software development encouraged more and more users to adopt Git.

Integrated Git Version Control System

Git is primarily a command-line system, much like most other UNIX/Linux systems. Don't fret. Xcode 4 has built-in support for Git, and the project navigator even shows the commit statuses of your file. However, the decision to use Git over any other version control system shouldn't be based solely on this. The main reasons I advocate using Git for your next iOS app are its cheap branching and its nonlinear development support.

Versions Editor

The versions editor is the third type of editor available in Xcode 4. (As mentioned previously in this chapter, the other two editor types are the standard editor and the assistant editor.) The integrated versions editor comes in handy when you want to visually analyze differences between two versions of a file. If your project uses Git (or SVN), you can compare a file with a previous revision from its repository. The versions editor allows you to pick any older version of a file by scrubbing through a timeline resembling the classical Time Machine UI. With OS X Lion, Apple might even consider adding local versions support to Xcode, and you would be able to compare local versions of files in addition to the versions in the repository.

Git Best Practices

Apple's AppStore is a walled garden, and if you're an active developer, Apple has probably rejected you at least once. Imagine a product development with, say, ten features. Out of these ten features, you develop four for version 1.0, another two for version 1.5, and the remaining four for version 2.0. You have submitted version 1.0 and are working on the fifth and sixth features. A couple of weeks into development, you get a reply from Apple that your app is rejected. Now, assume that the third feature violates some of Apple's policies and isn't allowed in its current state. In a traditional SCM system, you check out the old code, work on the fix, and submit the fix to Apple. You then come back to your code that has the fifth and sixth features added and painstakingly merge these bug fixes into your latest-and-greatest code. Although SVN and other source code control systems offer branching, it's quite hard to use, and as the size of projects grows, branching becomes an expensive operation (both timewise and disk-usagewise). With Git, this kind of merging, branching, and parallel development is very easy, mostly because of the way Git stores change sets.

> For a deeper introduction to Git, I recommend reading *Pro Git* or *Version Control with Git*. (See the "Further Reading" section at the end of the chapter for details.) The first book gives you an in-depth understanding about how Git works; the second helps you get started with and make the best use of Git in your project.

Follow these steps when using Git in your next iOS app:

1. Let your master branch reflect your latest-and-greatest code for the version available on the App Store.
2. For every new version you're working on, create a new branch.
3. For every major feature you implement, create a branch from the version branch.
4. Merge the feature branch with the version branch when the feature is complete.

5. Merge your version branch with the master whenever you submit your app to the App Store.

6. Optionally, tag your master branch after the app is approved.

When you follow these steps, you can easily fix bugs and issues with a particular version and merge your changes with the latest branch you're working on, all within couple of minutes.

For example, if Apple were to reject your app, all you would need to do is check out the master branch, make your fixes, resubmit to the App Store, check out your current working branch, and merge the changes you made in the master to it. With Git, nonlinear development gets really easy. Try using it. You will not regret it.

Schemes

The most powerful yet most confusing addition to Xcode 4 is schemes. In Xcode 3, there is a build configuration selection combo box, where you specify an active configuration, an active target, an active executable, the active architecture (instruction set), and the target device before running the app. Even the default set of options has an overwhelmingly high combination of selections, and for complicated project settings, choosing the right executable for the right target or device and instruction set becomes challenging. To top it off, Xcode 3 even allows choosing a wrong executable as active for a given target. Schemes have been introduced in Xcode 4 to help developers handle these issues easily. A scheme is a single entity that combines all of the above-mentioned settings. A scheme is a set of instructions for building a product. The product can be (in most cases will be) a collection of targets with its own build configurations. You can also use your existing `xcconfig` files for those targets. You learn about `xcconfig` configuration files later in this chapter.

Why Schemes?

The previous method of choosing four different options whenever you want to build something makes doing it right every time difficult. At times, you may have built and submitted the debug version of the app to the App Store or debugged the release configuration of the app, only to find your breakpoints weren't getting hit. With schemes in place, all you have to do is choose your scheme, and every other option is automatically applied. When you're building your product for debugging, you obviously don't want to strip off debug statements. On the other hand, when you're building for the App Store, you almost always want to optimize your build for performance and strip off debug statements. That holds good for Ad Hoc deployments as well. Wait! It doesn't end there. These schemes can also be shared among coworkers by committing into the repository.

Think of Schemes as Implementing Your Intentions

With schemes, you can automatically choose the correct configuration for a target by choosing a scheme that matches your *intent*. That is, if your intent is, "I want to debug this product," choose the Run scheme. If your intent is, "I want to submit this product to AppStore/Adhoc distribution," choose the Archive scheme.

With schemes, you select one option based on your intent. All your other settings are applied automatically. With some tweaks, you can customize the settings that are applied when a scheme is created. You will find out more about this in the next section.

Creating a Scheme

The easiest way to create a scheme is to let Xcode 4 automatically create one for you. When you open a project created on Xcode 3 in Xcode 4, it automatically creates a default scheme. Every scheme has its own unique settings panel that allows you to customize or tweak the default scheme setting. The following list discusses actions in a scheme.

- **Run**—The Run action builds the included targets using the debug build configuration. On the settings panel of this action, you can change the debugger you want to use (GDB or LLDB) and the build configuration, the default being Debug. The Run action's settings panel is where you specify command-line arguments, provide default data, or provide mock location data (using GPX files) to your app. You can also enable diagnostics-related arguments like Enabling Zombies or Guard Malloc from here (from Xcode 4.1 onward).

  To take it even further, you can duplicate the scheme and try different debuggers (and/or settings) on each.

- **Test**—The Test action runs your test targets. On the settings panel, you can customize which tests should be executed. Schemes are fully integrated with the OCUnit Objective-C testing framework and tests written will show up on the settings panel. Test failures show up on the issue and log navigators instead of the console, which means that navigating to the correct method that caused the test case to fail is now easier.

  By duplicating this scheme, you can create two test schemes: one testing your model classes and one testing, say, your helper methods.

- **Profile**—The Profile action builds your target and attaches it to Instruments. When you choose this, Instruments automatically launches and shows you the list of instruments available. You can edit the scheme to always launch the Time Profiler tool or the Leaks tool (or any other tool) automatically.

  You can duplicate the profile scheme so that you have two schemes: one launching Leaks and the other launching Time Profiler.

- **Analyze**—The Analyze action runs the Clang static analyzer on your code and warns you of potential memory leaks. There isn't much to customize here except the build configuration to use for this scheme.

- **Archive**—The Archive action is used for making `xcarchive` files (or `ipa` files) used for submitting to the App Store. Archives automatically appear on the Xcode organizer from which you can validate/submit to the App Store. With a dedicated Archive scheme, you're no longer required to create an Ad Hoc build configuration or an "App Store" build configuration for your product like you do in Xcode 3. These specific distribution configurations differ in most cases from the release configuration only by the signing certificate. Because the signing happens later, you can use the release configuration for archiving your apps. Signing it for submitting to the App Store or for Ad Hoc distribution is done through the Xcode organizer.

Sharing Your Schemes

By default, a scheme created by Xcode 4 is saved to the project bundle under the `xcuserdata` directory. Normally, this directory is excluded from repositories, which means schemes generated on your machine stay on your machine. In some cases, you might want to share schemes with coworkers. To do so, go to the Manage scheme options panel and select the Shared checkbox for every scheme you want to share. This is illustrated in Figure 2-7.

When you check the Shared option, your schemes are copied over to the `xcshareddata` directory. By adding this directory to your repository, you can share your custom schemes with coworkers.

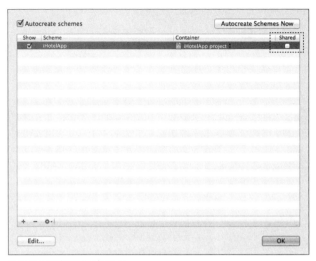

Figure 2-7 Sharing schemes

Schemes are actually a better way to customize your IDE or environment than Xcode 3's method of using multiple configurations. Give it a try and you will like it.

Build Configurations You Can Comment

In most projects, you depend on Xcode's build settings panel to edit/change your build settings. But this build settings panel has one major drawback: You cannot easily comment on a particular change you made on your project's configuration. Xcode provides an alternative way to do this by using xcconfig files.

Creating an xcconfig File

An xcconfig file is a plaintext file that contains build configurations for your target. Start by adding a Debug configuration file to your project. You can choose this from the templates in the new file wizard's Other section. Figure 2-8 illustrates this.

Name the file debug.xcconfig. Now open the build settings editor and select Basic and Levels as options. Copy these build settings to the configuration file you just created. You can select a row and use Cmd-C to copy and paste them in the configuration file.

Repeat these steps for the Release configurations in another file called release.xcconfig. Once you create this basic configuration, you can set all settings to default in the build settings editor.

Now, you need to tell Xcode to use this build configuration file instead of the specified build settings. To do so, select the project and then, in the project settings editor, choose the project again. In the Info panel, expand the configuration section and choose your config file.

That's it. You have now created a build configuration file that's readable and commentable that you can share with coworkers through your SCM. Ready to refactor this?

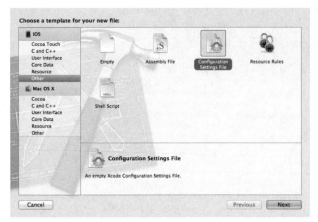

Figure 2-8 Adding a new configuration file

Refactoring the Build Configuration File

When you created the build configuration file, you probably noticed that many identical settings appear on both the debug and release configurations. You can easily avoid duplicating them by creating a `shared.xcconfig` file and copying those settings to it. Once you're done with the `shared.xcconfig` file, remove those entries from the `debug.xcconfig` and `release.xcconfig` files. Now use the `include` statement to add the `shared.xcconfig` entries to both files:

```
#include "shared.xcconfig"
```

This will automatically import all the shared settings into both the configuration files. Now when you run your app, everything should work. You can even add this to every new project you create or even to Xcode's new project templates.

Migrating Your Code to Objective-C ARC

ARC is easily the most important addition to iOS so far. Code compiled under ARC crashes less, and with ARC, it's very easy to avoid most of the common memory leaks and related crashes. __weak references can automatically ensure that your pointers are `nil`-ed as soon as they are deallocated. New technologies come in every day (both within and outside of Apple). But a majority of them don't end up getting adopted by developers because they require a steep learning curve. A technology, no matter how good it is, gets adopted only if the learning curve is not steep. To increase the adoption rate of ARC, Apple provides a tool built right into Xcode (starting from version 4.2) that helps you migrate your existing code base to ARC. Although ARC sounds great theoretically, it's not really that straightforward to use and a majority of open source code base (on sites like Github) hasn't yet been migrated to ARC.

In this section, I show you how to use the Convert to Objective-C ARC tool to convert your code base to ARC. Using the tool as such is straightforward. But dealing with the ARC-related compilation errors generated by the tool could be intimidating even for some prolific developers. A little understanding of how ARC works internally will help you understand these ARC-related compilation errors. You will read more details about ARC in Chapter 5. Chapter 28 covers some internals, including how ARC works under the hood.

Using the Convert to Objective-C ARC Tool

From Xcode's Edit menu, navigate to Refactor and click the Convert to Objective-C ARC option. Figure 2-9 illustrates this.

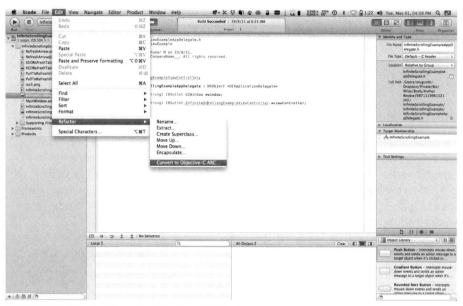

Figure 2-9 Invoking the Convert to Objective-C ARC tool

The Convert to Objective-C ARC tool opens with the first screen showing a list of targets in your project. ARC conversion is done for every target and not to individual classes or files. This means that, if you have already converted a target to ARC, converting another target within the same project (that is slightly different) is easier. Figure 2-10 shows the first step where you select the target and files within a target that need to be migrated to ARC.

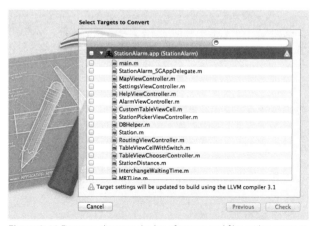

Figure 2-10 First step showing the list of targets and files within a target

Select the files that you want to migrate to ARC from here and click Check. The ARC migration tool performs a quick check, and if everything is fine, it converts your project and shows you a diff of the previous code to the new code. If you're okay with the changes made by ARC, click OK.

> **Superficially speaking, the ARC migration tool works by analyzing your project's code base and mostly removes calls to** `retain`**,** `release`**, and to your** `dealloc` **blocks. But ARC is a lot more than that. Chapter 5 covers ARC in depth and Chapter 28 covers the internals of ARC.**

Xcode 4 Organizer

Xcode 4 Organizer is a one-stop shop for anything related to Xcode that's not programming-specific. From Organizer, you can manage project repositories, perform SCM operations, and manage your application archives, provisioning profiles, and devices. Open the Organizer window by pressing Cmd-Shift-2 or clicking the rightmost button on your Xcode 4 toolbar. You will use the Xcode 4 Organizer mostly to access your application archives, submit your apps to the App Store, and manage your devices and provisioning profiles. Xcode 4 also has a new feature called Automatic Device Provisioning, discussed in the next section.

The first tab of Xcode 4 Organizer shows the list of devices and provisioning profiles currently loaded. Xcode 4 provides an easy way to export this list and import it on a new machine. If you ever want to migrate your developer settings to another computer, this is the place you should look for.

Automatic Device Provisioning

From the provisioning profiles list on the Devices tab, you can see a check box near the footer called Automatic Device Provisioning. When enabled, Xcode 4 can automatically download and install a developer certificate and a distribution certificate from your iOS developer program portal. Xcode 4 can also create a wild card provisioning profile (with an Implicit App ID) automatically, and that profile can be used for your apps that don't require an Explicit App ID.

> **Apps with any of the following features cannot use the implicit App ID and, hence, cannot depend on Xcode's Automatic Device Provisioning: Push Notification, Game Center, iCloud, and In App Purchases.**

Viewing Crash Logs and Console NSLog Statements

The Devices tab shows you a list of devices that have been connected to your development machine at least once. When you expand a device by clicking the disclosure triangle, you will be able to see Device logs and screenshots for that device. When a device is connected, you will see additional entries like Console logs, provisioning profiles installed on the device, and a list of applications provisioned. (This includes apps you run via Xcode or apps that you install via Ad Hoc distribution.)

Viewing an Application's Sandbox Data

With Xcode 4 (beginning with Xcode 4.2), you can view, delete, or add files to an app's sandbox inside a device. This makes debugging on the device easier. To access the device's sandbox, select the connected device from the left pane, choose Applications from the list, and choose the application for which you want to see the sandbox. Delete or add files from here or copy them locally to your computer.

Managing Repositories

Xcode 4 automatically adds the repositories for any project you have opened into the Repositories tab. The repositories section serves as a pretty good alternative for Git (or SVN) GUI access for most purposes.

If you're a "Unix-y" person and prefer to use the command line, I suggest you stay away from any GUI tools, and use them only for viewing diffs. A quick, lightweight tool I recommend is GitX. It has a command line tool to "pipe" Git diff output and shows you a visual diff. My workflow has always been like this:

```
git diff | gitx
```

Accessing Your Application Archives

You can access your application archives from the Xcode 4 Organizer and validate or submit your apps from there. In the previous section, you saw how to archive an application using the Archive scheme action. This archive can be accessed from the Archives tab of Xcode 4 Organizer.

Viewing Objective-C and SDK Documentation

Organizer also makes it easy to access the SDK documentation. The Documentation tab of Organizer shows the list of docsets installed. In most cases, that will be the latest two iOS SDKs, the latest two Mac OS X SDKs, and the current Xcode library.

Summary

In this chapter, you learned about the important features of Xcode 4 IDE. Then you learned about the LLVM compiler, including the new LLVM literals and the Migrate to Objective-C ARC tool. You learned the right way to use your version control system and learned about an under-used yet powerful way to comment your project's build settings using a separate configuration file. A major take-away from this chapter should be literals. The LLVM compiler is under heavy development right now, so there might be more literals and similar language and LLVM-related advancements coming in the near future.

Further Reading

Apple Documentation

The following documents are available in the iOS Developer Library at `developer.apple.com` or through the Xcode Documentation and API Reference.

Debugging with Xcode 4

Orientation to Xcode 4

Designing User Interfaces with Xcode 4

WWDC Sessions

The following session videos are available at `developer.apple.com`.

WWDC 2011, "Session 307: Moving to the Apple LLVM Compiler"

WWDC 2011, "Session 316: LLVM Technologies in Depth"

WWDC 2011, "Session 313: Mastering Schemes with Xcode 4"

WWDC 2012, "Session 402: Working Efficiently with Xcode"

Blogs

Cocoa Samurai. "Xcode 4 keyboard shortcuts now available!"
`http://cocoasamurai.blogspot.com/2011/03/xcode-4-keyboard-shortcuts-now.html`

Pilky.me. "Xcode 4: the super mega awesome review"
`http://pilky.me/view/15`

MK blog. (Mugunth Kumar). "Migrating your code to Objective-C ARC"
`http://blog.mugunthkumar.com/articles/migrating-your-code-to-objective-c-arc`

Cocoaphony: Mac and iPhone, on the brain. (Rob Napier). "Building the Build System—Part 1—Abandoning the Build Panel"
`http://robnapier.net/blog/build-system-1-build-panel-360`

Web Resources

GitX. (Mac OS X Git Client)
`http://gitx.frim.nl/`

LLVM. "Objective-C Literals"
`http://clang.llvm.org/docs/ObjectiveCLiterals.html`

The Pragmatic Studio. "Xcode 4 Shortcuts"
`http://pragmaticstudio.com/media/Xcode4Shortcuts.pdf`

Books

Chacon, Scott. *Pro Git* (Apress 2009; ISBN 978-1430218333).
`http://git-scm.com/book`

Loeliger, Jon. *Version Control with Git: Powerful Tools and Techniques for Collaborative Software Development* (O'Reilly Media 2009; ISBN 978-0596520120).
`http://oreilly.com/catalog/9780596520137`

Part II

Getting the Most Out of Everyday Tools

Chapter 3
Everyday Objective-C

This chapter covers many everyday best practices for Cocoa development, along with several underused features that more developers should be familiar with. Chapter 4 delves deeper into broad Cocoa patterns; here you focus on language features.

You begin by learning the critical Cocoa naming conventions that will improve your code's readability. Next you are introduced to memory management using Automatic Reference Counting (ARC). Then you find out how to best use properties and accessors to manage data in your objects. Finally, you learn about categories, extensions, and protocols, which are all commonly used throughout Cocoa.

By the end of this chapter, you should be very comfortable with the most important language features of Objective-C and feel confident that you're using the best practices of experienced Cocoa developers.

Naming Conventions

Throughout iOS, naming conventions are extremely important. If you understand how to read them correctly, the names of methods and functions throughout the iOS SDK tell you a great deal about how they are supposed to be called and what they do. Once you're used to the naming conventions, you can often guess what the name of a class or method is, making it much easier to find the documentation for it. This section touches on some of the most important naming convention rules and those that cause problems for developers with experience in other languages.

> The best source of information on Cocoa naming conventions is Apple's *Coding Guidelines for Cocoa*, which is available at `developer.apple.com`.

The first thing to know is that, in Cocoa, ease of reading is more important than ease of writing. Code spends much more of its life being read, maintained, and debugged than written. Cocoa naming conventions always favor the reader by striving for clarity over brevity. This is in stark contrast to C, which favors extremely terse naming. Because Objective-C is a dynamic language, the compiler provides far fewer safeguards than a static language such as C++. Good naming is a critical part of writing bug-free code.

The most important attribute of a good name is clarity. The names of methods should make it clear what types they accept and return. For instance, this method is extremely confusing:

```
- (void)add;  // Confusing
```

It looks like `add` should take a parameter, but it doesn't. Does it add some default object?

Names like these are much clearer:

```
- (void) addEmptyRecord;
- (void) addRecord: (Record *) record;
```

Now it's clear that `addRecord:` accepts a `Record`. The type of the object should match the name if there is any chance of confusion. For instance, this is a common mistake:

```
- (void) setURL: (NSString *) URL;   // Incorrect
```

It's incorrect because something called `setURL:` should accept an `NSURL`, not an `NSString`. If you need a string, then you need to add some kind of indicator to make this clear:

```
- (void) setURLString: (NSString *) string;
- (void) setURL: (NSURL *) URL;
```

This rule shouldn't be overused. It's better to have a property called `name` than `nameString`, as long as there is no `Name` class in your system that might confuse the reader.

Clear naming also means that you should avoid abbreviations in most cases. Use `backgroundColor` rather than `bgcolor`, and `stringValue` rather than `to_str`. There are exceptions to the use of abbreviations, particularly for things that are best known by their abbreviation. For example, `URL` is better than `uniformResourceLocator`. An easy way to determine whether an abbreviation is appropriate is to say the name out loud. You say "source" not "src." But most people say "URL" as either "u-ar-el" or "earl." No one says "uniform resource locator" in speech, so you shouldn't in code. There are a few abbreviations, such as `alloc`, `init`, `rect`, and `pboard`, that Cocoa uses for historical reasons that are considered acceptable. Apple has generally been moving away from even these abbreviations as it releases new frameworks.

Several kinds of variables are in a program: instance variables (ivars), static variables, automatic (stack) variables, and so on. It can be very difficult to understand code if you don't know what kind of variable you're looking at. Naming conventions should make the intent of a variable clear. After coding in many different styles with different teams, my recommendations are the following:

- Prefix static (package-scoped) variables with `s` and nonconstant global variables with `g`. Generally, you should avoid nonconstant globals. The following is a static declaration:

  ```
  static MYThing *sSharedInstance;
  ```

- Constants are named differently in Cocoa than in Core Foundation. In Core Foundation, constants are prefixed with a `k`. In Cocoa, they are not. File-local (static) constants should generally be prefixed with `k` in my opinion, but there is no hard-and-fast rule here. The following are examples of a file constant and a public constant:

  ```
  static const NSUInteger kMaximumNumberOfRows = 3;
  NSString * const MYSomethingHappenedNotification = @"SomethingHappened";
  ```

■ Method arguments are generally prefixed with an article such as *a*, *an*, or *the*. The last is less common and sometimes suggests a particularly important or unique object. Prefixing your arguments this way helps avoid confusing them with local variables and ivars. It is particularly helpful to avoid modifying them unintentionally.

■ Prefix instance variables (ivars) with an underscore. This is the standard that Xcode applies in recent versions and is consistent with automatic synthesis.

■ Classes should always begin with a capital letter. Methods and variables should always begin with a lowercase letter. All classes and methods should use camel case—never underscores—to separate words.

> **Cocoa and Core Foundation use slightly different naming conventions, but their basic approach is the same. For more information on Core Foundation naming, see Chapter 27.**

Cocoa naming is tightly coupled with memory management and object ownership. With ARC (discussed in the following section), memory management conventions are no longer as critical, but it is important to understand when working with non-ARC code. The naming convention is quite simple. An object can have one or more owners. When an object has no owners, it's destroyed (much like in garbage-collected systems). You take ownership whenever you use a method that begins with `alloc`, `new`, `copy`, or `mutableCopy`. You also take ownership whenever you call `retain`. When you're finished with an object, you relinquish ownership by calling `release` or `autorelease`. You use `autorelease` to delay the release of an object until the next time the autorelease pool drains (usually at the end of the event loop). This is mostly used for return values. You should always release objects that you own when you're done with them, and you should never release objects that you do not own.

The easiest way to implement this ownership system is to exclusively use properties to manage your ivars (except in `init` and `dealloc`), and to release all your properties in `dealloc`.

Even in ARC code, you need to be aware of this naming convention and avoid using `alloc`, `new`, `copy`, `mutableCopy`, `retain`, and `release` to mean anything other than their traditional meanings.

> **See *Advanced Memory Management Programming Guide* (`developer.apple.com`) for more details on manual memory management.**

Automatic Reference Counting

Automatic Reference Counting (ARC) greatly reduces the most common programmer error in Cocoa development: mismatching `retain` and `release`. ARC does not eliminate `retain` and `release`; it just makes them a compiler problem rather than a developer problem most of the time. In the vast majority of cases, this is a major win, but it's important to understand that `retain` and `release` are still going on. ARC is not the same thing as garbage collection. Consider the following code, which assigns a value to an ivar:

```
@property (nonatomic, readwrite, strong) NSString *title;
...
_title = [NSString stringWithFormat:@"Title"];
```

Without ARC, `_title` is underretained in the preceding code. The `NSString` assigned to it is autoreleased, so it will disappear at the end of the run loop, and the next time someone accesses `_title`, the program will crash. This kind of error is incredibly common and can be very difficult to debug. Moreover, if `_title` had a previous value, that old value has been leaked because it wasn't released.

Using ARC, the compiler automatically inserts extra code to create the equivalent of this:

```
id oldTitle = _title;
_title = [NSString stringWithFormat:@"Title"];
[_title retain];
[oldTitle release];
```

The calls to `release` and `retain` still happen, so there is a small overhead, and there may be a call to `dealloc` during the `release`. But generally this makes the code behave the way the programmer intended it to without creating an extra garbage collection step. Memory is reclaimed faster than with garbage collection, and decisions are made at compile time rather than at runtime, which generally improves overall performance. As with other compiler optimizations, the compiler is free to optimize memory management in various ways that would be impractical for the programmer to do by hand. ARC-generated memory management is often dramatically faster than the equivalent hand-coded memory management.

But this is not garbage collection. In particular, it cannot handle reference (retain) loops the way Snow Leopard garbage collection can. For example, the object graph in Figure 3-1 shows a retain loop between Object A and Object B:

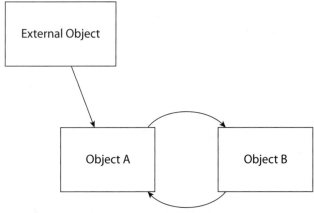

Figure 3-1 A retain loop

If the link from "External Object" to "Object A" is broken, then under Snow Leopard garbage collection, both Object A and Object B will be destroyed because they are orphaned from the program. Under ARC, Object A and Object B will not be destroyed because each still has a retain count greater than zero. So in iOS, you need to keep track of your strong relationships to avoid reference loops.

Property relationships have two main forms: `strong` and `weak`, which map to the former `retain` and `assign`. As long as there is a strong reference to an object, it will continue to exist. This is nearly identical to `shared_ptr` in C++, except that the code to manage the reference counts is injected by the compiler rather than determined at runtime with operator overloads.

Objective-C has always had the problem of reference loops, but they really don't come up that often in practice. Anywhere you would have used an `assign` property in the past, use a `weak` property under ARC, and you should be fine. Most reference loops are caused by delegates, and a `delegate` property should almost always be `weak`. Weak references have the advantage of automatically being set to `nil` when the referenced object is destroyed. This is a significant improvement over `assign` properties, which can point to freed memory.

> Prior to ARC, the default storage class for synthesized properties was `assign`. The default storage class is now `strong`. This can cause some confusion when converting code. I recommend always providing a storage class for properties.
>
> The other major cause of retain loops is blocks. See Chapter 22 for more information on blocks.

There are two major changes when switching to ARC for most code:

- Don't use `retain`, `release`, or `autorelease`. You can just delete these. ARC should do the right thing.
- If your `dealloc` only releases ivars, you don't need `dealloc`. This will be done automatically for you, and you can't call `release` in any case. If you still need `dealloc` to do other things (remove KVO observations, for instance), don't call `[super dealloc]`. This last change is surprising, but the compiler will give you errors if you forget.

As noted previously, ARC is not garbage collection. It is a compiler feature that injects calls to `retain` and `release` at appropriate places in the code. This means that it is fully interoperable with existing, manual memory management code, as long as all the code uses the correct naming conventions. For example, if you call a method named `copySomething`, ARC will expect the result of that method to have a +1 retain count. If needed, it will insert a balancing `release`. It doesn't matter to ARC whether that +1 retain count was created by ARC code inside of `copySomething` or by manual memory management inside of `copySomething`.

This breaks if you violate Cocoa's naming conventions. For instance, if you have a method that returns the copyright notice as an autoreleased string and call the method `copyRight`, then how ARC behaves depends on whether the calling and called code are both compiled with ARC.

ARC looks at the name `copyRight`, sees that it begins with `copy`, and so assumes that it returns a +1 retain count object. If `copyRight` is compiled with ARC and the calling code is compiled with ARC, everything will still work. ARC will inject an extra retain in `copyRight` because of its name, and it will inject an extra release in the calling code. It may be a little less efficient, but the code will neither crash nor leak.

If, however, the calling code is compiled with ARC, but `copyRight` is not, then the calling code will inject an extra release, and the code will crash. If the calling code is not compiled with ARC, but `copyRight` is, then ARC will inject an extra retain, and the code will leak.

The best solution to this problem is to follow Cocoa's naming conventions. In this example, you could name this method `copyright` and avoid the problem entirely. ARC determines the memory management rules based on whole camel case words in the method.

If renaming an incorrect method is impossible, you can add the attribute `NS_RETURNS_RETAINED` or `NS_RETURNS_NOT_RETAINED` to your method declaration to tell the compiler which memory management rule to use. These are defined in `NSObjCRuntime.h`.

ARC introduces four restrictions on your code so that it can properly add `retain` and `release` calls for you:

- **No calls to** `retain`, `release`, **or** `autorelease`—This is usually the easiest rule. Just delete them. It also means that you cannot override these methods, but you should almost never do that anyway. If you were overriding these methods to implement the Singleton pattern, see Chapter 4 for information about how to properly implement this pattern without overriding these methods.

- **No object pointers in C structs**—This seldom comes up, but if you have been storing an object in a C struct, you either need to store it in an object or cast it to `void*` (see the next rule for more information on casting to `void*`). C structs can be destroyed at any time by calling `free`, and this interferes with automatically tracking objects that are stored in them.

- **No casting between** `id` **and** `void*` **without a bridging cast**—This mostly impacts Core Foundation code. See Chapter 27 for full details on bridging casts and how to use them with ARC.

- **No** `NSAutoreleasePool`—Rather than creating your own autorelease pools by hand, just wrap any code you want to have its own pool in a `@autoreleasepool{}` block. If you had special code to control when you drained your pool, it is almost certainly unnecessary. `@autoreleasepool` is up to 20 times faster than `NSAutoreleasePool`.

Most code will have no problem with these rules. There is a new tool in Xcode under the Edit→Refactor menu called Convert to Objective-C ARC…. It will do the majority of the work for you.

ARC is perhaps the greatest advancement in Objective-C since the autorelease pool. If at all possible, you should convert your code to ARC. If you can't convert everything, convert as much as you can. It is faster, less buggy, and easier to write than manual memory management. Switch to ARC today.

Properties

Objective-C 2.0 introduced several interesting changes. A key improvement was non-fragile ivars. This allows classes to add ivars without recompiling their subclasses. This feature mostly affects framework developers like Apple rather than application developers, but it has some useful side effects. The most popular is synthesized properties.

With the most recent versions of iOS, the compiler will automatically synthesize ivars and the necessary methods for your properties. There is generally no need to use `@synthesize` anymore.

My recommendation is to switch entirely to properties and automatically synthesized ivars. Put your public properties in the header and your private properties in an extension in the `.m` file. So a full example might look like this:

MyClass.h

```
@interface MyClass : NSObject
@property (nonatomic, readwrite, weak) id delegate;
@property (nonatomic, readonly, strong) NSString *readonlyString;
@end
```

MyClass.m

```
@interface MyClass () // Private methods
@property (nonatomic, readwrite, strong) NSString *readonlyString;
@property (nonatomic, readwrite, strong) NSString *privateString;
@end
```

This automatically creates ivars `_delegate`, `_readonlyString`, and `_privateString`. Use these ivars only in `init`, `dealloc` (if you have one), and in overridden accessors.

Note how `readonlyString` is redefined in the class extension to be `readwrite`. This allows you to create a private setter.

Property Attributes

While I'm discussing properties, I want you to also consider the attributes you apply to your properties. Consider each category in turn.

- **Atomicity** (`atomic`, `nonatomic`; **LLVM 4 added the** `atomic` **attribute to match** `nonatomic`.)— This is an easy attribute to misunderstand. Its purpose is to make setting and getting the property thread-safe. That does not mean that the underlying object is thread-safe. For instance, if you declare an `NSMutableArray` property called `stuff` to be atomic, then `self.stuff` is thread-safe and `self.stuff=otherStuff` is thread-safe. But accessing the array with `objectAtIndex:` is not thread-safe. You will need additional locking to handle that. The `atomic` attribute is implemented similar to this:

```
[_propertyLock lock];
id result = [[value retain] autorelease];
[_propertyLock unlock];
return result;
```

The pattern of `retain/autorelease` ensures that the object will not be destroyed until the caller's autorelease pool drains. This protects the caller from other threads releasing the object in the middle of access. Managing the lock and calling `retain` and `autorelease` can be expensive (though atomic properties are much cheaper with ARC). If you will never access this property from another thread, or if you need more elaborate locking anyway, then this kind of atomicity is wasteful. It turns out that this is the case most of the time, and you usually want to use `nonatomic`. In the fairly small number of cases where this kind of atomicity is useful, there unfortunately is no way to call it out because there is no `atomic` attribute. It's helpful to add a comment in these cases to make it clear that you're making the property atomic on purpose.

ARC provides significant performance benefits to atomic properties, and best practices regarding `nonatomic` may change in the future.

- **Writability** (`readwrite`, `readonly`)—These should be fairly self-explanatory. If a property is `readonly`, then only a getter will be available. If it is `readwrite`, then both a setter and getter will be available. There is no `writeonly` attribute.

- **Setter semantics** (`weak`, `strong`, `copy`)—These should be fairly obvious, but there are some things to consider. First, you often should use `copy` for immutable classes such as `NSString` and `NSArray`. It's possible that there is a mutable subclass of your property's class. For instance, if you have an `NSString` property, you might be passed an `NSMutableString`. If that happens and you hold only a reference to the value (`strong`), your property might change behind your back as the caller mutates it. That often isn't what you want, and so you will note that most `NSString` properties use the `copy` semantic. The same is also usually true for collections such as `NSArray`. Copying immutable classes is generally very fast because it can almost always be implemented with `retain`.

Property Best Practices

Properties should represent the state of the object. Getters should have no externally visible side effects (they may have internal side effects such as caching, but those should be invisible to callers). Generally, they should be efficient to call and certainly should not block.

Private Ivars

Although I prefer properties for everything, some people prefer ivars, especially for private variables. You can declare ivars in the `@implementation` block like this:

```
@implementation Something {
    NSString *_name;
}
```

This syntax moves the private ivar out of the public header, which is good for encapsulation, and keeps the public header easier to read. ARC automatically retains and releases ivars, just like other variables. The default storage class is `strong`, but you can create weak ivars as shown here:

```
@implementation Something {
    __weak NSString *name;
}
```

Accessors

Avoid accessing ivars directly. Use accessors instead. I will discuss a few exceptions in a moment, but first I want to discuss the reasons for using accessors.

Prior to ARC, one of the most common causes of bugs was failure to use accessors. Developers would fail to retain and release their ivars correctly, and the program would leak or crash. Because ARC automatically manages retains and releases, some developers may believe that this rule is no longer important, but there are other reasons to use accessors.

- **Key-value observing**—Perhaps the most critical reason to use accessors is that properties can be observed. If you don't use accessors, you need to make calls to `willChangeValueForKey:` and `didChangeValueForKey:` every time you modify a property. Using the accessor will automatically call these when they are needed.

- **Side effects**—You or one of your subclasses may include side effects in the setter. There may be notifications posted or events registered with `NSUndoManager`. You shouldn't bypass these unless it's necessary. Similarly, you or a subclass may add caching to the getter that direct ivar access will bypass.

- **Locking**—If you introduce locking to a property in order to manage multithreaded code, direct ivar access will violate your locks and likely crash your program.

- **Consistency**—One could argue that you should just use accessors when you know you need them for one of the preceding reasons, but this makes the code very hard to maintain. It's better that every direct ivar access be suspicious and explained instead of having to constantly remember which ivars require accessors and which do not. This makes the code much easier to audit, review, and maintain. Accessors, particularly synthesized accessors, are highly optimized in Objective-C, and they are worth the overhead.

That said, you don't use accessors in a few places:

- **Inside of accessors**—Obviously, you cannot use an accessor within itself. Generally, you also don't want to use the getter inside of the setter either (this can create infinite loops in some patterns). An accessor may speak to its own ivar.

- **Dealloc**—ARC greatly reduces the need for `dealloc`, but it still comes up sometimes. It's best not to call external objects inside of `dealloc`. The object may be in an inconsistent state, and it's likely confusing to the observer to receive several notifications that properties are changing when what's really meant is that the entire object is being destroyed.

- **Initialization**—Similar to `dealloc`, the object may be in an inconsistent state during initialization, and you generally shouldn't fire notifications or have other side effects during this time. This is also a common place to initialize read-only variables like an `NSMutableArray`. This avoids declaring a property `readwrite` just so you can initialize it.

Accessors are highly optimized in Objective-C and provide important features for maintainability and flexibility. As a general rule, you refer to all properties, even your own, using their accessors.

Categories and Extensions

Categories allow you to add methods to an existing class at runtime. Any class, even Cocoa classes provided by Apple, can be extended with categories, and those new methods will be available to all instances of the class. This approach was inherited from Smalltalk and is somewhat similar to extension methods in C#.

Categories were designed to break up large classes into more manageable pieces, hence the name. If you look at large Foundation classes, you will find that sometimes they're broken into several pieces. For instance, NSArray includes the NSExtendedArray, NSArrayCreation, and NSDeprecated categories defined in NSArray.h, plus the NSArrayPathExtensions category defined in NSPathUtilities.h. Most of these are split up to make it simpler to implement in multiple files, but some categories, such as the UIStringDrawing category on NSString, exist specifically to allow different code to be loaded at runtime. On Mac, AppKit loads the NSStringDrawing category. On iOS, UIKit loads the UIStringDrawing category. This provides a more elegant way to split up the code than #ifdef does. On each platform, you simply compile the appropriate implementation (.m) files, and the functionality becomes available.

Prior to Objective-C 2.0, @protocol definitions could not include optional methods. Developers used categories as "informal protocols." The complier knows the methods defined in a category, but it will not generate a warning if the methods aren't implemented. This state of affairs made all the protocol's methods optional. I discuss this further in the section "Formal and Informal Protocols," later in this chapter, but for iOS I do not recommend this use of categories. Formal protocols now support optional methods directly.

Because the compiler will not check that you have implemented methods in the category, using categories solely to break up large classes has tradeoffs. An implementation file that's getting overly large is often an indication that you need to refactor your class to make it more focused rather than define categories to split it up. But if your class is correctly scoped, you may find splitting up the code with categories is convenient. On the other hand, using categories can scatter the methods into different files, which can be confusing, so use your best judgment.

Declaration of a category is straightforward. It looks like a class interface declaration with the name of the category in parentheses:

```
@interface NSMutableString (Capitalize)
- (void)capitalize;
@end
```

Capitalize is the name of the category. Note that no ivars are declared here. Categories cannot declare ivars, nor can they synthesize properties (which is the same thing). You'll see how to add category data in the section "Category Data Using Associative References." Categories can *declare* properties, because this is just another way of declaring methods. They just can't synthesize properties, because that creates an ivar. The Capitalize category doesn't require that capitalize actually be implemented anywhere. If capitalize isn't implemented and a caller attempts to invoke it, the system will raise an exception. The compiler gives you no protection here. If you do implement capitalize, then *by convention* it looks like this:

```
@implementation NSMutableString (Capitalize)
- (void)capitalize {
   [self setString:[self capitalizedString]];
}
@end
```

I say "by convention" because there is no requirement that this be defined in a category implementation or that the category implementation must have the same name as the category interface. However, if you provide an `@implementation` block named `Capitalize`, then it must implement all the methods from the `@interface` block named `Capitalize`. Adding the parentheses and category name after the class name allows you to continue adding methods in another compile unit (.m file). You can implement your category methods in the main `implementation` block, in a named category `implementation` block for the class, or not implement them at all.

Technically a category can override methods, but that's dangerous and not recommended. If two categories implement the same method, then it is undefined which one is used. If a class is later split into categories for maintenance reasons, your override could become undefined behavior, which is a maddening kind of bug to track down. Moreover, using this feature can make the code hard to understand. Category overrides also provide no way to call the original method. I recommend against using categories to override existing methods, except for debugging. Even for debugging, I prefer swizzling, which is covered in Chapter 28.

> Because of the possibility of collisions, it's often a good idea to add a prefix to your category methods if there is any chance that it might collide with Apple or with other categories in your project (such as third-party libraries). For example, you could use the name `MY_capitalize`, which would be safer. Cocoa generally doesn't use embedded underscores like this, but in this case, it's clearer than the alternatives. Whether you do this or not is somewhat a matter of your risk tolerance. I once added a category method `pop` on `NSMutableArray`. This collided with a private Apple category that the authors of `UINavigationController` had added. Since Apple and I implemented our versions of `pop` differently, the resulting bugs were extremely difficult to track down.

A very good use of categories is to provide utility methods to existing classes. When doing this, I recommend naming the header and implementation files using the name of the original class plus the name of the extension. For example, you might create a simple `MyExtensions` category on `NSDate`:

NSDate+MYExtensions.h

```
@interface NSDate (MYExtensions)
- (NSTimeInterval)timeIntervalUntilNow;
@end
```

NSDate+MYExtensions.m

```
@implementation NSDate (MYExtensions)
- (NSTimeInterval)timeIntervalUntilNow {
   return [self timeIntervalSinceNow];
}
@end
```

If you have only a few utility methods, it's convenient to put them together into a single category with a name such as `MYExtensions` (or whatever prefix you use for your code). Doing so makes it easy to drop your favorite extensions into each project. Of course, this is also code bloat, so be careful about how much you throw into a "utility" category. Objective-C can't do dead-code stripping as effectively as C or C++.

If you have a large group of related methods, particularly a collection that might not always be useful, it's a good idea to break those into their own category. Look at `UIStringDrawing.h` in UIKit for a good example.

+load

Categories are attached to classes at runtime. It's possible that the library that defines a category is dynamically loaded, so categories can be added quite late. (Although you can't write your own dynamic libraries in iOS, the system frameworks are dynamically loaded and include categories.) Objective-C provides a hook called `+load` that runs when the category is first attached. Like `+initialize`, you can use this to implement category-specific setup such as initializing static variables. You can't safely use `+initialize` in a category because the class may implement this already. If multiple categories implemented `+initialize`, the one that would run wouldn't be defined.

Hopefully, you're ready to ask the obvious question: "If categories can't use `+initialize` because they might collide with other categories, what if multiple categories implement `+load`?" This turns out to be one of the few really magical parts of the Objective-C runtime. The `+load` method is special-cased in the runtime so that every category may implement it and all the implementations will run. There are no guarantees on order, and you shouldn't try to call `+load` by hand.

`+load` is called regardless of whether the category is statically or dynamically loaded. It's called when the category is added to the runtime, which often is at program launch, before `main`, but could be much later.

Classes can have their own `+load` method (not defined in a category), and those will be called when the classes are added to the runtime. This is seldom useful unless you're dynamically adding classes.

You don't need to protect against `+load` running multiple times the way you do with `+initialize`. The `+load` message is only sent to classes that actually implement it, so you won't accidentally get calls from your subclasses the way you can in `+initialize`. Every `+load` will be called exactly once. You shouldn't call `[super load]`.

Category Data Using Associative References

Although categories can't create new ivars, they can do the next best thing: They can create associative references. Associative references allow you to attach key-value data to arbitrary objects.

Consider the case of a `Person` class. You'd like to use a category to add a new property called `emailAddress`. Maybe you use `Person` in other programs, and sometimes it makes sense to have an email address and sometimes it doesn't, so a category can be a good solution to avoid the overhead when you don't need it. Or maybe you don't own the `Person` class, and the maintainers won't add the property for you. In any case, how do you attack this problem? First, just for reference, take a look at the `Person` class:

```
@interface Person : NSObject
@property (nonatomic, readwrite, copy) NSString *name;
@end

@implementation Person
@end
```

Now you can add a new property, emailAddress, in a category using an associative reference:

```
#import <objc/runtime.h>
@interface Person (EmailAddress)
@property (nonatomic, readwrite, copy) NSString *emailAddress;
@end

@implementation Person (EmailAddress)

static char emailAddressKey;

- (NSString *)emailAddress {
  return objc_getAssociatedObject(self, &emailAddressKey);
}

- (void)setEmailAddress:(NSString *)emailAddress {
  objc_setAssociatedObject(self, &emailAddressKey,
                           emailAddress,
                           OBJC_ASSOCIATION_COPY);
}
@end
```

Note that associative references are based on the key's memory address, not its value. It does not matter what is stored in emailAddressKey; it only needs to have a unique address. That's why it's common to use an unassigned static char as the key.

Associative references have good memory management, correctly handling copy, assign, or retain semantics according to the parameter passed to objc_setAssociatedObject. They are correctly released when the related object is deallocated.

Associative references are a great way of attaching a relevant object to an alert panel or control. For example, you can attach a "represented object" to an alert panel, as shown in the following code. This code is available in the sample code for this chapter.

ViewController.m (AssocRef)

```
id interestingObject = ...;
UIAlertView *alert = [[UIAlertView alloc]
                       initWithTitle:@"Alert" message:nil
                       delegate:self
                       cancelButtonTitle:@"OK"
                       otherButtonTitles:nil];
objc_setAssociatedObject(alert, &kRepresentedObject,
```

(continued)

```
                  interestingObject,
         OBJC_ASSOCIATION_RETAIN_NONATOMIC);
[alert show];
```

Now, when the alert panel is dismissed, you can figure out why you cared:

```
- (void)alertView:(UIAlertView *)alertView
clickedButtonAtIndex:(NSInteger)buttonIndex {
  UIButton *sender = objc_getAssociatedObject(alertView,
                                   &kRepresentedObject);
  self.buttonLabel.text = [[sender titleLabel] text];
}
```

Many programs handle this with an ivar in the caller, such as `currentAlertObject`, but associative references are much cleaner and simpler. For those familiar with Mac development, this is similar to `representedObject`, but more flexible.

One limitation of associative references (or any other approach to adding data via a category) is that they don't integrate with `encodeWithCoder:`, so they're difficult to serialize via a category.

Class Extensions

Objective-C 2.0 adds a useful twist on categories, called *class extensions*. These are declared exactly like categories, except the name of the category is empty:

```
@interface MYObject ()
- (void)doSomething;
@end
```

Class extensions are a great way to declare private methods inside of your `.m` file. The difference between a category and an extension is that methods declared by an extension are exactly the same as methods declared in the main interface. The compiler will make sure you implement them all, and they will be added to the class at compile time rather than at runtime as categories are. You can even declare synthesized properties in extensions.

Formal and Informal Protocols

Protocols are an important part of Objective-C, and in Objective-C 2.0, formal protocols have become common. In Objective-C 1.0, there was no `@optional` tag for protocol methods, so all methods were mandatory. It's rare that this kind of protocol is useful. Often, you want some or all of the protocol to be optional. Because this wasn't possible in Objective-C 1.0, developers commonly used "informal protocols," and sometimes you'll still come across these.

An informal protocol is a category on `NSObject`. Categories tell the compiler that a method exists, but don't require that the method be implemented. This technique allowed developers to document the interface and prevent compiler warnings, while indicating that any child of `NSObject` could implement the methods. This isn't a great approach to defining an interface, but in Objective-C 1.0, it was the best there was.

With Objective-C 2.0, formal protocols can declare optional methods, and many informal protocols on Mac are migrating to formal protocols. Luckily, iOS has always used Objective-C 2.0, so formal protocols are the norm.

Most developers are familiar with how to declare that a class implements a formal protocol. You simply include the protocols in angle brackets after the superclass:

```
@interface MyAppDelegate : NSObject <UIApplicationDelegate,
                                     UITableViewDatasource>
```

Declaring a protocol is similarly easy:

```
@protocol UITableViewDataSource <NSObject>

@required
- (NSInteger)tableView:(UITableView *)tableView
  numberOfRowsInSection:(NSInteger)section;
- (UITableViewCell *)tableView:(UITableView *)tableView
        cellForRowAtIndexPath:(NSIndexPath *)indexPath;

@optional
- (NSInteger)numberOfSectionsInTableView:(UITableView *)tv;
- (NSString *)tableView:(UITableView *)tableView
  titleForHeaderInSection:(NSInteger)section;
...
```

Note some important points in this example. First, protocols can inherit just like classes. The UITableViewDataSource protocol inherits from the <NSObject> protocol. Your protocols should almost always inherit from <NSObject>, just as your classes inherit from NSObject.

> NSObject **is split into both a class and a protocol. This is primarily to support** NSProxy**, which inherits from the protocol, but not the class.**

For delegate protocols, the delegating object is always the first parameter. This is important because it allows a single delegate to manage multiple delegating objects. For instance, one controller could be the delegate for multiple UIAlertView instances. Note the slight difference in naming convention when there are parameters other than the delegating object. If no other parameters exist, the class name comes last (numberOfSectionsInTableView:). If there are other parameters, the class name comes first as its own parameter (tableView:numberOfRowsInSection:).

Once you've created your protocol, you will often need a property to hold it. The typical type for this property is id<Protocol>:

```
@property(nonatomic, readwrite, weak) id<MyDelegate> delegate;
```

This means "any object that conforms to the MyDelegate protocol." It's possible to declare both a specific class and a protocol, and it's possible to declare multiple protocols in the type:

```
@property(nonatomic, readwrite, weak) MyClass* <MyDelegate,
                            UITableViewDelegate> delegate;
```

This indicates that `delegate` must be a subclass of `MyClass` and must conform to both the `<MyDelegate>` and `<UITableViewDelegate>` protocols.

Protocols are an excellent alternative to subclassing in many cases. A single object can conform to multiple protocols without suffering the problems of multiple inheritance (as found in C++). If you are considering an abstract class, a protocol is often the better choice. Protocols are extremely common in well-designed Cocoa applications.

Summary

Much of good Objective-C is "by convention" rather than enforced by the compiler. This chapter covers several of the important techniques you'll use every day to get your programs to the next level. Conforming to Cocoa's naming conventions will greatly improve the reliability and maintainability of your code, and give you key-value coding and observing for free. Correct use of properties will make memory management easy, especially since the addition of ARC. And good use of categories and protocols will keep your code easy to understand and extend.

Further Reading

Apple Documentation

The following documents are available in the iOS Developer Library at `developer.apple.com` or through the Xcode Documentation and API Reference.

Coding Guidelines for Cocoa

Programming With Objective-C

Other Resources

Cocoa with Love. (Matt Gallagher), "Method names in Objective-C"
`cocoawithlove.com/2009/06/method-names-in-objective-c.html`

CocoaDevCentral. (Scott Stevenson), "Cocoa Style for Objective-C,"
`cocoadevcentral.com/articles/000082.php`

Chapter 4
Hold On Loosely: Cocoa Design Patterns

If you're like most iOS developers, Objective-C is not your first language. You probably have a background in other object-oriented languages such as Java, C++, or C#. You may have done development in C. None of these languages really prepare you for how to think in Objective-C.

In the beginning, there was Simula, and Simula had two children: C++ from Bell Labs and Smalltalk from Xerox PARC. From C++ sprang Java, which tried to make things easier. Microsoft wrote Java.NET and called it C#. Today, most developers are trained in this branch of Simula. Its patterns include generic programming, static typing, customization through subclassing, method calling, and strong object ownership.

Objective-C and Cocoa come from the Smalltalk fork. NeXT developed a framework called NeXTSTEP. It was written in Objective-C and implemented many of Smalltalk's patterns. When Apple brought NeXTSTEP to the Mac, Apple renamed it Cocoa, although the NS prefix remains to this day. Cocoa has very different patterns, and this is what sometimes gives new developers trouble. Common Cocoa patterns include protocols, dynamic typing, customization through delegation, message passing, and shared object ownership.

I'm not going to give a computer science history lesson here, but it's important to understand that Objective-C is not Java and it's not C++. It's really Smalltalk. Because few developers learn Smalltalk, most need to adjust their thinking to get the most out of Objective-C.

In this chapter, I use the terms Objective-C and Cocoa interchangeably. Technically, Objective-C is a language, and Cocoa is a collection of frameworks implemented in Objective-C. In principle, you could use Objective-C without Cocoa, but in practice, this is never done. In the following sections, you find out about the major Cocoa design patterns and how best to apply them in your programs.

The pattern names used in this chapter come from the book *Design Patterns* (Addison-Wesley Professional 1994. ISBN: 978-0201633610) by Eric Gamma, Richard Helm, Ralph Johnson, and John Vlissides—sometimes called "The Gang of Four." Apple maps its patterns to the *Design Pattern* names in the chapter, "Cocoa Design Patterns," in the *Cocoa Fundamentals Guide* (see the "Further Reading" section at the end of this chapter).

Understanding Model-View-Controller

The most important pattern in Smalltalk and Cocoa is called *model-view-controller (MVC)*. This is an approach to assigning responsibilities within a program. *Model* classes are responsible for representing information. *View* classes are responsible for interfacing with the user. *Controller* classes are responsible for coordinating between models and views.

> There are subtle differences between how Smalltalk and Cocoa implement MVC. This chapter
> discusses only how Cocoa uses MVC.

Using Model Classes

A good model class encapsulates a piece of data in a presentation-independent way. A classic example of
a good model class is `Person`. A `Person` might have a name, an address, a birth date, and an image. The
`Person` class, or related model classes, would encapsulate storing and retrieving related information from a
data source, but would have no display or editing features. The same `Person` class should be easily reusable
on an iPhone, an iPad, a Mac, or a command-line program. Model classes should reference only other model
classes. They should never reference views or controllers. A model class might have a delegate that happens to
be a controller, but it should implement this using a protocol so that it does not need to reference the specific
controller class.

Model class names are generally simple nouns like `Person`, `Dog`, and `Record`. You often include a two- or
three-letter prefix, such as `RNPerson`, to identify them as your code and prevent collisions.

Model classes can be mutable or immutable. An *immutable* class cannot change once it's created. `NSString` is
a good example of this. A *mutable* class like `NSMutableString` can change after it's created. In this context,
"change" refers only to changes that are visible outside the object. It doesn't matter if internal data structures
like caches change.

Immutable objects offer many advantages. They can save time and memory. Immutable objects can implement
`copy` by calling `retain`. Because it's impossible for the object to change, you don't have to make a real copy.
Immutable objects are inherently thread-safe without locking, which makes them much faster and safer to
access in multithreaded code. Because everything is configured at initialization time, it's much easier to ensure
the object is always in a consistent state. You should use immutable model classes unless there is a reason to
make them mutable.

Model classes are often the most testable and reusable classes in the system. Designing them well is one of
the best ways to improve the overall quality of your code. Historically, Apple sample code has not included
well-designed model classes, which confuses new developers who are led to believe that controllers (or
worse, views) are supposed to hold data. More recent sample code from Apple has improved, and the
example project, `The Elements`, includes good examples of model classes. Look at `AtomicElement` and
`PeriodicElements`. (See the "Further Reading" section at the end of this chapter.)

Using View Classes

View classes are responsible for interfacing with the user. They present information and accept user events.
(This is the biggest deviation from Smalltalk MVC, where controller classes are responsible for user events.)
View classes should not reference controller classes. As with model classes, view classes may have a delegate
that happens to be a controller, but they shouldn't reference the controller class directly. They also shouldn't
reference other views, except their immediate superview and subviews. Views may reference model classes, but
generally only the specific model object that they are displaying. For instance, a `PersonView` object might
display a single `Person` object. It's easier to reuse view objects that do not reference custom model objects.

For instance, a `UITableViewCell` is highly reusable because it displays only strings and images. There is sometimes a trade-off between reusability and ease-of-use in view objects, and finding the right balance is an important part of your architecture. In my experience, specialized views that handle a specific model class are often very useful for application writers. Framework writers, such as the UIKit team, need to emphasize reusability.

Model-specific view class names often append `View` to the model class, such as `PersonView` or `RecordView`. You should do this only if the view is a subclass of `UIView`. Some kinds of view classes have special names. Reusable views are generally called *cells,* such as `UITableViewCell` on iOS or `NSCell` on Mac. Lightweight, hardware-optimized view classes are generally called *layers,* such as `CALayer` or `CGLayer`. Whether or not they are subclasses of `UIView` or `NSView`, they are still MVC view classes.

Views are responsible for accepting events from users, but not for processing them. When a user touches a view, the view may respond by alerting a delegate that it has been touched, but it should not perform logic or modify other views. For example, pressing a Delete button should simply tell a delegate that the Delete button has been pressed. It should not tell the model classes to delete the data, nor tell the table view to remove the data from the screen. Those functions are the responsibility of a controller.

Using Controller Classes

Between the models and the views lie the controllers, which implement most of the application-specific logic. Most controllers coordinate between model classes and view classes. For example, `UITableViewController` coordinates between the data model and the `UITableView`.

Some controllers coordinate between model objects or between view objects. These sometimes have names ending in `Manager`, such as `CALayoutManager` and `CTFontManager`. It's common for managers to be singletons.

Controllers are often the least-reusable parts of a program, which is why it's so critical not to allow view and model classes to reference them directly. Even controllers should avoid referencing other controllers directly. In this context, "directly" means referring to specific classes. It's fine to refer to protocols that are implemented by a controller. For instance, `UITableView` references `<UITableViewDelegate>`, but should not reference `MYTableViewController`.

A common mistake is to allow many objects to reference the application delegate directly. For example, you may want to access a global object. A common, but incorrect, solution is to add this global object as a property on the application delegate and access it as shown here:

```
// Do not do this
MyAppDelegate *appDelegate =
  (MyAppDelegate*)[[UIApplication sharedApplication] delegate];
Something *something = [appDelegate something];
// Do not do this
```

It's very difficult to reuse code that uses this pattern. It relies on `MYAppDelegate`, which is hard to move to other programs that have their own application delegate. The better way to access global objects is the Singleton pattern, discussed later in this chapter.

The model-view-controller pattern is very effective at improving code reuse. Applying it properly to your programs helps them fit into the Cocoa framework and simplify development.

Understanding Delegates and Data Sources

A *delegate* is a helper object that manages the behavior of another object. For example, `UITableView` needs to know how tall each row should be. `UITableView` has a `rowHeight` property, but this isn't sufficient for all problems. What if the first row should be taller than the other rows? Apple might have added a `firstRowHeight` property for that case. Then it might have added `lastRowHeight` and `evenRowHeight` properties. `UITableView` would become much more complicated, and still would be limited to uses that Apple had specifically designed it for.

Instead, `UITableView` takes a delegate, which can be any object class that conforms to the `<UITableViewDelegate>` protocol. Every time `UITableView` is ready to draw a row, it asks its delegate how tall that row should be, allowing you to implement arbitrary logic for row height. It could be based on the data in that row, or a user configuration option, or any other criterion that is appropriate for your application. Delegation makes customization extremely flexible.

Some objects have a special kind of delegate called a *data source*. `UITableView` has a data source protocol called `<UITableViewDataSource>`. Generally a delegate is responsible for appearance and behavior, whereas a data source is responsible for the data to be displayed. Splitting the responsibilities this way can be useful in some cases, but most of the time the delegate and the data source are the same object. This object is generally the controller. For instance, `UITableViewController` conforms to both `<UITableViewDelegate>` and `<UITableViewDataSource>`.

As a general rule, objects do not retain their delegates. If you create a class with a `delegate` property, it should almost always be declared `weak`. In most cases, an object's delegate is also its controller, and the controller almost always retains the original object. If the object retained its delegate, you would have a retain loop and would leak memory. There are exceptions to this rule. For example, `NSURLConnection` retains its delegate, but only while the connection is loading. After that, `NSURLConnection` releases its delegate, avoiding a permanent retain loop.

Delegates are often observers (see "Working with the Observer Pattern" later in this chapter). It's common for objects to have delegate methods that parallel their notifications. For example, `UIApplication` sends its delegate `applicationWillTerminate:`. It also posts the notification `UIApplicationWillTerminateNotification`.

Configuring your objects using delegation is a form of the Strategy pattern. The Strategy pattern encapsulates an algorithm and allows you to change how an object behaves by attaching different strategy (algorithm) objects. A delegate is a kind of Strategy object that encapsulates the algorithms determining the behavior of another object. For instance, a table view's delegate implements an algorithm that determines how high the table view's rows should be. Delegation reduces the need for subclassing by moving customization logic into helper objects, improving reusability and simplifying your code by moving complex customization logic out of the main program flow. Before adding configuration properties to your classes, consider adding a delegate instead.

Working with the Command Pattern

The Command pattern encapsulates a request as an object. Rather than calling a method directly, you package the method call into an object and dispatch it, possibly at a later time. This can provide significant flexibility and allows requests to be queued, redirected, logged, and serialized. It also supports undoing operations by storing the inverse of the commands. Cocoa implements the Command pattern using target-action and `NSInvocation`. In this section, you discover how to use `NSInvocation` to create more complex dispatch objects such as trampolines.

Using Target-Action

The simplest form of the Command pattern in Cocoa is called *target-action*. This isn't a full implementation of the Command pattern because it doesn't encapsulate the request into a separate object, but it allows similar flexibility.

`UIControl` is an excellent example of target-action. You configure a `UIControl` by calling `addTarget:action:forControlEvents:`, which establishes a *target*, which is the object to send a message, an *action*, which is the message to send, and a set of events that will trigger the message. The action selector must conform to a particular signature. In the case of `UIControl`, the signature must be in one of the following forms:

```
- (void)action;
- (void)action:(id)sender;
- (void)action:(id)sender forEvent:(UIEvent *)event;
```

`UIControl` can then dispatch its `action` like this:

```
[target performSelector:action
            withObject:self
            withObject:event];
```

Because of how Objective-C message passing works, this use of `performSelector:...` works whether `action` takes one, two, or no parameters. (See Chapter 28 for details of how Objective-C message passing is implemented.)

Target-action is very common in Objective-C. Controls, timers, toolbars, gesture recognizers, `IBAction`, notifications, and other parts of Cocoa rely on this pattern.

Target-action is similar to delegation. The main difference is that in target-action, the selector is configurable, whereas in delegation, the selector is defined by a protocol. It's easier for a single object to be the target of several `NSTimer` objects than it is to be the delegate of several `UITableView` objects. To listen to multiple `NSTimer` objects, you only need to configure them with different actions:

```
[NSTimer scheduledTimerWithTimeInterval:1 target:self
            selector:@selector(firstTimerFired:) ...];
[NSTimer scheduledTimerWithTimeInterval:1 target:self
            selector:@selector(secondTimerFired:) ...];
```

To listen to multiple table views, you need to check which table view sent the request:

```
- (NSInteger)numberOfSectionsInTableView:(UITableView*)tv {
  if (tv == self.tableView1) {
    return [self.dataset1 count];
  }
  else if (tv == self.tableView2) {
    return [self.dataset2 count];
  }
  else {
    NSAssert(NO, @"Bad tv: %@", tv);
    return 0;
  }
}
```

Each delegate method must include this `if` logic, which can become very cumbersome. For this reason, multiple instances of a class generally do not share the same delegate.

On the other hand, delegation allows you to verify at compile time that the required methods are implemented. The compiler cannot verify that the target of an `NSTimer` implements a given action.

> While the compiler cannot determine if a target implements a given action, you can check for simple typos by turning on the Undeclared Selector warning (GCC_WARN_UNDECLARED_SELECTOR, `-Wundeclared-selector`). This generates a warning if an `@selector(...)` expression references an unknown selector.

Target-action is generally not the best tool for new code. If you have a small number of callbacks, blocks are often a better solution. See Chapter 23 for more information on blocks. If you have a large number of callbacks, delegation is generally best.

Using Method Signatures and Invocations

`NSInvocation` is a traditional implementation of the Command pattern. It bundles a target, a selector, a method signature, and all the parameters into an object that can be stored and invoked at a later time. When the invocation is invoked, it will send the message, and the Objective-C runtime will find the correct method implementation to execute.

A *method implementation* (`IMP`) is a function pointer to a C function with the following signature:

```
id function(id self, SEL _cmd, ...)
```

Every method implementation takes two parameters, `self` and `_cmd`. The first parameter is the `self` pointer that you're familiar with. The second parameter, `_cmd`, is the selector that was sent to this object. This is a reserved symbol in the language and is accessed exactly like `self`. For more details on how to work with method implementations, see Chapter 28.

> Although the `IMP` typedef suggests that every Objective-C method returns an `id`, obviously there are many Objective-C methods that return other types such as integers or floating-point numbers, and many Objective-C methods return nothing at all. The actual return type is defined by the message signature, discussed later in this section, not the `IMP` typedef.

`NSInvocation` includes a target and a selector. As discussed in the earlier section "Using Target-Action," a target is the object to send the message to, and the selector is the message to send. A selector is roughly the name of a method. We say "roughly" because selectors don't have to map exactly to methods. A selector is just a name, like `initWithBytes:length:encoding:`. A selector isn't bound to any particular class or any particular return value or parameter types. It isn't even specifically a class or instance selector. You can think of a selector as a string. So `-[NSString length]` and `-[NSData length]` have the same selector, even though they map to different method's implementations.

`NSInvocation` also includes a method signature (`NSMethodSignature`), which encapsulates the return type and the parameter types of a method. An `NSMethodSignature` does not include the name of a method, only the return value and the parameters. Here is how you can create one by hand:

```
NSMethodSignature *sig =
        [NSMethodSignature signatureWithObjCTypes:"@@:*"];
```

This is the signature for `-[NSString initWithUTF8String:]`. The first character (`@`) indicates that the return value is an `id`. To the message passing system, all Objective-C objects are the same. It can't tell the difference between an `NSString` and an `NSArray`. The next two characters (`@:`) indicate that this method takes an `id` and a `SEL`. As discussed previously, every Objective-C method takes these as its first two parameters. They're implicitly passed as `self` and `_cmd`. Finally, the last character (`*`) indicates that the first "real" parameter is a character string (`char*`).

> If you do work with type encoding directly, you can use `@encode(type)` to get the string that represents that type rather than hard-coding the letter. For example, `@encode(id)` is the string "@".

You should seldom call `signatureWithObjCTypes:`. We do it here only to show that it's possible to build a method signature by hand. The way you generally get a method signature is to ask a class or instance for it. Before you do that, you need to consider whether the method is an instance method or a class method. The method `-init` is an instance method and is marked with a leading hyphen (–). The method `+alloc` is a class method and is marked with a leading plus (+). You can request instance method signatures from instances and class method signatures from classes using `methodSignatureForSelector:`. If you want the instance method signature from a class, you use `instanceMethodSignatureForSelector:`. The following example demonstrates this for `+alloc` and `-init`.

```
SEL initSEL = @selector(init);
SEL allocSEL = @selector(alloc);
NSMethodSignature *initSig, *allocSig;

// Instance method signature from instance
initSig = [@"String" methodSignatureForSelector:initSEL];

// Instance method signature from class
initSig = [NSString instanceMethodSignatureForSelector:initSEL];

// Class method signature from class
allocSig = [NSString methodSignatureForSelector:allocSEL];
```

If you compare `initSig` and `allocSig`, you will discover that they are the same. They each take no additional parameters (besides `self` and `_cmd`) and return an `id`. This is all that matters to the message signature.

Now that you have a selector and a signature, you can combine them with a target and parameter values to construct an `NSInvocation`. An `NSInvocation` bundles everything needed to pass a message. Here is how you create an invocation of the message `[set addObject:stuff]` and invoke it:

```
NSMutableSet *set = [NSMutableSet set];
NSString *stuff = @"Stuff";
SEL selector = @selector(addObject:);
NSMethodSignature *sig = [set methodSignatureForSelector:selector];

NSInvocation *invocation =
    [NSInvocation invocationWithMethodSignature:sig];
[invocation setTarget:set];
[invocation setSelector:selector];
// Place the first argument at index 2.
[invocation setArgument:&stuff atIndex:2];
[invocation invoke];
```

Note that the first argument is placed at index 2. As discussed previously, index 0 is the target (`self`), and index 1 is the selector (`_cmd`). `NSInvocation` sets these automatically. Also note that you must pass a pointer to the argument, not the argument itself.

Invocations are extremely flexible, but they're not fast. Creating an invocation is hundreds of times slower than passing a message. Invoking an invocation is efficient, however, and invocations can be reused. They can be dispatched to different targets using `invokeWithTarget:` or `setTarget:`. You can also change their parameters between uses. Much of the cost of creating an invocation is in `methodSignatureForSelector:`, so caching this result can significantly improve performance.

Invocations do not retain their object arguments by default, nor do they make a copy of C string arguments. To store the invocation for later use, you call `retainArguments` on it. This method retains all object arguments and copies all C string arguments. When the invocation is released, it releases the objects and frees its copies of the C strings. Invocations do not provide any handling for pointers other than Objective-C objects and C strings. If you're passing raw pointers to an invocation, you're responsible for managing the memory yourself.

> If you use an invocation to create an NSTimer, **such as by using** timerWithTimeInterval:invoc ation:repeats:**, the timer automatically calls** retainArguments **on the invocation.**

Invocations are a key part of the Objective-C message dispatching system. This integration with the message dispatching system makes them central to creating trampolines and undo management.

Using Trampolines

A *trampoline* "bounces" a message from one object to another. This technique allows a proxy object to move messages to another thread, cache results, coalesce duplicate messages, or any other intermediary processing you'd like. Trampolines generally use forwardInvocation: to handle arbitrary messages. If an object does not respond to a selector, before Objective-C throws an error, it creates an NSInvocation and passes it to the object's forwardInvocation:. You can use this to forward the message in any way that you'd like. For full details, see Chapter 28.

In this example, you create a trampoline called RNObserverManager. Any message sent to the trampoline will be forwarded to registered observers that respond to that selector. This provides functionality similar to NSNotification, but is easier to use and faster if there are many observers.

Here is the public interface for RNObserverManager:

RNObserverManager.h (ObserverTrampoline)

```
#import <objc/runtime.h>
@interface RNObserverManager: NSObject

- (id)initWithProtocol:(Protocol *)protocol
            observers:(NSSet *)observers;
- (void)addObserver:(id)observer;
- (void)removeObserver:(id)observer;

@end
```

You initialize this trampoline with a protocol and an initial set of observers. You can then add or remove observers. Any method defined in the protocol will be forwarded to all the current observers if they implement it.

Here is the skeleton implementation for RNObserverManager, without the trampoline piece. Everything should be fairly obvious.

RNObserverManager.m (ObserverTrampoline)

```
@interface RNObserverManager()
@property (nonatomic, readonly, strong)
                            NSMutableSet *observers;
@property (nonatomic, readonly, strong) Protocol *protocol;
```

(continued)

```
@end

@implementation RNObserverManager
- (id)initWithProtocol:(Protocol *)protocol
              observers:(NSSet *)observers {
  if ((self = [super init])) {
    _protocol = protocol;
    _observers = [NSMutableSet setWithSet:observers];
  }
  return self;
}

- (void)addObserver:(id)observer {
   NSAssert([observer conformsToProtocol:self.protocol],
          @"Observer must conform to protocol.");
  [self.observers addObject:observer];
}

- (void)removeObserver:(id)observer {
  [self.observers removeObject:observer];
}
@end
```

Now you override `methodSignatureForSelector:`. The Objective-C message dispatcher uses this method to construct an `NSInvocation` for unknown selectors. You override it to return method signatures for methods defined in `protocol`, using `protocol_getMethodDescription`. You need to get the method signature from the protocol rather than from the observers because the method may be optional, and the observers might not implement it.

```
- (NSMethodSignature *)methodSignatureForSelector:(SEL)sel
{
  // Check the trampoline itself
  NSMethodSignature *
  result = [super methodSignatureForSelector:sel];
  if (result) {
    return result;
  }

  // Look for a required method
  struct objc_method_description desc =
            protocol_getMethodDescription(self.protocol,
                                          sel, YES, YES);
  if (desc.name == NULL) {
    // Couldn't find it. Maybe it's optional
    desc = protocol_getMethodDescription(self.protocol,
                                         sel, NO, YES);
  }

  if (desc.name == NULL) {
    // Couldn't find it. Raise NSInvalidArgumentException
    [self doesNotRecognizeSelector: sel];
    return nil;
```

```
    }

    return [NSMethodSignature
                    signatureWithObjCTypes:desc.types];
}
```

Finally, you override `forwardInvocation:` to forward the invocation to the observers that respond to the selector:

```
- (void)forwardInvocation:(NSInvocation *)invocation {
  SEL selector = [invocation selector];
  for (id responder in self.observers) {
    if ([responder respondsToSelector:selector]) {
      [invocation setTarget:responder];
      [invocation invoke];
    }
  }
}
```

To use this trampoline, you create an instance, set the observers, and then send messages to it as the following code shows. Variables that hold a trampoline should generally be of type `id` so that you can send any message to it without generating a compiler warning.

```
@protocol MyProtocol <NSObject>
- (void)doSomething;
@end

...

id observerManager = [[RNObserverManager alloc]
                    initWithProtocol:@protocol(MyProtocol)
                            observers:observers];
[observerManager doSomething];
```

Passing a message to this trampoline is similar to posting a notification. You can use this technique to solve a variety of problems. For example, you can create a proxy trampoline that forwards all messages to the main thread as shown here:

RNMainThreadTrampoline.h (ObserverTrampoline)

```
@interface RNMainThreadTrampoline : NSObject
@property (nonatomic, readwrite, strong) id target;
- (id)initWithTarget:(id)aTarget;
@end
```

RNMainThreadTrampoline.m (ObserverTrampoline)

```
@implementation RNMainThreadTrampoline

- (id)initWithTarget:(id)aTarget {
```

(continued)

```
    if ((self = [super init])) {
      _target = aTarget;
    }
    return self;
  }

  - (NSMethodSignature *)methodSignatureForSelector:(SEL)sel
  {
    return [self.target methodSignatureForSelector:sel];
  }

  - (void)forwardInvocation:(NSInvocation *)invocation {
    [invocation setTarget:self.target];
    [invocation retainArguments];
    [invocation performSelectorOnMainThread:@selector(invoke)
                                 withObject:nil
                              waitUntilDone:NO];
  }
  @end
```

`forwardInvocation:` can transparently coalesce duplicate messages, add logging, forward messages to other machines, and perform a wide variety of other functions. See Chapter 28 for more discussion, including how to couple with `NSProxy`.

Using Undo

The Command pattern is central to undo management. By storing Command objects (`NSInvocation`) in a stack, you can provide arbitrary undo and redo functionality.

Before performing an action that the user should be able to undo, you pass its inverse to `NSUndoManager`. A convenient way to do this is with `prepareWithInvocationTarget:`. For example:

```
  - (void)setString:(NSString *)aString {
    // Make sure there is really a change
    if (! [aString isEqualToString:_string]) {
      // Send the undo action to the trampoline
      [[self.undoManager prepareWithInvocationTarget:self]
        setString:_string];
      // Perform the action
      _string = aString;
    }
  }
```

When you call `prepareWithInvocationTarget:`, the undo manager returns a trampoline that you can send arbitrary messages to. These are converted into `NSInvocation` objects and stored on a stack. When the user wants to undo an operation, the undo manager just invokes the last command on the stack.

The Command pattern is used throughout Cocoa and is a useful tool for your architectures. It helps separate request dispatching from the requests themselves, improving code reusability and flexibility.

Working with the Observer Pattern

The Observer pattern allows an object to notify many observers of changes in its state, without requiring that the observed object have special knowledge of the observers. The Observer pattern comes in many forms in Cocoa, including `NSNotification`, delegate observations, and key-value observing (KVO). It encourages weak coupling between objects, which makes components more reusable and robust.

Delegate observations are discussed in "Understanding Delegates and Data Sources" earlier in this chapter. KVO is discussed fully in Chapter 22. The rest of this section focuses on `NSNotification`.

Most Cocoa developers have encountered `NSNotificationCenter`, which provides loose coupling by allowing one object to register to be notified of events defined by string names. Notifications can be simpler to implement and understand than KVO. Here are examples of how to use it well.

Poster.h

```
// Define a string constant for the notification
extern NSString * const PosterDidSomethingNotification;
```

Poster.m

```
NSString * const PosterDidSomethingNotification =
                        @"PosterDidSomethingNotification";

...

// Include the poster as the object in the notification
[[NSNotificationCenter defaultCenter]
   postNotificationName:PosterDidSomethingNotification
               object:self];
```

Observer.m

```
// Import Poster.h to get the string constant
#import "Poster.h"

...

   // Register to receive a notification
   [[NSNotificationCenter defaultCenter] addObserver:self
     selector:@selector(posterDidSomething:)
     name:PosterDidSomethingNotification object:nil];

...

- (void) posterDidSomething:(NSNotification *)note {
   // Handle the notification here
}
```

(continued)

```
- (void)dealloc {
    // Always remove your observations
    [[NSNotificationCenter defaultCenter]
       removeObserver:self];
    [super dealloc];
}
```

Notice the name `PosterDidSomethingNotification`. It begins with the class of the poster, which should always be the class of the `object`. It then follows a "will" or "did" pattern. This naming convention is very similar to delegate methods and that's intentional. The ending `Notification` is traditional for notification names so that they can be distinguished from other string constants like keys or paths.

This example uses a string constant for the notification name, which is critical for avoiding typos. Notification string constants do not traditionally begin with a `k` as some constants do. I recommend that the value of the string constant match the name of the string constant as shown in this example. This makes obvious which constant is being used when you see the value in debug logs.

The placement of `const` is important when declaring string constants. This declaration is correct:

```
extern NSString * const RNFooDidCompleteNotification;
```

This declaration is incorrect:

```
extern const NSString * RNFooDidCompleteNotification;
```

The former is a constant pointer to an immutable string. The latter is a changeable pointer to an immutable string. `NSString` is always immutable because it is an immutable class. So `NSString * const` is useful. `const NSString *` is useless. This rule is easier to remember if you read the declaration from right to left: `const` **pointer to** `NSString`.

As I mentioned earlier, the beginning of the notification name should always be the class of `object`. In this case, the class is `Poster`. The `object` is also almost always `self` (the object posting the notification). For consistency, the notification should always include an `object`, even if it is a singleton.

The observer should consider carefully whether to observe a specific object or `nil` (all notifications with a given name, regardless of the value of `object`). Observing a specific object can be cleaner and ensures that the observer won't receive notifications from instances that it's unaware of. A class that has a single instance today may have additional instances tomorrow.

If you observe a specific instance, it should generally be something you `retain` in an ivar. Observing something does not `retain` it, and the object you're observing could deallocate. Continuing to observe a deallocated object won't cause a crash; you just won't receive notifications from that object anymore. But it's sloppy and likely indicates a flaw in your design. It also uses unneeded slots in the notification table, which is bad for performance.

Although observing an object that deallocates won't cause a crash, notifying a deallocated observer will, so you should always call `removeObserver:` in your `dealloc` if any part of your object calls `addObserver:…`. Make a habit of this. It's one of the most common and preventable causes of crashes in code that uses notifications.

Calling `addObserver:selector:name:object:` multiple times with the same parameters causes you to receive multiple callbacks, which is almost never what you want. Generally, it's easiest to start observing notifications in `init` and stop in `dealloc`. But what if you want to watch notifications from one of your properties, and that property can change? This example shows how to write `setPoster:` so that it properly adds and removes observations for a `poster` property:

```
- (void)setPoster:(Poster *)aPoster {
  NSNotificationCenter *nc =
                   [NSNotificationCenter defaultCenter];

  if (_poster != nil) {
    // Remove all observations for the old value
    [nc removeObserver:self name:nil object:_poster];
  }

  _poster = aPoster;

  if (_poster != nil) {
    // Add the new observation
    [nc addObserver:self
          selector:@selector(anEventDidHappen:)
              name:PosterDidSomethingNotification
            object:_poster];
  }
}
```

The checks for `nil` are very important here. Passing `nil` as the `object` or the `name` means "any object" or "any notification."

While observing specific instances is cleaner and protects you against surprises when new objects are added to the system, there are reasons to avoid it. First, you may not really care which object is posting the notification. The object may not actually exist when you want to start observing notifications it might post, or the object instance may change over time.

There are also performance considerations when observing notifications. Every time a notification is posted, `NSNotificationCenter` has to search through the list of all registered observers to determine which observers to notify. The time required to search this list is proportional to the total number of observations registered in the `NSNotificationCenter`. When the total number of observations in the program reaches a few hundred, the time to search this list can become noticeable on an iPhone, particularly older models. The time required to call `removeObserver:` is similarly proportional to the total number of observations, which can cause serious performance problems if you have a large number of observations and post many notifications or remove observers often.

What if you want to observe a notification from a large number of objects, but not necessarily every object that might post that notification? For instance, you might be interested in changes to music tracks, but only the tracks in your current playlist. You could observe each track individually, but that can be very expensive. A better technique is to observe `nil` and check in the callback whether you were actually interested, as shown here:

```
// Observe all objects, whether in your tracklist or not
[[NSNotificationCenter defaultCenter]
  addObserver:self selector:@selector(trackDidChange:)
  name:TrackDidChangeNotification object:nil];

...

- (void)trackDidChange:(NSNotification *)note {
  // Verify that you cared about this track
  if ([self.tracks containsObject:[note object]]) {
    ...
  }
}
```

This technique reduces the number of observations, but adds an extra check during the callback. Whether this tradeoff is faster or slower depends on the situation, but it's generally better than creating hundreds of observations.

Posting notifications is synchronous, which trips up many developers who expect the notification to execute on another thread or otherwise run asynchronously. When you call `postNotification:`, observers are notified one at a time before returning. The order of notification is not guaranteed.

Notifications are a critical part of many Cocoa programs. You just need to keep the preceding issues in mind, and they'll be a very useful part of your architecture.

Working with the Singleton Pattern

The Singleton pattern is in many ways just a global variable. It provides a global way to access a specific object. The Singleton pattern is common throughout Cocoa. In most cases, you can identify it by a class method that begins with `shared`, such as `+sharedAccelerometer`, `+sharedApplication`, and `+sharedURLCache`. Some singleton access methods have other prefixes, such as `+[NSNotificationCenter defaultCenter]` and `+[NSUserDefaults standardUserDefaults]`. These are generally older classes inherited from NeXTSTEP. Most new frameworks use the `shared` prefix followed by their class name (without its namespace prefix).

The Singleton pattern is one of the most misused patterns in Cocoa because of some unfortunate sample code published by Apple. In the *Cocoa Fundamentals Guide*, Apple includes an implementation of the Singleton pattern that overrides the major memory management methods, `allocWithZone:`, `copyWithZone:`, `retain`, `retainCount`, `release`, and `autorelease`. Using Apple's example, multiple calls to `[[Singleton alloc] init]` return the same object, which is almost never needed or appropriate. Apple's explanation to this code indicates that it's useful only in cases where it's mandatory that there be only one instance of the class, which is seldom the case. Most of the time, it's merely convenient to have one instance of the class that is easily accessible. Many classes, such as `NSNotificationCenter`, work perfectly well if multiple instances exist. Unfortunately, many developers do not carefully read the explanation and incorrectly copy this example.

Sometimes a strict singleton is appropriate. For example, if a class manages a unique shared resource, it may be impossible to have more than one instance. In this case, it's often better to treat the creation of multiple instances as a programming error with NSAssert rather than transparently returning a shared instance. You will see how to implement this kind of assertion later in this section.

If you are creating a transparently strict singleton, make sure that it's an implementation detail and not something the caller must know. For instance, the class should be immutable. If the caller has requested distinct instances using +alloc, it's very confusing if changes to one modify the other.

In the vast majority of cases, use a shared singleton rather than a strict singleton. A shared singleton is just a specific instance that is easy to fetch with a class method, and is generally stored in a static variable. There are many ways to implement a shared singleton, but my recommendation is to use Grand Central Dispatch (GCD):

```
+ (MYSingleton *)sharedSingleton {
  static dispatch_once_t pred;
  static MYSingleton *instance = nil;
  dispatch_once(&pred, ^{instance = [[self alloc] init];});
  return instance;
}
```

This approach is easy to write, it's fast, and it's thread-safe. Other approaches achieve thread safety by adding an @synchronize in +sharedSingleton, but this approach adds a significant performance penalty every time +sharedSingleton is called. It is also possible to use +initialize, but the GCD approach is simpler.

Although a shared singleton is usually the best approach, sometimes you do require a strict singleton. For example, you may have a singleton that manages the connection to the server, and the server protocol may forbid multiple simultaneous connections from the same device. As a general rule, you first try to redesign the protocol so that it doesn't have this restriction, but there are cases where doing so is impractical and a strictly enforced singleton is the best approach.

In most cases, the best way to implement a strict singleton is with the same pattern as a shared singleton, but treat calls to init as a programming error with NSAssert, as shown here:

```
+ (MYSingleton *)sharedSingleton {
  static dispatch_once_t pred;
  static MYSingleton *instance = nil;
  dispatch_once(&pred, ^{instance = [[self alloc] initSingleton];});
  return instance;
}

- (id)init {
  // Forbid calls to -init or +new
  NSAssert(NO, @"Cannot create instance of Singleton");

  // You can return nil or [self initSingleton] here,
  // depending on how you prefer to fail.
  return nil;
```

(continued)

```
}

// Real (private) init method
- (id)initSingleton {
  self = [super init];
  if ((self = [super init])) {
    // Init code
  }
  return self;
}
```

This approach's advantage is that it prevents callers from believing they're creating multiple instances when that is forbidden. Frameworks should avoid silently fixing programming errors. Doing so just makes it harder to track down bugs.

As discussed earlier in the section "Understanding Model-View-Controller," developers often use the application delegate to store global variables like this:

```
// Do not do this
MYAppDelegate *appDelegate =
        (MYAppDelegate*)[[UIApplication sharedApplication] delegate];
Something *something = [appDelegate something];
// Do not do this
```

In almost all cases, globals like this are better implemented with a Something singleton like the following:

```
Something *something = [Something sharedSomething];
```

This way, when you copy the Something class to another project, it's self-contained. You don't have to extract bits of the application delegate along with it. If the application delegate is storing configuration information, it's best to move that into NSUserDefaults or a singleton Configuration object.

The Singleton pattern is one of the most common patterns in well-designed Cocoa applications. Don't overuse it. If an object is used in only a few places, just pass the object where it is needed. But for objects that have application-wide scope, a Singleton is a very good way to maintain loose coupling and improve code reusability.

Summary

This chapter explored the most pervasive patterns in Cocoa, particularly Strategy, Observer, Command, and Singleton. You discovered how several patterns combine to facilitate Cocoa's central architecture: model-view-controller. Cocoa uses design patterns focused on loose coupling and code reusability. Understanding these patterns will help you anticipate how Apple frameworks are structured and improve your code's integration with iOS. The patterns Apple uses in iOS are well-established and have been studied and refined for years throughout industry and academia. Correctly applying these patterns will improve the quality and reusability of your own programs.

Further Reading

Apple Documentation

The following documents are available in the iOS Developer Library at `developer.apple.com` or through the Xcode Documentation and API Reference.

"Cocoa Design Patterns," *Cocoa Fundamentals Guide*. This entire document is valuable to understanding Cocoa, but the section, "Cocoa Design Patterns," focuses on how Cocoa applies the well-established software patterns.

The Elements (Sample Code). Historically, Apple sample code has not demonstrated good design or coding practices. The focus has typically been to show how a specific feature works, and the sample code typically ignores Apple's recommendations and common best practices. Apple appears to have changed its approach to sample code, and some recent examples are well-designed and written. *The Elements* is a good example that developers can use to model their own projects.

Notification Programming Topics. Explains the Observer pattern implemented with NSNotification.

Undo Architecture. Explains how to use NSUndoManager using the Command pattern.

Other Resources

Gamma, Erich *et al. Design Patterns: Elements of Reusable Object-Oriented Software* (Addison-Wesley Professional, 1994. ISBN: 978-0201633610). This book is a collection of well-known design patterns, explained in practical terms with code examples in C++ and Smalltalk. It should be part of every developer's library. Erich Gamma and his coauthors did not invent these patterns, and *Design Patterns* is not an exhaustive list of all patterns. This book attempts to catalog patterns that the authors found in common use among developers and to provide a framework by which developers can apply known solutions to their unique problems.

AgentM, "Elegant Delegation," *Borkware Rants*. AgentM provides a somewhat different `MDelegateManager` class than my `RNObserverManager`. It was designed for Objective-C 1.0, so it does not rely on @protocol, but it's still worth studying.
`borkware.com/rants/agentm/elegant-delegation`

Burbeck, Steve. "Applications Programming in Smalltalk-80™: How to use Model-View-Controller (MVC)." (1987, 1992). This is the definitive paper defining the MVC pattern in Smalltalk. NeXTSTEP (and later Cocoa) modified the pattern somewhat, but the Smalltalk approach is still the foundation of MVC.
`st-www.cs.illinois.edu/users/smarch/st-docs/mvc.html`

## Chapter 5

Memory Management with Objective-C ARC

The shift from the GCC compiler to the LLVM compiler gave Apple control over certain language and compiler-specific features. One arguably important feature is *Automatic Reference Counting* (ARC). For the most part, ARC "just works." On a new project, things are easy, but when you try to migrate an existing code base to use Objective-C ARC, you will probably find that the road is long and winding.

In Chapter 2, I showed you how to use Xcode's built-in Objective-C ARC migration tool. In this chapter, I show you how to migrate your code base to ARC and solve the most commonly faced errors by finding out how ARC works. Let's get started.

Introduction to Objective-C ARC

In a nutshell, ARC is a memory management model where the compiler automatically inserts `retains` and `releases` appropriately for you. This means that you don't have to worry about managing memory and focus or writing the next great app. You're relieved from worrying about leaks, for the most part.

A Brief History

Prior to ARC, Apple platform developers used manual reference counting memory management on iOS. On Mac, developers had a choice. They could choose either garbage collection memory management or manual reference counting memory management. Garbage collection is mostly a high-level language (C# or Java) feature that *non-deterministically* manages memory for you. A programming language is designed either for writing high-performance applications or for writing software that behaves like a daemon process, running mostly in the background, servicing requests. Java, C#, and other languages fall into the latter category, whereas derivatives of C, namely C++ and Objective-C, are languages that were designed to write native code where high-performance code was crucial. High performance can be achieved only when memory management is *deterministic*. Garbage collection, although it sounds good theoretically, fares poorly when implemented for a language like C mainly because it's non-deterministic. Non-deterministic memory management means that as a developer, you will not know when a memory you allocate will be released. Apple devices like the iPhone and iPad have very limited memory on board, which means that apps should be designed to run within their resource constraints. Deterministic memory management was, in fact, one of the design goals of iOS, which is why iOS supported only a reference counting memory management model.

Manual Versus Automatic Reference Counting

Until iOS 5, the memory management model in iOS was manual reference counting where the developer kept track of the number of references to every allocated memory block. Good coding practices such as using properties (instead of using ivars directly) to read and write a variable (and such) eases manual memory management to some extent. But even for a highly skilled developer, manual memory management is still overkill. Many developers continue to depend on autorelease pools. For example, compare the following two code blocks:

Manual Memory Management Code without autorelease

```
NSDictionary *myDict = [[NSDictionary alloc] init];
// Use the dictionary here
. . .
// more code
. . .

// I don't need my dictionary any more, let's deallocate the memory
[myDict release];
```

Manual Memory Management Code with autorelease

```
//autoreleased object
NSDictionary *myDict = [NSDictionary dictionary];

// use the dictionary here
. . .
// more code
. . .
// I don't need my dictionary any more, but let's leave it to the
// autorelease pool to release the dictionary for us
```

In the first block of code, you create a dictionary and release it when you no longer need it. This kind of code is highly deterministic and has better performance. You know exactly when the memory associated with the dictionary is released. In the second block of code, you're at the mercy of the autorelease pool to drain off the memory allocated during that run loop. Although this code might not show up as a huge performance problem in most cases, code like this slowly accumulates and makes the app heavily dependent on autorelease pools. Even objects that you know can be deallocated earlier linger around until the autorelease pool drains them off, making the memory management model slightly non-deterministic.

What Is ARC?

ARC is a compiler-level feature that automatically inserts `retains` and `releases` for you, which means that when you use ARC, you don't have to raise the minimum deployment target. Though ARC was announced with iOS 5, your app compiled with ARC can run on iOS 4 devices as well. This also means that every app you write today should use the ARC compiler, and there is no reason to do otherwise.

In the first code block, just shown, after allocating memory, you don't have to write the `[dict release]` statement. ARC inserts it automatically for you. In fact, when you compile the file with the ARC flag turned on, you will get a compile time error when you call `release` or `retain`. Within a project, you can mix and match

ARC code with non-ARC code. You do so by using the two compiler flags, -fno-objc-arc and -fobjc-arc. In the next section, you find out when to use which flag. You can even mix and match ARC code and non-ARC code within the same class!

Using Non-ARC Code within a File Compiled with ARC

```
#if !__has_feature(objc_arc)
// Non ARC code goes here
-(void) release {
  // release your variables here
}
#endif
```

The Clang language extension __has_feature **function-like macro is even more powerful. You can use it with any supported Clang language feature to conditionally compile code.**

Integrating Non-ARC Third-Party Code into Your ARC Project

When you use a third-party framework that isn't compiled with ARC, you compile files from that library using the -fno-objc-arc compiler switch. (Refer to Chapter 2 to find out how to add a compiler switch to a file.) In Chapter 2, you learned how to use Xcode's Convert to Objective-C ARC tool. When you use that tool to migrate, you can choose a list of files that must be excluded from migration (refer to Figure 2-10 in Chapter 2). All these excluded files will have the -fno-objc-arc compiler switch automatically added by the migration tool. If you don't add this switch, the ARC compiler is going to throw errors for all retain/release statements in your third-party library.

Integrating ARC Code into Your Non-ARC Project

Sometimes you will find a third-party library that is compiled using ARC. Unlike the contrary, when you use this code in your non-ARC project, you will not get a compiler time error. You will instead be leaking memory. To avoid this situation, compile the ARC ready third-party source code with -fobjc-arc.

ARC Code in a Framework

If you're writing a framework, either for publishing it as open source or for sharing it with other teams within your company, you need to warn developers who use your code that the files should be compiled with ARC. Accidentally using ARC code in a non-ARC project leaks memory, and it's your responsibility to warn developers. You do so using a macro as shown here:

```
#if ! __has_feature(objc_arc)
#error This file is ARC only.
#endif
```

The preceding code block should be familiar. It's the same Clang language extension explained in the previous section.

Zeroing Weak References

Earlier in this chapter, I said that ARC is a compile time feature, But there's more. Starting from iOS 5 and Mac OS X 10.7 (Lion), the runtime supports something called *zeroing weak references*. Weak references are used to avoid retain cycles by allowing you to hold onto a pointer without bumping the retain count. But when the memory (that is pointed to) gets deallocated, the weak reference will be pointing to garbage. Accessing that pointer after the memory is deallocated crashes your app. Zeroing weak references solves this problem by "nil-ing" out pointers when the memory gets deallocated.

Ownership Qualifiers

ARC works by inserting retains and releases automatically for you. But the "automatic" code generator needs some hints to know when exactly it should release the allocated memory. ARC uses the ownership qualifiers you specify to determine the lifecycle of a pointer. Ownership qualifiers are arguably the most esoteric concept in ARC. After you understand ownership qualifiers, you will start thinking in terms of object graphs rather than retains and releases. ARC supports four kinds of ownership qualifiers:

- `__strong`
- `__weak`
- `__unsafe_unretained`
- `__autoreleasing`

The first three can also be used in property declarations. When an ownership qualifier is used with a property, you don't prefix it with the double underscore (__). The default ownership qualifier (when you don't specify one) is `__strong`.

> In Xcode 4.2/LLVM 3, the default ownership qualifier for stack variables was `__strong` and for properties it was `__assign`. From Xcode 4.3/LLVM 3.1 and higher the default ownership qualifier for properties was also changed to `__strong`. The migration wizard in Xcode 4.3 was updated to reflect these changes.

__strong

You specify strong ownership using either the qualifier

```
__strong NSString* myObject = nil;
```

or

```
@property(nonatomic, strong) NSObject *myObject;
```

When declaring properties, you use the second syntax. A strong ownership is synonymously equivalent to a "retain." It bumps up the retain count. Thinking in ARC, you use strong ownership for holding references to anything you "own." Strong ownership includes a view owning its subviews, a parent controller owning its child view controllers, and contained objects. As I already mentioned, `__strong` is the default ownership qualifier.

You are not obliged to write the keywords __strong or strong to specify strong ownership. The preceding two lines of code can henceforth be written as follows.

```
NSString* myObject = nil;
```

or

```
@property(nonatomic) NSObject *myObject;
```

__unsafe_unretained

When you don't own a pointer but want to hold a reference to it, you use the _unsafe_unretained ownership qualifier. The following lines show use of this qualifier:

```
__unsafe_unretained UIView* mySubview;
```

or

```
@property(nonatomic, unsafe_unretained) UIView* mySubview;
```

A common use of this qualifier is to maintain references to a subview in a view controller. Remember that when you create a view and add it to another view using the addSubview: method, the super view "owns" the subview. Sometimes, in your view controller, you may want to own another reference to this subview.

Figure 5-1 illustrates a frequently used ownership qualifier pattern. A UITextField is normally added as a subview of a UIView, and the UIViewController maintains a reference to it without owning it.

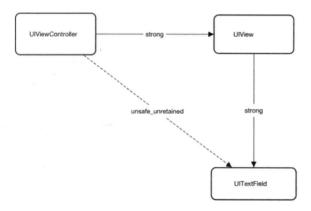

Figure 5-1 Ownership of a UITextField

You can use __unsafe_unretained in many similar cases including maintaining a reference to a delegate. When used with properties, __unsafe_unretained is synonymous to the assign modifier that you used prior to ARC.

__weak

The __weak modifier behaves exactly like the __unsafe_unretained modifier, except that when the pointer that is being pointed to gets deallocated, the weak pointer becomes nil. Of course, this requires runtime support. As such, weak references are supported from iOS 5 and higher.

> Unlike the rest of ARC, weak references are not a compile-time setting. The deployment runtime needs to support zeroing weak references. If your minimum deployment target is lower than iOS 5.0, you cannot use weak references. The Xcode Convert to Objective-C ARC migration wizard will convert your "assign" modifiers to __unsafe_unretained instead of __weak if your deployment target doesn't support zeroing weak references.

The use of a weak reference is similar to __unsafe_unretained. You use a weak reference in places where you normally use __unsafe_unretained on runtimes that support zeroing weak references.

__auto_releasing

__auto_releasing is an ownership qualifier that you use when you want to return an auto-released variable from a method. NSError objects that are allocated in a method and returned to the caller via a pointer to a pointer use __auto_releasing ownership. The one important difference between this qualifier and the other three is that __auto_releasing cannot be used with properties.

> Ownership qualifiers, unlike const modifiers, are position-independent. So __weak NSObject* myObject; and NSObject __weak *myObject; mean the same. This is because ARC ownership qualifiers apply to the pointer and not the value pointed to (unlike a const modifier). The LLVM compiler can henceforth deduce your intent.

ARC nils Declared Variables

Another important benefit of using ARC is that all uninitialized variables are "nil-ed" by default. This means that a declaration like

```
NSObject *myObject1, *myObject2;
```

points to nil when compiled with ARC. However, note that, unlike other high-level programming languages, ARC doesn't automatically set a scalar variable to zero.

This means that the variables declared in the following lines will not equate to zero:

```
int a;
int b;
```

Objective-C Naming Conventions

ARC honors the Objective-C method family and performs memory management automatically. In Objective-C, all methods that start with `alloc`, `copy`, `mutableCopy`, and `new` create a new object and bump up the retain count. In other words, you need to deallocate the memory associated with this object when you no longer need it. When you use ARC, it will release any Objective-C pointer returned by a method that is expected to return a retainable object pointer type in the calling method. However, there is a caveat.

If you write a method called `newPersonName`, in the `Person` object, ARC assumes that the method returns a newly allocated object. Everything goes well when the calling code and the method, `newPersonName`, are compiled using the ARC compiler (or using a non-ARC compiler). But if this method is in a class that is not compiled with ARC and the calling code is compiled with ARC, you will crash. Conversely, if the `newPersonName` method is compiled with ARC but your calling code isn't, the allocated memory will leak.

Overriding the Default Behavior

Although technically you can't override LLVM's behavior, you can change a method's family using the Clang source annotation, `NS_RETURNS_RETAINED` and `NS_RETURNS_NOT_RETAINED`. Your `newPersonName` method can be annotated as shown here to tell the ARC compiler not to assume that this method, despite starting with `new`, returns an unretained object pointer.

```
-(NSString*) newPersonName NS_RETURNS_NON_RETAINED;
```

Another way to tell the ARC compiler that this method returns a unretained object pointer is to rename the method to something like `-(NSString*) personName;`. But if your implementation is in a static library and you don't have access to the source code, you will not be able to rename methods. The two Clang source annotations we showed can be used in these cases.

Toll-Free Bridging

Unlike the Objective-C libraries, older C-based or Core Foundation class libraries (`CF*` methods) that you use in Objective-C don't follow an established set of naming conventions. This means that the ARC compiler cannot, with 100 percent certainty, release memory when it's done. Prior to ARC, you can bridge between a `CF*` object to a `NS*` object with a cast (also known as *toll-free bridging*). That is, you can cast a `CFString*` to an `NSString*` just by typecasting. With ARC, you can't do this anymore, at least without specifying an ownership transfer modifier.

ARC allows the following ownership transfer modifiers:

- `__bridge`
- `__bridge_retained`
- `__bridge_transfer`

__bridge

The first modifier, `__bridge`, is a plain cast that tells ARC not to bump retain count or change ownership. The designers of the LLVM compiler didn't make `__bridge` the default ownership transfer modifier, because it would be preposterous to make such a bold decision. Because Core Foundation methods don't have a proper naming convention, the compiler cannot make an educated decision about how to transfer the ownership of a C pointer type without you specifying one explicitly.

__bridge_retained

You use the second modifier when you want to transfer the C pointer type by bumping up the retain count. You use this modifier when you want to return a retained pointer from an Objective-C method that creates a Core Foundation object and releases the object using a `CFRelease` method. You normally return a retained pointer if your Objective-C method belongs to the `NSRETURNS_RETAINED` family.

__bridge_transfer

You use the last modifier when you want to transfer a Core Foundation pointer type to an Objective-C pointer by bumping the retain count. You use this modifier when your Core Foundation method creates an object and you want the object's memory management to be handled by ARC. You're essentially "transferring" ownership to ARC.

> Two macros, `CFBridgingRetain` and `CGBridgingRelease`, can be used in place of this typecast. You find out more about these Core Foundation macros and ARC in Chapter 27 and Chapter 28. Chapter 27 also covers toll-free bridging in detail.

ARC Internals

ARC, as you know already, inserts retain and releases automatically for you. The ARC compiler has two parts, the front-end compiler and the optimizer.

Front-End Compiler

The front-end compiler works by inserting a matching release for every "owned" object. An object is owned if its ownership qualifier is `__strong`. Objects created within a method scope are torn down by a release inserted automatically at the end of the method. Objects owned by the class (ivar/properties) are torn down in the `dealloc` method. In fact, you don't have to write a `dealloc` method or call the super class's `dealloc`. ARC does all these automatically for you. Moreover, this compiler-generated code even performs better than a `release` written by you because the compiler can make some assumptions. Under ARC, since no class can override the `release` method, there is no reason to invoke it. Instead, ARC optimizes the call by directly calling the `objc_release`. The same applies to `retain` as well. ARC optimizes the call by emitting an `objc_retain` rather than a retain message.

ARC Optimizer

Although the front-end compiler sounds superior, the emitted code may at times contain duplicate calls to `retain` and `release`. The ARC optimizer takes care of removing unwanted retains and releases which ensures that the generated code runs as fast (or faster) than an equivalent manual reference-counted code.

You will find out more about the internal workings of ARC in Chapter 28.

Common ARC Migration Errors

Earlier in this chapter, you read about ARC and how to handle toll-free bridging under ARC and how ARC works internally. Following are some of the common migration errors that you will encounter while migrating to ARC.

- **Casting an Objective-C pointer type to a C pointer type (and vice versa)**—Because ARC doesn't automatically do toll-free bridging, when you convert an Objective-C pointer to a C pointer, you have to use an ownership transfer qualifier. You will encounter this error when trying to compile a source code that relies heavily on Core Foundation library.

- **Receiver type doesn't declare the method with selector**—ARC relies on the method family to determine how to release/retain the memory returned by a method. If you call a method on an object, ARC mandates that the method be defined. Prior to ARC, you can still call the method even if it is not defined or declared. Calling such a method results in a runtime exception. Object doesn't respond to this selector. With ARC, this is no longer the case. The ARC compiler flags a compile time error when it encounters this situation.

> LLVM 3.0 mandates that you define all private methods inside a category. Calling a private method that isn't defined but declared after the calling code results in a compile time error. (Calling a private method that isn't defined but declared *before* the calling code works). LLVM 3.1 is a two-pass compiler that discovers methods declared within the same file, which means that you can omit private method definitions in the implementation file. Nevertheless, for the sake of writing clean code, I recommend defining all private methods or at least the ones that you will call (excluding IBActions) in the implementation file within a category. In fact, Xcode 4.3 and higher automatically create an empty private category for you (For example: The default templates like the `UIViewController` subclass have an empty private category).

- **performSelector may cause a leak because its selector is unknown**—This error again roots down to the same fact that the ARC compiler needs to know the method family of every method call. This is not a problem if you invoke a selector using its name. When you call a selector defined in a `SEL` variable, ARC cannot deduce if the selector method belongs to the family of methods that "creates" objects (`NS_RETURNS_RETAINED`) or the family of methods that returns an auto-released object (`NS_RETURNS_NOT_RETAINED`). The compiler emits this to warn you to do the memory management manually.

Clang doesn't have a source annotation that helps to imply the method family of a selector. Purists may opt to suppress this warning forcefully using a pragma.

```
#pragma clang diagnostic push
#pragma clang diagnostic ignored "-Warc-performSelector-leaks"
    [self performSelector:self.mySelector];
  #pragma clang diagnostic pop
```

Note that this pragma only suppresses the warning and doesn't fix the leak if your method returns a created/retained object.

- **Implicit conversion of an Objective-C pointer to 'void *' is disallowed with ARC**—With ARC, you can no longer cast a void pointer to an `id` and vice versa. ARC depends on ownership qualifiers to determine when to release the allocated memory. When you do a simple cast between id and a void pointer, you need to use a qualifier so that the ARC compiler knows how to treat the pointers. You do this using a bridge cast that you learn in one of the earlier sections, "Toll-Free Bridging." Here is an example:

```
id selfPointerAsId = (__bridge void*)self;
```

▪ **ARC forbids Objective-C objects in structs or unions**—ARC mandates that you use an Objective-C object rather than a struct when your struct has an Objective-C object. Structs don't have a release method (or a destructor in C++ terms). As such, when you want to release a struct, you should also ensure that you release the memory allocated for objects within a struct. ARC works only when ownership qualifiers are clear. You own the object, and the object owns all of its sub-objects. An Objective-C object has init and release/dealloc methods that take care of setting it up and tearing it down. This means that when you "own" an object (think strong ownership), you don't have to worry about ownership of the sub-objects. With a struct, it becomes impossible to implement such ownerships. The backbone of ARC, the ownership qualifier, breaks when a struct contains an object and this is the reason why ARC forbids Objective-C objects within a struct.

However, note that a struct for representing a mathematical complex number like the one shown here is perfectly valid.

```
typedef struct _ComplexNumber {

double realValue;
double imaginaryValue;
} ComplexNumber;
```

Under ARC, you can't use an object/pointer within a struct. Everything else is valid.

▪ **'NSAutoreleasePool' is unavailable: not available in automatic reference counting mode**—ARC deprecates NSAutoreleasePool in favor of a more powerful @autoreleasepool block. In fact, the Convert To Objective-C ARC tool converts your NSAutoreleasePool block in the main.m to an @autoreleasepool block like this:

```
@autoreleasepool {
    return UIApplicationMain(argc, argv, nil,
    NSStringFromClass([AppDelegate class]));
}
```

The @autoreleasepool block is much faster (up to six times faster) than the previous NSAutoreleasePool block. In fact, even if you don't use ARC, you should at least use @autoreleasepool. The @autoreleasepool block is an LLVM feature and can in fact be used independently of ARC.

Workarounds When Using ARC

ARC has its own share of disadvantages, with the biggest being retain cycles. To add to this disadvantage, the LLVM static analyzer doesn't detect a retain cycle.

Retain Cycles

A retain cycle occurs when two objects have mutual ownership of each other. A classical real-life scenario is when a child view controller owns a strong pointer to the parent view controller, as shown in Figure 5-2.

Figure 5-2 Retain cycles happen when you own a reference to your own parent

To avoid such retain cycles, use the __weak (or __unsafe_unretained) ownership qualifier. A weak reference ensures that the pointer's retain count isn't bumped up, as shown in Figure 5-3.

Figure 5-3 Avoiding retain cycles using a weak reference to your own parent

If you have learnt to think in terms of ownerships rather than release/retains, you can avoid retain cycles in most cases, except, well, blocks. Blocks and ARC hate each other, to some extent.

Blocks and ARC

A block is an ad hoc piece of code that captures the context. That is, any variable that can be accessed outside the block can be accessed within the block. Capturing the context yields surprisingly new design patterns and Chapter 22 explains them in detail. When a block "captures" the context, it makes a copy of every scalar variable in its scope. Every Objective-C object within the scope is retained. A common mistake that happens is when a block is owned by "self," and within the block, you update ivars.

Code that Causes a Retain Cycle

```
self.myBlock = ^(NSString* returnedString) {

self.labelControl.text = returnedString;
};
```

The preceding code shows a classical pitfall when using blocks with ARC. The block, which is retained by `self`, captures `self` (retain) again. This causes a retain cycle. Very subtle, yet very dangerous.

To avoid this situation, capture an unretained reference to `self` within a block. Prior to ARC, you used the __block keyword along with __unsafe_unretained to make an unretained copy of a reference and used that reference within a block.

Using __block to Avoid Retain Cycles (Non-ARC)

```
__block id safeSelf = self;
self.myBlock = ^(NSString* returnedString) {

safeSelf.labelControl.text = returnedString;
};
```

ARC changes the semantics of __block, and you shouldn't use it. Under ARC, __block references are retained rather than copied, which means that the preceding code still causes a retain cycle under ARC. The correct practice is to use a __weak (or unsafe_unretained) reference, as shown in the following code:

Using __weak to Avoid Retain Cycles (ARC)

```
__weak id safeSelf = self; //iOS 5+
//__unsafe_unretained id safeSelf = self; //iOS 4+

self.myBlock = ^(NSString* returnedString) {

safeSelf.labelControl.text = returnedString;
}
```

Summary

In this chapter, you found out about the ins and outs of ARC. I later explained the migration process and the various scenarios to avoid when using ARC.

Lastly, you read about the drawbacks and retain cycles associated with ARC. Fret not; within a few months of using ARC, you will become proficient in writing code that avoids retain cycles. Despite those minor drawbacks, I still recommend that you use ARC in all future projects.

Further Reading

Apple Documentation

The following document is available in the iOS Developer Library at developer.apple.com or through the Xcode Documentation and API Reference.

Memory Management Programming Guide

WWDC Sessions

The following session videos are available at developer.apple.com.

WWDC 2011, "Session 322: Objective-C Advancements in Depth"

WWDC 2011, "Session 323: Introduction to Automatic Reference Counting"

WWDC 2011, "Session 308: Blocks and Grand Central Dispatch in Practice"

WWDC 2012, "Session 406: Adopting Automatic Reference Counting"

Blogs

MK blog. (Mugunth Kumar). "Migrating your code to Objective-C ARC"
```
http://blog.mugunthkumar.com/articles/migrating-your-code-to-objective-c-
arc
```

mikeash.com. "Friday Q&A 2010-07-16: Zeroing Weak References in Objective-C"
```
http://mikeash.com/pyblog/friday-qa-2010-07-16-zeroing-weak-references-in-
objective-c.html
```

LLVM. "Automatic Reference Counting"
```
http://clang.llvm.org/docs/AutomaticReferenceCounting.html
```

Chapter 6

Getting Table Views Right

Table views are arguably the most frequently used control on the iOS platform. Most of the quality apps on the App Store use table views not just for showing a hierarchical list of data but also for laying out complex structured, scrollable views. Table views are used as cheap substitutes for creating vertically scrollable views even if the content they display is not a list of data. For example, in the built-in contacts app, the contacts list is a `UITableView` and so is the view for adding a new contact. Additionally, new interaction patterns have been introduced by third-party application developers and have been quite commonly used on other apps as well.

iOS has been around for five years, so this chapter assumes that you are well versed with concepts like `UITableViewDelegate` and `UITableViewDataSource`.

> **If you are not familiar with** `UITableViewDelegates` **and** `UITableViewDataSource`, **read Chapter 8 in** *Beginning iPhone Development: Exploring the iPhone SDK* **by Dave Mark and Jeff Lamarche (Apress 2009, ISBN 978-1430216261) before finishing this chapter.**

This chapter focuses on the advanced aspects of table views and shows you how to create complex (yet common) UIs such as pull-to-refresh and infinite scrolling lists. It also briefly explains how to use table view row animations to create accordions or options drawers (a UI that shows available toolbar elements just below the table view cell that is acted upon) and several other interesting UI paradigms.

After exploring new user interaction paradigms, you find out about the best practices for writing cleaner `UITableViewController` code (code that's easy to modify later). With that, it's time to get started.

UITableView Class Hierarchy

A `UITableView` is a subclass of `UIScrollView` that allows users to scroll through a list of `UITableViewCells` (which are in turn a subclass of `UIView`).

`UITableView` and `UIScrollView` share several things in common. For a heavily customized view that is not a list of data, you can use a `UIScrollView` and populate it with `UIView` or `UIControl` subclasses, but there are certain advantages to using a `UITableView` in this case. First, when possible, use of a higher-level abstraction is always advisable. Second, a `UITableView` takes care of several subtle functionalities automatically. One of them is the capability of easily dequeing and reusing `UITableViewCells`, which improves performance and reduces memory consumption. Another is its elegant and easy way for populating content through its data source and for receiving feedback on actions through the delegate. If you use a custom `UIScrollView`, you must do these two manually. Although it's not difficult to implement a scrollable view using `UIScrollView`, you probably will not gain any advantage by doing it yourself.

Understanding Table Views

A UITableView is normally used in conjunction with several other classes such as UITableView Controller and UITableViewCell and protocols UITableViewDelegate and UITableViewDataSource. This section briefly discusses the functionalities of each of these classes/protocols.

UITableViewController

A UITableViewController is a subclass of a UIViewController that performs some additional functions related to table view loading. If you are initializing a UITableViewController from a nib file, the nib loading mechanism loads the archived table view from the nib. If not, it creates an unconfigured table view. In both cases, you can access the table view using the tableView property of the UITableViewController.

Additionally, a UITableViewController reloads the table and clears the cell selection, as the table view is about to appear for the first time (viewWillAppear). It then, in viewDidAppear, flashes the scroll indicators to indicate that the view is scrollable. You can override these methods and provide custom implementations as well.

The UITableViewController also handles the delegates and data source for your table. For table views created without a nib file, the delegate and data source become the table view controller. For table views created with a nib file, the delegate and data source are set from that file.

My recommendation is to use a separate UITableViewController for every table view you use within your view. Using multiple UITableViewControllers makes it easy to understand (and modify) the code later in the project's lifecycle. You find out how to use multiple UITableViewControllers within a single view/nib file later in this chapter.

UITableViewCell

UITableViewCell is a subclass of UIView that adds certain properties and functionalities to a UIView that are useful when used inside a UITableView. You don't need to manually add custom elements because a UITableViewCell has often-used elements such as a title, a subtitle, and a thumbnail image exposed via the properties textLabel, detailedTextLabel, and imageView. When you create a UITableViewCell, you specify the kind of cell you need by specifying a UITableViewCellStyle. The style decides what custom elements the newly created cell supports. For example, a cell created with the UITableViewCellStyleDefault style doesn't support the imageView property. The second most important functionality provided by a UITableViewCell is the capability to maintain distinct selected and highlighted states and the capability to animate between these states.

In most cases, you will be using a custom subclass of a UITableViewCell in your app. The next section discusses the different ways of creating a UITableViewCell and the pros and cons of using it.

Speed Up Your Table View Scrolling

You might already know how to create a custom table view cell that scrolls butter-smooth like Tweetie (Twitter for iPhone). Loren Brichter has open-sourced his custom table view cell and explained in his blog how to do it (see the

"Further Reading" section at the end of this chapter). In this section, you develop a table view and populate it with cells created using different techniques, including Loren's, and you analyze performance using Instruments. When you finish this section, you will understand *why* Loren's method makes your table view scroll smoothly. You also find out *how* to troubleshoot and find performance bottlenecks if your table views aren't scrolling as fast as they should. Once you know the "how" behind a technique, you can apply that technique elsewhere.

A Word on Performance and Interface Builder

Whenever you talk about performance, the first thing you hear from most iOS developers is, "Don't use Interface Builder." Using Interface Builder (IB) to build interfaces is a controversial topic in the iOS developer community. Veteran Mac developers, or those who have switched from developing native Windows apps (using VB or C#), understand what IB does and why it should be used. Some web developers, on the other hand, often correlate IB to web-authoring tools and thus assume that IB slows down the app and degrades performance. My advice is to never pay attention to any advice about improving the performance of your app without measuring it. Tools such as Instruments can help you with performance measurement; later in this chapter, you find out how to use them.

Keep in mind that IB is not a code generator. It is an editor that generates XML-based archives of your view. In most cases, nib files do not lower the performance compared to an equivalently coded UI. (I illustrate this later in this chapter.) Additionally, using a nib file helps you isolate your "view" to a separate file, which keeps your controller free of view-related code (especially in your `viewDidLoad` method). That's a cleaner way to implement and adhere to the MVC design pattern.

To Use or Not to Use Interface Builder?

Having said that, the only place where I recommend using a coded UI over IB is for high-performance `UITable ViewCells`. The iOS rendering mechanism slows down when your `UITableViewCell` has many subviews. As of this writing, based on my profiling (explained later in this chapter), only the latest, dual-core A5-powered iPad 2 and the new iPad gives acceptable scrolling performance for a fairly customized `UITableViewCell` (probably because of the super fast graphics processor). By "acceptable scrolling performance," I mean getting at least 60 frames per second when scrolling the table. You learn how to measure this later in this chapter. The old iPhone 3G was slowest at 25 fps; other devices fall somewhere between 60 and 25 fps.

When you profile the app using Instruments, you will also notice that the performance hit when scrolling a table view isn't caused by unarchiving nib files, but by rendering multiple subviews. Hence, a coded UI doesn't mean moving your `addSubView:` methods to the `initWithStyle:reuseIdentifier` method but rather overriding the `drawRect` method and directly drawing your content instead of using subviews. Avoiding subviews (especially subviews that have transparency and blend with other views beneath) improves performance.

In the next section, you will first write a table view with a thousand rows and measure the scrolling performance using Instruments. You also will use Instruments to identify areas with alpha-blended layers that are time-consuming to render. You gradually improve the performance by avoiding subviews and measuring performance in each step.

To complete the example, I recommend having an iOS device provisioned for development, because performance-related profiling is best done when the app runs on a real device. Because the iOS Simulator runs on a Mac that has a super-fast processor, your scrolling is going to be very smooth, and you will obviously get 60 fps.

UITableView with Subviews in a Custom UITableViewCell

Download the code from the book's website and extract the `Chapter 6/TableViewPerformance` folder. This is a simple view-based iPhone application created using Xcode without storyboards, but with ARC enabled. Open `TableViewPerformanceViewController.xib`. You will see a `UITableView` in it. This is the table view that you will populate with different kinds of cells in this example.

Open the file `CustomCell.xib`. You will see a title label, a subtitle label, and a label for displaying the timestamp. This interface builder nib file will look like Figure 6-1.

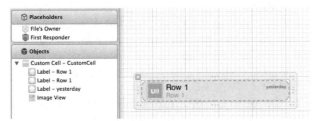

Figure 6-1 Custom cell nib

Now open `TableViewPerformanceViewController.m` and add the following code for the `TableViewDataSource`.

```
- (NSInteger)numberOfSectionsInTableView:(UITableView *)tableView {
return 1;
}

- (NSInteger)tableView:(UITableView *)tableView
         numberOfRowsInSection:(NSInteger)section {
return 1000;
}

// table with normal XIB based cells
- (UITableViewCell *)tableView:(UITableView *)tableView
         cellForRowAtIndexPath:(NSIndexPath *)indexPath {
    static NSString *CellIdentifier = @"CustomCell";

    CustomCell *cell = (CustomCell*)[tableView
dequeueReusableCellWithIdentifier: CellIdentifier];
    if (cell == nil) {
    NSArray *nib = [[NSBundle mainBundle]
loadNibNamed:@"CustomCell" owner:self options:nil];
    cell = (CustomCell*)[nib objectAtIndex:0];
  }

  cell.titleLabel.text = [NSString stringWithFormat:@"Row %d",
  indexPath.row];
  cell.subTitleLabel.text = [NSString stringWithFormat:@"Row %d",
  indexPath.row];
  cell.timeTitleLabel.text = @"yesterday";
  cell.thumbnailImage.image = [UIImage imageNamed:@"ios5"];
```

```
    cell.selectionStyle = UITableViewCellSelectionStyleNone;
    return cell;
}
```

Nothing fancy here. The sample code just sets some arbitrary values to the cells. Now profile this app in Instruments. Choose the device as your target and click and hold the Play button and choose Profile to profile the app. Choose the Core Animation trace template, as shown in Figure 6-2.

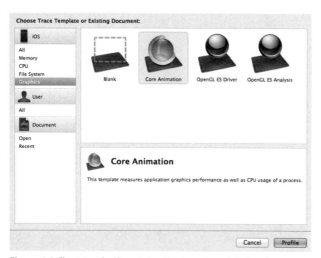

Figure 6-2 Choosing the Core Animation trace template from Instruments

Expand the debug options panel in Instruments by clicking View → Detail in Instruments. (This will be selected and showing up by default.) Select the Color Blended Layers check box, as shown in Figure 6-3.

Figure 6-3 Select Color Blended Layers

The app is now running on your device. Because you turned on Color Blended Layers, your iPhone screen will look similar to Figure 6-4. Now scroll your table view and look at the frames per second measurement on

Instruments. Depending on your debug device's processor and GPU speed, this measurement may vary. We got somewhere around 38 fps to 45 fps on the iPhone 4 running a beta of iOS 6.

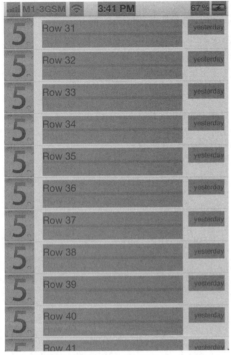

Figure 6-4 iOS device screen using custom cells

With Color Blended Layers, iOS shows transparent layers in red and opaque layers in green. In Figure 6-4, most of the areas around the labels in the custom cell are transparent and blended. Try scrolling the list and observe the fps. On an old iPhone 3G, we were getting close to 25 fps (which is no way near butter-smooth scrolling). Transparent layers (shown in red) have a computational cost to render. The system has to blend the transparent layer with the layer beneath it to compute its color and then draw it. The rendering speed can be drastically improved by avoiding transparent layers and replacing them with opaque equivalents. In the next section, we show you a couple of techniques that will improve the frame rate.

First, replace the custom table view cell with the system's default `UITableViewCell`. Replace the table view data source code with the code in the following section. (This is commented in the sample code)

UITableView with a Default UITableViewCell

Replace the code in `cellForRowAtIndexPath:` in the controller file with this:

```
- (UITableViewCell *)tableView:(UITableView *)tableView
 cellForRowAtIndexPath:(NSIndexPath *)indexPath {

    static NSString *CellIdentifier = @"Cell";
    UITableViewCell *cell = [tableView
```

```
dequeueReusableCellWithIdentifier:CellIdentifier];
    if (cell == nil) {
        cell = [[UITableViewCell alloc]
initWithStyle:UITableViewCellStyleSubtitle
reuseIdentifier:CellIdentifier]];

    }
cell.textLabel.text = [NSString stringWithFormat:
@"Row %d", indexPath.row];
    cell.detailTextLabel.text = [NSString stringWithFormat:
@"Row %d", indexPath.row];
    cell.imageView.image = [UIImage imageNamed:@"ios5"];
    return cell;
}
```

Instead of using your custom cell in this code, you use UIKit's built-in `UITableViewCell` with style `UITableViewCellStyleSubtitle`. Now profile the app again. When you turn on Color Blended Layers, your iPhone screen will look like Figure 6-5.

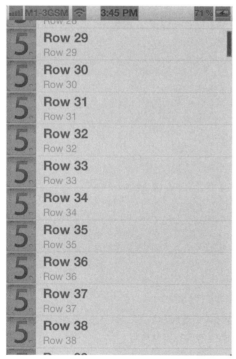

Figure 6-5 iOS device screen using built-in cells

Observe that the transparent layers are all gone except for a few near the images. When you scroll the list, you find that the performance is slightly better and feels smoother than previously. Observe that the fps measurement hits 60. When you hit 60 fps, you can technically stop improving the scrolling performance, but in this case, only the latest iPhone 4 and iPhone 4S were able to reach 60 fps while scrolling. The iPhone 3G and 3GS were much slower.

Moreover, with built-in cells, you're limited to just four styles, and in normal cases that just might not be enough. In the next method, we show you how to use a custom cell that uses Core Graphics methods to draw the image and text directly on the cell without using subviews.

UITableView with a Custom Drawn UITableViewCell

Loren Brichter of Tweetie (now known as Twitter for iPhone) wrote about butter-smooth scrolling in Tweetie. In this example, you use Loren's technique to create a custom cell for your `UITableView`.

The file `CustomDrawnCell.m` in the source code shows how to draw directly in the `drawRect` method. Make changes to the table view controller to use this cell instead of the default cell. When you run this code on your device and turn on Color Blended Layers, you see something like Figure 6-6.

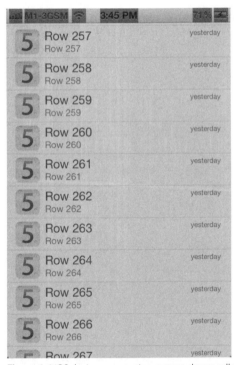

Figure 6-6 iOS device screen using custom-drawn cells

With your custom-drawn cells, every part of the table view cell is opaque, and your table view scrolling is fast and smooth. I was getting 60 fps on nearly every device, including the oldest one, iPhone 3G.

The only problem with this method is that the code you write to draw the content gets annoyingly difficult to read (although it's not difficult to write). Whatever technique you use, try to make your cells as opaque as possible.

Now that you know why Loren's method is fast, you can easily troubleshoot your apps for any performance bottlenecks. In the next section, you briefly look at what could slow down `UITableViewCell` rendering.

Things to Avoid in the UITableViewCell Rendering Method

Always avoid allocating resources while drawing, including allocating objects like `NSDateFormatter`, `UIFont`, or anything that you need while drawing. I recommend you do your allocation in a class-level initialize method and store it in a static variable. Use this static variable for every instance of your cell.

If you still find that the performance is low, use Instruments' Time Profiler on your project and look for bottlenecks. Now that you know how to use Instruments to measure your table view scrolling performance, you can easily improve your app's performance when you find bottlenecks.

Custom Nonrepeating Cells

As I showed you before, table views are used not just for showing a list of data but also for complex and structured scrollable layouts. If your table view structure has a nonrepeating pattern of cells, you can add the custom cell into the same nib file as the table view and connect `IBOutlets`, as illustrated in Figure 6-7. This way, you can just return a pointer to this `IBOutlet` in `cellForRowAtIndexPath:`.

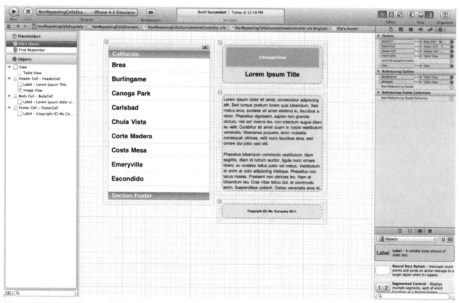

Figure 6-7 The different objects in the nib file and their connections

The following code shows how to return these objects from the `UITableViewDataSource` methods.

UITableViewDataSource Methods

```
-(CGFloat) tableView:(UITableView*) tableView
heightForRowAtIndexPath:(NSIndexPath *)indexPath {

    switch (indexPath.row) {

        case 0:
```

(continued)

```
            return self.headerCell.frame.size.height;
            break;
        case 1:
            return self.bodyCell.frame.size.height;
            break;
        case 2:
            return self.footerCell.frame.size.height;
            break;

        default:
            return 0;
            break;
    }
}

- (UITableViewCell *)tableView:(UITableView *)tableView
  cellForRowAtIndexPath:(NSIndexPath *)indexPath {

    switch (indexPath.row) {

        case 0:
            return self.headerCell;
            break;
        case 1:
            return self.bodyCell;
            break;
        case 2:
            return self.footerCell;
            break;

        default:
            return nil;
            break;
    }
}
```

You can get the code from the `Chapter 6/NonRepeatingCellsExample` folder on the book's website. Note that `UIKit` objects don't conform to `NSCopying` or `NSMutableCopying` protocols and hence cannot be copied or cloned. That means that if you need two body cells—say one in row 1 and another in row 2—you have to load them from their nib files every time you need them. But fret not; the nib file-loading methods are optimized for performance and once loaded, nibs are cached.

> You can use a similar technique as in the previous example for creating custom table view headers and footers. Just create custom table header/footer views within the same nib file and drag them to the `UITableView` in IB. Your view gets added as a header or footer depending on where you dropped it.

Table View Best Practices: Writing Clean Code with Lean Controllers

If you have been doing iOS development for quite a while, you know that your controller's `cellForRowAtIndexPath:` can easily get messy and unmanageable as your project evolves. When you use the model-view-controller paradigm in your software project (not just iOS), strive to make your controller as lean as possible. Keeping the controller lean is arguably the easiest way to keep your code readable and manageable. The next section discusses briefly how to refactor your code adhering to these ideas.

Data Binding Guidelines

When you're writing a table view controller subclass, the bulk of your code is written in the `UITableViewDelegate` and `UITableViewDataSource` methods. Focusing on how to write these methods clearly solves the problem. The `cellForRowAtIndexPath` method often contains code that sets values for every individual UI element of the cell. The best way to set the values for individual UI elements in the cell is to move this code elsewhere. Now, where should it be moved? That depends on the kind of custom cell you're using. Based on your app's functionality, your table views need to be bound with associated data.

This technique, often called *data binding*, is a bit underrepresented on iOS, at least when compared to Mac. The best way to bind data is to pass your data model object to the custom table view cell and let it bind the data.

We classify table view cells as three types based on how we normally associate data with them. The first type, a subclass of `UITableViewCell`, is a custom cell designed to display a specific kind of data, which in most cases is closely tied to the specifics of the app. An example of an RSS Reader app is a `FeedCell` that displays a feed.

The second type is designed and developed in a generic way similar to Apple's `UITableViewCell` implementation. You create your cells by specifying a style, and these cells can be used in other classes or projects for displaying many different types of data models. For example, you could create generic cells such as `MyTableViewSwitchCell` for displaying a title text, an on/off `UISwitch` or `MyTableViewInputCell` for displaying a title text, and a `UITextField` for data entry.

The third type of cell is a native `UITableViewCell` provided by the `UIKit` framework. In any of these three cases, as far as possible, try to move the data-binding code to the cell itself.

The first case is straightforward. Write a method within the `FeedCell` that accepts your model object as a parameter and set the individual UI elements to the values in the Feed model object. That is, move your data-binding code to the `FeedCell`, the subclass of `UITableViewCell`. For example, in the case of a RSS Reader app, the `FeedCell` public method will look similar to the following:

Bind Method in Your FeedCell

```
-(void) bind:(Feed*) feedToBeDisplayed {
self.titleLabel.text = feedToBeDisplayed.text;
self.timeStampLabel.text = feedToBeDisplayed.modifiedDateString;
 . . .
}
```

Instead of this code being written in the view controller's data source method, `cellForRowAtIndexPath:`, it's moved to the `UITableViewCell` subclass. This means that if the format of the cell needs to be changed at a later stage, such as when adding an author name field to your Feed model object and `FeedCell`, you can do so in one place.

When you use the system default `UITableViewCell` for displaying your data, I recommend adding this bind method to a category class on `UITableViewCell`.

The second case is when you have multiple models using the same `UITableViewCell`. I recommend creating multiple category classes, one for each model; for example, create `UITableViewCell+Feed.h/m` for displaying feeds and, say, `UITableViewCell+Subscription.h/m` for displaying subscriptions on the same cell. Be careful when naming the bind method. When a category contains a duplicated method name, it overrides the previously defined method, and there is no defined order in which this overriding happens. I recommend naming them `bind<ModelClassName>`, which is readable and understandable. For example, the names `bindFeed:(Feed*)` and `bindSubscription:(Subscription*)` follow this convention.

The third case is when you have a generic custom table cell like the `MyTableViewSwitchCell`. In this case, too, you can apply the previous technique. Add category methods on your generic custom table view cell.

More often than not, you would be reusing the same `FeedCell` in multiple tables and in multiple view controllers. Moving the data-binding code out of the table view controller (or any generic view controller) will reduce the clutter on the controllers and make it easy to maintain your code.

Multiple UITableViewControllers Inside a Single UIViewController

In most user interfaces, you will often use multiple table views within the same `UIViewController`. Figure 6-8 shows a project with multiple table views within a single `UIViewController`.

Figure 6-8 Interface Builder showing a project with multiple UITableViews within a single view

Spaghetti code starts creeping in when both the table's data source and delegate are set to the file's owner—the parent `UIViewController`. The second stage of "spaghettiness" creeps in when you add `UISearchDisplayController` to both these tables. Now your `cellForRowAtIndexPath:` method will look similar to the following code listing.

Sample cellForRowAtIndexPath

```
-(UITableViewCell *)tableView:(UITableView *)tableView
   cellForRowAtIndexPath:(NSIndexPath *)indexPath {
  if(tableView == self.firstTable)  {
    //return first table's cell
  else if(tableView == self.secondTable) {
    //return second table's cell
}
  else if(tableView ==
    self.firstSearchDisplayController.searchTableView) {
       //return first table's search cell
}
  else if(tableView ==
    self.secondSearchDisplayController.searchTableView) {
       //return second table's search cell
}
}
```

Obviously, there should be a better way, right? As it happens, there is. Instead of setting the delegate and data source to the file's owner, create custom `UITableViewController` subclasses for each table and set the delegate and data source to its own controller, as illustrated in Figure 6-9.

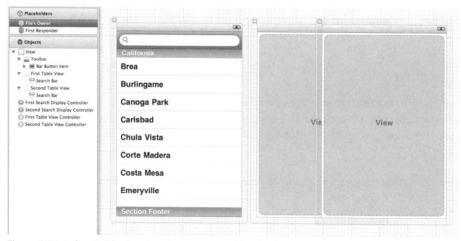

Figure 6-9 Interface Builder showing a better way to add multiple UITableView subclasses within a single view

Create custom subclasses called `FirstTableViewController` and `SecondTableViewController` and move the `cellForRowAtIndexPath` methods in the file's owner to these two classes. You will reduce the number of `if` statements used by half. You can do something similar to this to isolate the search display controller's delegate as well if the code for it gets long and unmanageable. You might end up creating more files and more classes, but that's just fine.

> The first rule of thumb for refactoring is to revisit your code to check whether the `if` statements you're using are truly for a logical branching and not for class-based switching. (In the `cellForRowAtIndexPath:` method).
>
> The second rule is to check if you are using an `if` condition to branch code for different kinds of tables, such as in the `cellForRowAtIndexPath:` method. As I showed you previously, code like this should be refactored and solved elegantly using object-oriented techniques. Every class-based switching like this can be solved in an object-oriented way. This refactoring technique holds good for any language, not just Objective-C.

Adhering to these two refactoring techniques will reduce much of the code in your controller class. Remember that your controller should act as a mediator among your models and UI elements defined at that level and not at the subclass level. In other words, a view controller can set the property of a UI element defined in its scope but not that of a UI element that is inside a subclass. For example,

```
self.textLabel.text = NSLocalizedString(@"Hello", @"")
```

is okay, but

```
self.customView.textLabel.text = NSLocalizedString(@"Hello", @"")
```

should be avoided. The recommended way is to move this code into the `customView`'s class. Apply these techniques and start writing cleaner and leaner controllers.

Summary

In this chapter, I showed you how to use Instruments to measure and improve the scrolling performance of your app. You also found out how to create a scrollable UI using table views even if the data displayed is not a list. Later in this chapter, you discovered some important refactoring techniques to keep your controller code cleaner.

Later in this book, we cover more table view gimmicks, including advanced table view customization and storyboarding in Chapters 12 and 21, respectively.

Further Reading

Apple Documentation

The following documents are available in the iOS Developer Library at `developer.apple.com` or through the Xcode Documentation and API Reference.

What's New in iOS 6

TableView Programming Guide iOS Developer Documentation

TableViewSuite

UIViewController Programming Guide

Other Resources

Fast scrolling in Tweetie with UITableView
`http://www.blog.atebits.com/2008/12/fast-scrolling-in-tweetie-with-uitableview/`

## Chapter 7

Great at Any Angle: Collection Views and Auto Layout

Until now (before iOS 6), developers used `UITableView` for almost any type of display that shows a collection of items. Though Apple has been using a UI that looks similar to the collection views in photo apps for quite a long time (since the original iPhone), that UI wasn't available for use by third-party developers. We depended on third-party frameworks like three20, or we hacked our own to display a collection list. For the most part they worked well, but implementing animation when items are added or deleted or implementing a custom layout like cover flow was difficult. Fret not. iOS 6 introduces a brand new controller for iOS just for displaying a collection of items. Collection view controllers are a completely new building block UI that's similar to table view controllers. In this chapter, I introduce you to collection views in general, and you find out how to display a list of items using a collection view.

Later in this chapter, I introduce you to one of Cocoa's features, Auto Layout. Auto Layout debuted for the Mac last year with Mac OS 10.7 SDK. Apple is introducing Auto Layout for iOS apps with iOS 6. In this chapter, you discover Auto Layout, focusing specifically on the layout options that are impossible to do now without writing code. I assume that you have experience customizing table views. If not, read Chapter 6 and then come back to this chapter. I also assume that you know the basics of storyboards and segues. If you don't, see Chapter 21 of this book. You don't need to know about complex custom transition effects with storyboards, just a basic understanding will do.

After you complete this chapter, you'll be proficient with two new UI concepts that will help you push the limits on your apps and take them to the next level. With that, it's time to get started.

Collection Views

iOS 6 introduces a new controller, `UICollectionViewController`. Collection views provide a much more elegant way to display items in a grid than was previously possible using `UIKit`. Collection views were available on the Mac OS X SDK (`NSCollectionView`); however the iOS 6 collection view (`UICollectionView`) is very different from the Mac equivalent. In fact, iOS 6's `UICollectionViewController`/`UICollectionView` is more similar to `UITableViewController`/ `UITableView`, and if you know how to use `UITableViewController`, you'll be at home with `UICollectionViewController`.

In this chapter, I explain the different classes that you need to know about, and I walk you though an app that displays a directory of images in a collection view. As I walk you through this, I'll be comparing

`UICollectionViewController` with the `UITableViewController` that you know well by now (because learning by association helps us better remember).

Classes and Protocols

In this section I describe the most important classes and protocols that you need to know when you implement a collection view.

UICollectionViewController

The first and most important class is the `UICollectionViewController`. This class functions similarly to `UITableViewController`. It manages the collection view, stores the data required, and handles the data source and delegate protocol.

UICollectionViewCell

This is similar to the good old `UITableViewCell`. You normally don't have to create a `UICollection ViewCell`. You'll be calling the `dequeueCellWithReuseIdentifier:indexPath:` method to get one from the collection view. Collection views behave like table views inside a storyboard. You create `UICollectionViewCell` types (like prototype table view cells) in Interface Builder (IB) inside the `UICollectionView` object instead. Every `UICollectionViewCell` is expected to have a `Cell Identifier`, failing which you'll get a compiler warning. The `UICollectionViewController` uses this `CellIdentifier` to enqueue and dequeue cells. The `UICollectionViewCell` is also responsible for maintaining and updating itself for selected and highlighted states. I show you how to do this later in this chapter.

The following code snippet shows how to get a `UICollectionViewCell`:

```
MKPhotoCell *cell = (MKPhotoCell*)
[collectionView dequeueReusableCellWithReuseIdentifier:
@"MKPhotoCell" forIndexPath:indexPath];
```

The code is straight from the sample code for this chapter that I walk you through later.

> If you're not using Interface Builder or storyboards, you can call the `registerClass:forCellWith ReuseIdentifier:` in `collectionView` to register a nib file.

UICollectionViewDataSource

You've probably already deduced by now that this should be similar to `UITableViewDataSource`. Yes, you're right. The data source protocol has methods that should be implemented by the `UICollectionViewController`'s subclass. In the first example, you will implement some of the methods in the data source to display photos in the collection view.

UICollectionViewDelegate

The delegate protocol has methods that should be implemented if you want to handle selection or highlighting events in your collection views. In addition, the `UICollectionViewCell` can also show context-sensitive menus like the Cut/Copy/Paste menu. The action handlers for these methods are also passed to the delegate.

Example

You start by creating a single view application in Xcode. In the second panel, select Use Storyboards and Use Automatic Reference Counting, Choose iPad as the target device and click Next, and choose a location to save your project.

Editing the Storyboard

Open the `MainStoryboard.storyboard` file and delete the only `ViewController` inside it. Drag a `UICollectionViewController` from the object library. Ensure that this controller is set as the initial view controller for the storyboard.

Now, open the only view controller's header file in your project. Change the base class from `UIView Controller` to `UICollectionViewController` and implement the protocols `UICollection ViewDataSource` and `UICollectionViewDelegate`. Go back to the storyboard and change the class type of the collection view controller to `MKViewController` (or whatever, depending on your class prefix).

Build and run your app. If you followed the steps properly, you'll see a black screen in your iOS simulator.

We could have used a blank application template and added a storyboard with a collection view as well. But that requires changes to App Delegate and your `Info.plist` because the Xcode's blank application template doesn't use storyboards.

Adding Your First Collection View Cell

Well, the application's output wasn't impressive, right? Lets' spice it up. Go ahead and add a class that is a subclass of `UICollectionViewCell`. In the sample code, I'm calling it `MKPhotoCell`. Open your storyboard and select the only collection view cell inside your collection view controller. Change the class to `MKPhotoCell`. This is shown in Figure 7-1.

Open the Attributes Inspector (the fourth tab) in the Utilities and set the Indentifier to `MKPhotoCell`. This step is very important. You will use this identifier to dequeue cells later in code. Add a blank `UIView` as the subview of your collection view cell and change the background color to red (so you can see it).

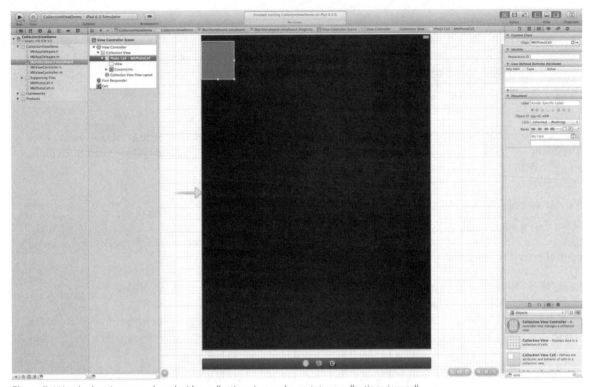

Figure 7-1 Xcode showing a storyboard with a collection view and a prototype collection view cell

Implementing Your Data Source

Now implement the data source as shown here:

```
-(NSInteger)numberOfSectionsInCollectionView:(UICollectionView *)
  collectionView {

  return 1;
}

- (NSInteger)collectionView:(UICollectionView *)
collectionView numberOfItemsInSection:(NSInteger)section {

  return 100;
}

// The cell that is returned must be retrieved from a call to
// -dequeueReusableCellWithReuseIdentifier:forIndexPath:
- (UICollectionViewCell *)collectionView:(UICollectionView *)
  collectionView cellForItemAtIndexPath:(NSIndexPath *)indexPath {

  MKPhotoCell *cell = (MKPhotoCell*)
   [collectionView dequeueReusableCellWithReuseIdentifier:@"MKPhotoCell"
   forIndexPath:indexPath];
```

```
      return cell;
   }
```

Now, build and run the app. You'll see a grid of 100 cells, each painted in red. Impressive. All it took was a few lines of code. What is even more impressive is that the grid rotates and aligns itself when you turn your iPad to landscape.

Using the Sample Photos

You'll now replace the red-colored subview with something more interesting. You're going to display photos from a directory. Copy some photos (about 50) to your project. You can use the sample photos from the example code.

Remove the red-colored subview you added in the previous section and add a UIImageView and a UILabel to the UICollectionViewCell. Add outlets in your UICollectionViewCell subclass and connect them appropriately in Interface Builder.

Preparing Your Data Source

Prepare your data source by iterating through the files in your directory. Add this to your viewDidLoad method in the collection view controller subclass. You can find it in the MKViewController class in the example code.

```
      self.photosList = [[NSFileManager defaultManager]
        contentsOfDirectoryAtPath:[self photosDirectory] error:nil];
```

The photosDirectory method is defined as

```
   -(NSString*) photosDirectory {
     return [[[NSBundle mainBundle] resourcePath]
   stringByAppendingPathComponent:@"Photos"];
   }
```

Now, update your data source methods to return data based on this information. Your previous code was returning 1 section and 100 items in that section. Change the value 100 to the number of photos you added to your project in the previous section. This is the size of the photosList array.

UICollectionViewDataSource Methods (MKViewController.m)
```
- (UICollectionViewCell *)collectionView:(UICollectionView *)
   collectionView cellForItemAtIndexPath:(NSIndexPath *)indexPath {

   MKPhotoCell *cell = (MKPhotoCell*) [collectionView
   dequeueReusableCellWithReuseIdentifier:@"MKPhotoCell"
   forIndexPath:indexPath];

   NSString *photoName = [self.photosList objectAtIndex:indexPath.row];
   NSString *photoFilePath = [[self photosDirectory]
    stringByAppendingPathComponent:photoName];
   cell.nameLabel.text =[photoName stringByDeletingPathExtension];
   UIImage *image = [UIImage imageWithContentsOfFile:photoFilePath];
   UIGraphicsBeginImageContext(CGSizeMake(128.0f, 128.0f));
   [image drawInRect:CGRectMake(0, 0, 128.0f, 128.0f)];
```

```
    cell.photoView.image = UIGraphicsGetImageFromCurrentImageContext();
    UIGraphicsEndImageContext();

    return cell;
}
```

Build and run the app now. You'll see the photos neatly organized in rows and columns. Note, however, that the app creates a `UIImage` from a file in the `collectionView:cellForItemAtIndexPath:` method. This is going to hurt performance.

Improving Performance

You can improve the performance by using a background GCD queue to create the images and optionally cache them for performance. Both these methods are implemented in the sample code. For an in-depth understanding on GCD, read Chapter 13 in this book.

Supporting Landscape and Portrait Photos

The previous example was good, but not great. Both portrait and landscape images were cropped to 128×128 and looked a bit pixelated. The next task is to create two cells, one for landscape images and another for portrait images. Because both portrait and landscape images differ only by the image orientation, you don't need an additional `UICollectionViewCell` subclass.

Create another `UICollectionViewCell` in your storyboard and change the class to `MKPhotoCell`. Change the image view's size so that the orientation is portrait and the older cells' orientation is landscape. You can use 180×120 for the landscape cell and 120×180 for the portrait cell. Change the `CellIdentifier` to something like `MKPhotoCellLandscape` and `MKPhotoCellPortrait`. You'll be dequeuing one of these based on the image size.

When you're done, your storyboard should look like Figure 7-2.

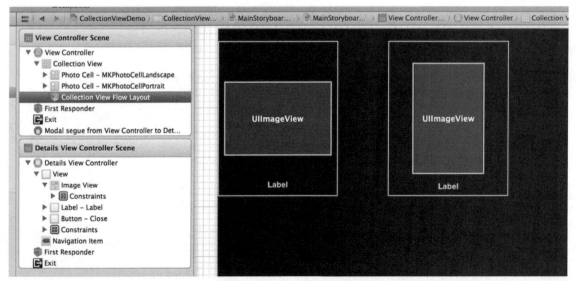

Figure 7-2 Storyboard showing two cells, one for landscape images and another for portrait images

Determining the Orientation

You now have to determine which cell to dequeue based on the image's orientation. You can get the image orientation by inspecting the size property of the `UIImage` after you create it. If the image is wider than its height, it's a landscape photo; if not it's in portrait orientation. Of course, if you're going to compute this in your `collectionView:cellForItemAtIndexPath:` method, your scrolling speed will be affected. The sample code precomputes the size and hence the orientation and stores the orientation in a separate array when the view loads, all on a background GCD queue. The following code listing shows the modified `collectionView:cellForItemAtIndexPath:`.

collectionView:cellForItemAtIndexPath: Method in MKViewController.m

```
- (UICollectionViewCell *)collectionView:(UICollectionView *)
  collectionView cellForItemAtIndexPath:(NSIndexPath *)indexPath {

  static NSString *CellIdentifierLandscape = @"MKPhotoCellLandscape";
  static NSString *CellIdentifierPortrait = @"MKPhotoCellPortrait";

  int orientation = [[self.photoOrientation objectAtIndex:
   indexPath.row] integerValue];

  MKPhotoCell *cell = (MKPhotoCell*)
    [collectionView dequeueReusableCellWithReuseIdentifier:
                             orientation == PhotoOrientationLandscape ?
                             CellIdentifierLandscape:CellIdentifierPortrait
                             forIndexPath:indexPath];

  NSString *photoName = [self.photosList objectAtIndex:indexPath.row];
  NSString *photoFilePath = [[self photosDirectory]
    stringByAppendingPathComponent:photoName];
  cell.nameLabel.text =[photoName stringByDeletingPathExtension];

  __block UIImage* thumbImage = [self.photosCache objectForKey:photoName];
  cell.photoView.image = thumbImage;

  if(!thumbImage) {
    dispatch_async(dispatch_get_global_queue
    (DISPATCH_QUEUE_PRIORITY_HIGH, 0), ^{

      UIImage *image = [UIImage imageWithContentsOfFile:photoFilePath];
      if(orientation == PhotoOrientationPortrait) {
        UIGraphicsBeginImageContext(CGSizeMake(180.0f, 120.0f));
        [image drawInRect:CGRectMake(0, 0, 120.0f, 180.0f)];
        thumbImage = UIGraphicsGetImageFromCurrentImageContext();
        UIGraphicsEndImageContext();
      } else {

        UIGraphicsBeginImageContext(CGSizeMake(120.0f, 180.0f));
        [image drawInRect:CGRectMake(0, 0, 180.0f, 120.0f)];
        thumbImage = UIGraphicsGetImageFromCurrentImageContext();
```

(continued)

```
        UIGraphicsEndImageContext();
    }

    dispatch_async(dispatch_get_main_queue(), ^{

        [self.photosCache setObject:thumbImage forKey:photoName];
        cell.photoView.image = thumbImage;
    });
  });
}

  return cell;
}
```

That completes the data source. But you're not done yet. Go ahead and implement the delegate method. The delegate method will tell you which collection view was tapped, and you'll display the photo in a separate form sheet.

Adding a Selected State to the Cell

First, add a `selectedBackgroundView` to your cell:

```
-(void) awakeFromNib {

    self.selectedBackgroundView = [[UIView alloc] initWithFrame:self.frame];
    self.selectedBackgroundView.backgroundColor = [UIColor
colorWithWhite:0.3
    alpha:0.5];

    [super awakeFromNib];
}
```

Now, implement the following three delegate methods and return YES for the delegate that asks if the collection view should select/highlight the said item.

UICollectionViewDelegate methods in MKViewController.m

```
- (BOOL)collectionView:(UICollectionView *)collectionView
    shouldHighlightItemAtIndexPath:(NSIndexPath *)indexPath {

  return YES;
}

- (BOOL)collectionView:(UICollectionView *)collectionView
    shouldSelectItemAtIndexPath:(NSIndexPath *)indexPath {

  return YES;
}
```

Now, build and run the app. Tap on a photo. You'll see a selection as shown in Figure 7-3.

Figure 7-3 Screenshot of iOS Simulator showing cell selection

Handling Taps on Your Collection View Cell

That was easy, right? Now, show a detailed view when you tap on this item. Create a details controller in your storyboard. Add necessary outlets in your controller subclass (a `UIImageView` and a `UILabel`) and a property to pass the photo to be displayed from the collection view.

Now, create a segue by Ctrl-clicking the collection view controller and dragging it to your details controller.

Select the segue and set the identifier of your segue to something like MainSegue. Also change the style to "modal" and presentation to "Form Sheet." You can leave the transition style to the default value or change it to whatever you prefer.

Implement the last delegate method, `collectionView:didSelectItemAtIndexPath:`.

collectionView:didSelectItemAtIndexPath: in MKViewController.m

```
-(void) collectionView:(UICollectionView *)collectionView
   didSelectItemAtIndexPath:(NSIndexPath *)indexPath {

   [self performSegueWithIdentifier:@"MainSegue" sender:indexPath];
}
```

Implement the `prepareForSegue` method.

prepareForSegue Method in MKViewController.m

```
-(void) prepareForSegue:(UIStoryboardSegue *)segue sender:(id)sender {

   NSIndexPath *selectedIndexPath = sender;
   NSString *photoName = [self.photosList
   objectAtIndex:selectedIndexPath.row];

   MKDetailsViewController *controller = segue.destinationViewController;
   controller.photoPath = [[self photosDirectory]
     stringByAppendingPathComponent:photoName];
}
```

Build and run the app and tap on a photo. You'll see the Details view showing the photo you tapped.

That completes handling the delegate. Wait! That's not the end of the road! You can also add multiple sections to the collection view and set decoration views and backgrounds to them. I'll show you how to do this now.

Adding a "Header" and a "Footer"

Well, a collection view doesn't call them headers and footers. When you open the `UICollectionViewDataSource` header file, you probably won't find any methods that can be used to add these elements. A collection view calls these views supplementary views, and there are two kinds of supplementary views: `UICollectionElementKindSectionHeader` and `UICollectionElementKindSectionFooter`.

You add a supplementary view in your storyboard by selecting the `CollectionView` and enabling the "Section Header" or "Section Footer" and dragging a `UICollectionReusableView` from the object browser to your collection view. The next important step is to set an identifier to your supplementary view. You can do this by selecting the newly dragged `UICollectionReusableView` and setting the identifier on the Utilities panel.

Next, you add methods (in fact, just one method) to your `UICollectionViewController` subclass.

Collection View Datasource Method to Provide a Footer

```
- (UICollectionReusableView *)collectionView:(UICollectionView
    *)collectionView
            viewForSupplementaryElementOfKind:(NSString *)kind
                            atIndexPath:(NSIndexPath *)indexPath {


   static NSString *SupplementaryViewIdentifier =
   @"SupplementaryViewIdentifier";

   return [collectionView

dequeueReusableSupplementaryViewOfKind:UICollectionElementKindSectionFooter
      withReuseIdentifier:SupplementaryViewIdentifier
      forIndexPath:indexPath];
}
```

This is slightly different from a table view where you create a section header in the data source method and return it. The collection view controller takes it to the next level by adding dequeue support to supplementary views as well.

Collection View Layout and Advanced Customization

The most important differentiator of a collection view and a table view is that a collection view doesn't know how to lay out its items. It delegates the layout mechanics to a `UICollectionViewLayout` subclass. The default layout is a *flow* layout provided by the `UICollectionViewFlowLayout` class. This class allows you to tweak various settings through `UICollectionViewDelegateFlowLayout` protocol. For instance, you can change the `scrollDirection` property from vertical (the default) to horizontal and instantly get a horizontally scrolling collection view.

The most important part of `UICollectionViewLayout` is that by subclassing it, you can create your own custom layouts. For example, a "CoverFlowLayout". `UICollectionViewFlowLayout` happens to be one such subclass that is provided by default from Apple. For this reason, collection view calls the "headers" and "footers" supplementary views. It's a header if you're using the built-in flow layout. It could be something else if you're using a custom layout like, say, a cover flow layout.

Collection views are powerful because the layout is handled independent of the class, and this alone means that you'll probably see a huge array of custom UI components, including springboard imitations, custom home screens, visually rich layouts like cover flow, pinterest-style UIs, and such. If you're making an app that requires a rich layout, upgrade your minimum target to iOS 6 just for using collection views. Use it and try making simple applications using collection views, and you'll gradually become a pro.

With that, it's time to start the next important layout technology, Cocoa Auto Layout.

Cocoa Auto Layout

Cocoa Auto Layout was introduced last year with Mac OS X SDK. Apple has a long-standing history of testing new SDK features on Mac SDKs and bringing them to iOS the following year. For example, Apple introduced `SceneKit.framework` in the Mac OS X 10.8 SDK, and this might just find its way to iOS in the next release (iOS 7).

Auto Layout is a *constraint-based layout* engine that *automatically* adjusts the frames based on constraints set on your object. The layout model used prior to iOS 6 is called the "springs and struts" model. Although that worked efficiently for the most part, you still had to write code to lay out your subviews when trying to do custom layout during a rotation. Also, because Auto Layout is constraint-based and not frame-based, you can size your UI elements based on content automatically. This means that translating your UI to a new language just got easier. You don't have to create multiple XIBs, one for every language anymore. Constraints are again instances of the `NSLayoutConstraint` class.

> Auto Layout is supported only on iOS 6. This means that if you're intending to support iOS 5 as a deployment target, well, you guessed it, you can't use Auto Layout. Xcode 4.4.1, the latest at the time of this writing, creates every nib file with Auto Layout turned on by default, even if your minimum deployment target is not iOS 6. Using Auto Layout in unsupported operating systems will crash the whole app when you navigate to that view controller. Before submitting your apps, check every nib file and ensure that you haven't—even accidentally—turned on Auto Layout for your iOS 5 apps.

Using Auto Layout

Whenever Apple introduces a new technology, you can use it in one of the following ways. You can either convert the whole application to use the new technology, or migrate it partially, or write new code using the new technology. With iOS 5, you followed one of these ways for ARC conversion and storyboards. In iOS 6, this applies to Auto Layout. You can migrate an existing application completely to Auto Layout, use it only for new UI elements, or do a partial conversion.

I recommend doing a complete rewrite, even if you've written lots of boilerplate layout code. With Auto Layout, you can almost completely get rid of it. Moreover, just by enabling Auto Layout in your nib file, Interface Builder automatically converts your auto-resize masks to equivalent Auto Layout constraints.

Getting Started with Auto Layout

Thinking in terms of Auto Layout might be difficult if you are a "pro" at using the struts and springs method. But for a newbie to iOS, Auto Layout might be easy. My first recommendation is to "unlearn" the struts and springs method.

Why do I say that? The human brain is an associative machine. When you actually lay out your UI elements, your mental model (or your designer's mental model) is indeed "thinking" in terms of visual constraints. This happens not only when you are designing user interfaces; you do it subconsciously when you're shifting furniture in your living room. You do it when you park your car in a parking lot. You do it when you hang a photo frame on your wall. In fact, when you were doing UI layout using the old springs and struts method, you were converting the relative, constraint-based layout in your mind to what springs and struts can offer, and you were writing code to do layout that cannot be done automatically. With Auto Layout, you can express how you think—by applying visual constraints.

Here's what your designer may normally say:

- "Place the button 10 pixels from the bottom."
- "Labels should have equal spacing between them."
- "The image should be centered horizontally at all times."
- "The distance between the button and the image should always be 100 pixels."
- "The button should always be at least 44 pixels tall."
- and so on . . .

Before Auto Layout, you were manually converting these design criteria to auto-resize mask constraints, and for criteria that cannot be easily done with auto-resize masks, you wrote layout code in the `layoutSubviews` method. With Auto Layout, most of these constraints can be expressed naturally.

For example, a constraint like "Labels should have equal spacing between them" is a constraint that cannot be implemented with the springs and struts method. You have to write layout code by calculating the frame sizes of the superview and calculating the frames of all the labels. The word "them" in the above-mentioned constraint makes the constraint relative to another label within the same view (a sibling view instead of the super view).

Relative Layout Constraints

Say hello to the Auto Layout's relative layout constraint. In fact, one of the most powerful features of Auto Layout is the capability to descriptively define the constraints of a UI element with respect to the parent and with respect to its sibling. A button can have a constraint that positions itself based on the image above it. With springs and struts, you were able to control the anchor points and frames of an element based only on its superview.

Here's an example. I'm going to make an app that displays nine images in a grid fashion as shown in Figure 7-4.

You can get the source code for this (AutoLayoutDemo) from the book's website. Open the storyboard and turn off Auto Layout. Run the app and rotate your simulator (or device if you're running on your device). Notice that in landscape mode, the images in the center row are not aligned the way you want them to be. This is because the springs and struts model allows you to anchor your elements relative only to the superview, whereas here the images in the center row are constrained to be equally spaced from the top row and bottom row images.

Now turn Auto Layout on. Run the app again and rotate the simulator. Notice that the images lay out properly on any orientation without you having to write a single line of code! When you turn on Auto Layout, Xcode by default converts the springs and struts-based resizing mask constraints to equivalent auto layout constraints, and in some cases (like this example), it works without you having to write layout code, including for layout animation.

Figure 7-4 Interface Builder showing a 3×3 grid of images

Well, that was simple. But that example didn't really show you the power of Auto Layout. Here's another one. Assume that you want to add a button just below each image view and set criteria that always places the buttons at the bottom of the image view regardless of its orientation, as illustrated in Figure 7-5.

If you were using the springs and struts model, this wouldn't be an easy constraint. You have to calculate the position of your button every time the layout changes (in your `layoutSubviews` method) and animate it to the new position. Not for the weak-hearted. Moreover, layout code like this is not portable across apps, and adding a new UI element to this view requires a huge amount of changes.

Auto Layout makes this easy. Add a button. Select the button and the image that you want the button to position it to. You'll now add vertical and horizontal spacing constraints to the button with respect to the image. This is shown in Figure 7-6.

Figure 7-5 Button placement below the image views

When you add these two constraints, as shown in Figure 7-6, and run the app, you'll see that the button position remains the same with respect to the image in both orientations without you having to again lay out the UI by code.

The next important feature of Auto Layout is the ability to express your layout constraint by code using a syntactic language that Apple calls Visual Format Language.

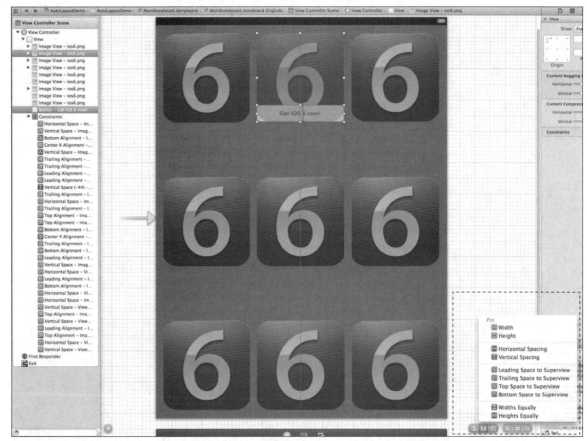

Figure 7-6 Screenshot of Interface Builder showing how to add vertical and horizontal spacing constraints using Auto Layout

Visual Format Language

In this section, you learn about Visual Format Language and how to lay out UI by code using that language, the Auto Layout way, instead of calculating frames and repositioning them. So far you've been working with Interface Builder to lay out your subviews. Visual Format Language will be helpful if you're using code to create and lay out the UI elements.

Adding a layout constraint to a view is easy. Create an `NSLayoutConstraint` instance and add it as a constraint to the view. You can create an `NSLayoutConstraint` instance using the Visual Format Language or using the class method.

Adding a Constraint to a View

```
[self.view addConstraint:
[NSLayoutConstraint constraintWithItem:self.myLabel
                      attribute:NSLayoutAttributeRight
                      relatedBy:NSLayoutRelationEqual
                         toItem:self.myButton
```

```
                  attribute:NSLayoutAttributeLeft
               multiplier:10.0
                  constant:100.0]];
```

Adding a Constraint Using Visual Format Language

```
NSDictionary *viewsDictionary = NSDictionaryOfVariableBindings(self.
myLabel, self.myButton);
NSArray *constraints = [NSLayoutConstraint
                        constraintsWithVisualFormat:@"[myLabel]-100-
                        [myButton]"
                        options:0 metrics:nil
                        views:viewsDictionary];
[self.view addConstrints:constraints];
```

The second method is more expressive and lets you specify a constraint using an *ASCII Art* style. In the preceding example, you created a constraint that ensures that the space between the label and the button is always 100 pixels. The following is the visual format language syntax for specifying this constraint:

```
[myLabel]-100-[myButton]
```

Well, that was easy. Now, take a look at a more complicated example. The Visual Format Language is powerful and expressive, almost like regular expressions, yet readable. You can connect the labels and buttons to the superview using the following syntax:

```
|-[myLabel]-100-[myButton]-|
```

You can add a nested constraint to the button as follows:

```
|-[myLabel]-100-[myButton (>=30)]-|
```

This constraint will ensure that myButton will be at least 30 pixels at all orientations.

You can even add multiple constraints to the same button:

```
|-[myLabel]-100-[myButton (>=30, <=50)]-|
```

This constraint will ensure that myButton will be at least 30 pixels but not more than 50 pixels at all orientations.

Now, when you add conflicting constraints like the one here

```
|-[myLabel]-100-[myButton (>=30, ==50)]-|
```

Auto Layout will gracefully try to satisfy by ignoring the conflicting constraint. But when your UI layout cannot be performed without ambiguities, Auto Layout will crash. I show you two methods to handle debugging errors in the next couple of sections.

Debugging Layout Errors

When Auto Layout throws exceptions, you see something like the following on your console. You can simulate this crash from the example code for this chapter. Open the AutoLayoutDemo code and go to the Details view by tapping the Get iOS 6 Now button. Rotate your iPad to see this crash.

```
Unable to simultaneously satisfy constraints.
    Probably at least one of the constraints in the following list is one
you
    don't want. Try this: (1) look at each constraint and try to figure out
    which you don't expect; (2) find the code that added the unwanted
    constraint or constraints and fix it. (Note: If you're seeing
    NSAutoresizingMaskLayoutConstraints that you don't understand, refer to
    the documentation for the UIView property
    translatesAutoresizingMaskIntoConstraints)
(
    "<NSAutoresizingMaskLayoutConstraint:0x762ed60 h=--& v=--&
     H:[UIView:0x762b2b0(748)]>",
    "<NSLayoutConstraint:0x762af50 V:[UIImageView:0x762b310(400)]>",
    "<NSLayoutConstraint:0x762adc0 V:|-(304)-[UIImageView:0x762b310]
     (Names: '|':UIView:0x762b2b0 )>",
    "<NSLayoutConstraint:0x762ad80 V:[UIImageView:0x762b310]-(300)-|
     (Names: '|':UIView:0x762b2b0 )>"
)

Will attempt to recover by breaking constraint
<NSLayoutConstraint:0x762af50 V:[UIImageView:0x762b310(400)]>

Break on objc_exception_throw to catch this in the debugger.
The methods in the UIConstraintBasedLayoutDebugging category on UIView
listed in <UIKit/UIView.h> may also be helpful.
```

The crash will list of the constraints Auto Layout is trying to satisfy. In the preceding case, there's a constraint that the image height should be at least 400 pixels tall and 300 pixels from the top and bottom edge of the superview. When the iPad was in portrait mode, the layout engine was able to lay out the UI with all the constraints, but when you rotated it to landscape mode, it crashed because in landscape mode, the height of the iPad is much less and can't satisfy the constraints provided. The log also says that it will try to recover by breaking a constraint.

You might not see crashes like this when you use Interface Builder because when you try to add a conflicting constraint, Interface Builder automatically adds several other constraints to counterbalance yours. In most cases, you will see such crashes when you use the Visual Format Language or manual code-based layout,.

The next common case for this crash is when your view automatically translated auto-resizing masks to constraints. When that happens, you'll see a lot more constraints (including many that weren't directly created by you) in the crash log. Debugging will be easier if you turn off the option that translates resizing masks to constraints on a per-view basis. You do so by calling the method

```
[self.view setTranslatesAutoresizingMaskIntoConstraints:NO];
```

While troubleshooting crashes for which there are log entries you don't understand, set `translates AutoResizingMaskIntoConstraints` to `NO`, one-by-one, for every view that you think might be offending, and when you spot the offending view, remove the translated constraints and write your own.

Summary

In this chapter, you learned about two powerful techniques that will revolutionize the way you write custom controls and layouts. Auto Layout, by itself, may not be a compelling reason to develop for only iOS 6, but if your app is data-centric (an RSS reader or a Twitter client, for example), collection views should be a compelling reason to support only iOS 6 in your apps. Auto Layout might be interesting down the line, say after five or six months when most users have already moved to iOS 6. To end users, Auto Layout may not offer benefits, but for developers, using Auto Layout means deleting tons of layout code from your controllers. When iOS 6 is out of NDA, you'll be seeing more and more third-party, open source controls (or controllers) and layout classes using collection views that are easier to use and understand. Collection views are a new building block element like table views and you should start learning them to take your apps to the next level.

Further Reading

Apple Documentation

The following documents are available in the iOS Developer Library at `developer.apple.com` or through the Xcode Documentation and API Reference.

UICollectionViewController Class Reference

UICollectionViewLayout Class Reference

Cocoa Auto Layout Guide

WWDC Sessions

WWDC 2012, "Session 205: Introducing Collection Views"

WWDC 2012, "Session 219: Advanced Collection Views and Building Custom Layouts"

WWDC 2012, "Session 202: Introduction to Auto Layout for iOS and OS X"

WWDC 2012, "Session 228: Best Practices for Mastering Auto Layout"

WWDC 2012, "Session 232: Auto Layout by Example"

Chapter 8
Better Drawing

Your users expect a beautiful, engaging, and intuitive interface. It's up to you to deliver. No matter how powerful your features, if your interface seems "clunky," you're going to have a hard time making the sale. This is about more than just pretty colors and flashy animations. A truly beautiful and elegant user interface is a key part of a user-centric application. Keeping your focus on delighting your user is the key to building exceptional applications.

One of the tools you need in order to create an exceptional user interface is custom drawing. In this chapter, you learn the mechanics of drawing in iOS, with focus on flexibility and performance. This chapter doesn't cover iOS UI design. For information on how to design iOS interfaces, start with Apple's *iOS Human Interface Guidelines* and *iOS Application Programming Guide*, available in the iOS developer documentation.

In this chapter, you find out about the several drawing systems in iOS, with a focus on UIKit and Core Graphics. By the end of this chapter, you will have a strong grasp of the UIKit drawing cycle, drawing coordinate systems, graphic contexts, paths, and transforms. You'll know how to optimize your drawing speed through correct view configuration, caching, pixel alignment, and use of layers. You'll be able to avoid bloating your application bundle with avoidable prerendered graphics.

With the right tools, you can achieve your goal of a beautiful, engaging, and intuitive interface, while maintaining high performance, low memory usage, and small application size.

iOS's Many Drawing Systems

iOS has several major drawing systems: UIKit, Core Graphics (Quartz), Core Animation, Core Image, and OpenGL ES. Each is useful for a different kind of problem.

- **UIKit**—This is the highest-level interface, and the only interface in Objective-C. It provides easy access to layout, compositing, drawing, fonts, images, animation, and more. You can recognize UIKit elements by the prefix UI, such as UIView and UIBezierPath. UIKit also extends NSString to simplify drawing text with methods like drawInRect:withFont:.

- **Core Graphics (also called Quartz 2D)**—The primary drawing system underlying UIKit, Core Graphics is what you use most frequently to draw custom views. Core Graphics is highly integrated with UIView and other parts of UIKit. Core Graphics data structures and functions can be identified by the prefix CG.

- **Core Animation**—This provides powerful two- and three-dimensional animation services. It is also highly integrated into UIView. Chapter 8 covers Core Animation in detail.

■ **Core Image**—A Mac technology first available in iOS 5, Core Image provides very fast image filtering such as cropping, sharpening, warping, and just about any other transformation you can imagine.

■ **OpenGL ES**—Most useful for writing high-performance games—particularly 3D games—Open GL ES is a subset of the OpenGL drawing language. For other applications on iOS, Core Animation is generally a better choice. OpenGL ES is portable between most platforms. A discussion of OpenGL ES is beyond the scope of this book, but many good books are available on the subject.

UIKit and the View Drawing Cycle

When you change the frame or visibility of a view, draw a line, or change the color of an object, the change is not immediately displayed on the screen. This sometimes confuses developers who incorrectly write code like this:

```
progressView.hidden = NO; // This line does nothing
[self doSomethingTimeConsuming];
progressView.hidden = YES;
```

It's important to understand that the first line (`progressView.hidden = NO`) does absolutely nothing useful. This code does not cause the progress view to be displayed while the time-consuming operation is in progress. No matter how long this method runs, you will never see the view displayed. Figure 8-1 shows what actually happens in the drawing loop.

All drawing occurs on the main thread, so as long as your code is running on the main thread, nothing can be drawn. That is one of the reasons you should never execute a long-running operation on the main thread. Not only does it prevent drawing updates but it also prevents event handling (such as responding to touches). As long as your code is running on the main thread, your application is effectively "hung" to the user. This isn't noticeable as long as you make sure that your main thread routines return quickly.

You may now be thinking, "Well, I'll just run my drawing commands on a background thread." You generally can't do that because drawing to the current UIKit context isn't thread-safe. Any attempt to modify a view on a background thread leads to undefined behavior, including drawing corruption and crashes. (See the section "Caching and Background Drawing," later in the chapter, for more information on how you can draw in the background.)

This behavior is not a problem to be overcome. The consolidation of drawing events is one part of iOS's capability to render complex drawings on limited hardware. As you see throughout this chapter, much of UIKit is dedicated to avoiding unnecessary drawing, and this consolidation is one of the first steps.

So how do you start and stop an activity indicator for a long-running operation? You use dispatch or operation queues to put your expensive work in the background, while making all of your UIKit calls on the main thread, as shown in the following code.

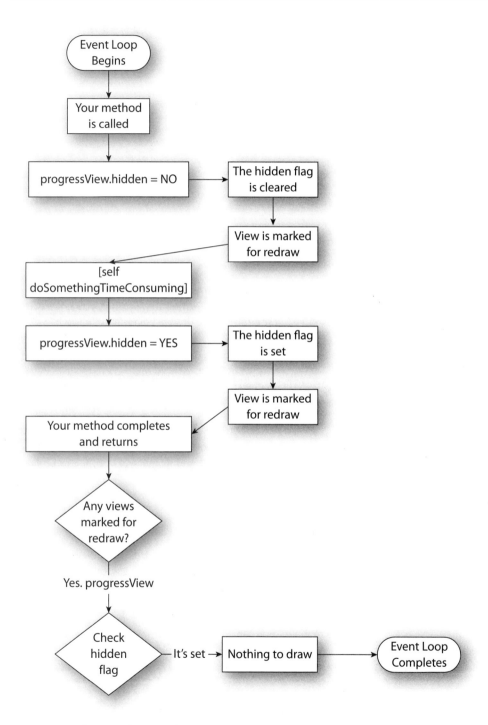

Figure 8-1 How the Cocoa drawing cycle consolidates changes

ViewController.m (TimeConsuming)

```
- (IBAction)doSomething:(id)sender {
  [sender setEnabled:NO];
  [self.activity startAnimating];

  dispatch_queue_t bgQueue = dispatch_get_global_queue(
                    DISPATCH_QUEUE_PRIORITY_DEFAULT, 0);

  dispatch_async(bgQueue, ^{
    [self somethingTimeConsuming];

    dispatch_async(dispatch_get_main_queue(), ^{
      [self.activity stopAnimating];
      [sender setEnabled:YES];
    });
  });
}
```

When the `IBAction` is called, you start animating the activity indicator. You then put a call to `somethingTimeConsuming` on the default background dispatch queue. When that finishes, you put a call to `stopAnimating` on the main dispatch queue. Dispatch and operation queues are covered in Chapter 13.

To summarize:

- iOS consolidates all drawing requests during the run loop, and draws them all at once.

- You must not block the main thread to do complex processing.

- You must not draw into the main view graphics context except on the main thread. You need to check each UIKit method to ensure it does not have a main thread requirement. Some UIKit methods can be used on background threads as long as you're not drawing into the main view context. See "CGLayer," later in this chapter, for examples.

View Drawing Versus View Layout

`UIView` separates the layout ("rearranging") of subviews from drawing (or "display"). This is important for maximizing performance because layout is generally cheaper than drawing. Layout is cheap because `UIView` caches drawing operations onto GPU-optimized bitmaps. These bitmaps can be moved around, shown, hidden, rotated, and otherwise transformed and composited very inexpensively using the GPU.

When you call `setNeedsDisplay` on a view, it is marked "dirty" and will be redrawn during the next drawing cycle. Don't call it unless the content of the view has really changed. Most UIKit views automatically manage redrawing when their data is changed, so you generally don't need to call `setNeedsDisplay` except on custom views.

When a view's subviews need to be rearranged because of an orientation change or scrolling, UIKit calls `setNeedsLayout`. This, in turn, calls `layoutSubviews` on the affected views. By overriding `layoutSubviews`, you can make your application much smoother during rotation and scrolling events. You can rearrange your subviews' frames without necessarily having to redraw them, and you can hide or show views based on orientation. You can also call `setNeedsLayout` if your data changes in ways that only need layout updates rather than drawing.

Custom View Drawing

Views can provide their content by including subviews, including layers, or implementing `drawRect:`. Typically if you implement `drawRect:`, you don't mix this with layers or subviews, although it's legal and sometimes useful to do so. Most custom drawing is done with UIKit or Core Graphics, although OpenGL ES has become easier to integrate when needed.

2D drawing generally breaks down into several operations:

- Lines
- Paths (filled or outlined shapes)
- Text
- Images
- Gradients

2D drawing does not include manipulation of individual pixels because that is destination-dependent. You can achieve this with a bitmap context, but not directly with UIKit or Core Graphics functions.

Both UIKit and Core Graphics use a "painter" drawing model. This means that each command is drawn in sequence, overlaying previous drawings during that event loop. Order is very important in this model, and you must draw back-to-front. Each time `drawRect:` is called, it's your responsibility to draw the entire area requested. The drawing "canvas" is not preserved between calls to `drawRect:`.

Drawing with UIKit

In the "old days" before iPad, most custom drawing had to be done with Core Graphics because there was no way to draw arbitrary shapes with UIKit. In iPhoneOS 3.2, Apple added `UIBezierPath` and made it much easier to draw entirely in Objective-C. UIKit still lacks support for lines, gradients, shading, and some advanced features like controlling anti-aliasing and precise color management. Even so, UIKit is now a very convenient way to manage the most common custom drawing needs.

The simplest way to draw rectangles is with `UIRectFrame` or `UIRectFill`, as shown in the following code:

```
- (void)drawRect:(CGRect)rect {
  [[UIColor redColor] setFill];
  UIRectFill(CGRectMake(10, 10, 100, 100));
}
```

Notice how you first set the pen color using –[`UIColor setFill`]. Drawing is done into a graphics context provided by the system before calling `drawRect:`. That context includes a lot of information including stroke color, fill color, text color, font, transform, and more. At any given time, there is just one stroke pen and one fill pen, and their colors are used to draw everything. The "Managing Graphics Contexts" section, later in this chapter, covers how to save and restore contexts, but for now just note that drawing commands are order-dependent, and that includes commands that change the pens.

> The graphics context provided to `drawRect:` is specifically a view graphics context. There are other types of graphics contexts, including PDF and bitmap contexts. All of them use the same drawing techniques, but a view graphics context is optimized for drawing onto the screen. This distinction will be important when I discuss `CGLayer`.

Paths

UIKit includes much more powerful drawing commands than its rectangle functions. It can draw arbitrary curves and lines using `UIBezierPath`. A Bézier curve is a mathematical way of expressing a line or curve using a small number of control points. Most of the time, you don't need to worry about the math because `UIBezierPath` has simple methods to handle the most common paths: lines, arcs, rectangles (optionally rounded), and ovals. With these, you can quickly draw most shapes needed for UI elements. The following code is an example of a simple shape scaled to fill the view, as shown in Figure 8-2. You draw this several ways in the upcoming examples.

FlowerView.m (Paths)

```
- (void)drawRect:(CGRect)rect {
  CGSize size = self.bounds.size;
  CGFloat margin = 10;
  CGFloat radius = rint(MIN(size.height - margin,
                           size.width - margin) / 4);

  CGFloat xOffset, yOffset;
  CGFloat offset = rint((size.height - size.width) / 2);
  if (offset > 0) {
    xOffset = rint(margin / 2);
    yOffset = offset;
  }
  else {
    xOffset = -offset;
    yOffset = rint(margin / 2);
  }

  [[UIColor redColor] setFill];
  UIBezierPath *path = [UIBezierPath bezierPath];
  [path addArcWithCenter:CGPointMake(radius * 2 + xOffset,
                                     radius + yOffset)
                  radius:radius
              startAngle:-M_PI
                endAngle:0
               clockwise:YES];
  [path addArcWithCenter:CGPointMake(radius * 3 + xOffset,
                                     radius * 2 + yOffset)
                  radius:radius
```

```
                startAngle:-M_PI_2
                  endAngle:M_PI_2
                 clockwise:YES];
   [path addArcWithCenter:CGPointMake(radius * 2 + xOffset,
                                      radius * 3 + yOffset)
                    radius:radius
                startAngle:0
                  endAngle:M_PI
                 clockwise:YES];
   [path addArcWithCenter:CGPointMake(radius + xOffset,
                                      radius * 2 + yOffset)
                    radius:radius
                startAngle:M_PI_2
                  endAngle:-M_PI_2
                 clockwise:YES];
   [path closePath];
   [path fill];
}
```

Figure 8-2 Output of FlowerView

`FlowerView` creates a path made up of a series of arcs and fills it with red. Creating a path doesn't cause anything to be drawn. A `UIBezierPath` is just a sequence of curves, like an `NSString` is a sequence of characters. Only when you call `fill` is the curve drawn into the current context.

Note the use of the `M_PI` (π) and `M_PI_2` (π⁄₂) constants. Arcs are described in radians, so π and fractions of π are important. `math.h` defines many such constants that you should use rather than recomputing them. Arcs measure their angles clockwise, with 0 radians pointing to the right, π⁄₂ radians pointing down, π (or -π) radians pointing left, and -π⁄₂ radians pointing up. You can use 3π⁄₂ for up if you prefer, but I find `–M_PI_2` easier to visualize than `3*M_PI_2`. If radians give you a headache, you can make a function out of it:

```
CGFloat RadiansFromDegrees(CGFloat d) {
  return d * M_PI / 180;
}
```

Generally, I recommend just getting used to radians rather than doing so much math, but if you need unusual angles, it can be easier to work in degrees.

When calculating `radius` and `offset`, you use `rint` (round to closest integer) to ensure that you're point-aligned (and therefore pixel-aligned). That helps improve drawing performance and avoids blurry edges. Most of the time, that's what you want, but in cases where an arc meets a line, it can lead to off-by-one drawing errors. Usually, the best approach is to move the line so that all the values are integers, as discussed in the following section.

Understanding Coordinates

There are subtle interactions among coordinates, points, and pixels that can lead to poor drawing performance and blurry lines and text. Consider the following code:

```
CGContextSetLineWidth(context, 3.);

// Draw 3pt horizontal line from {10,100} to {200,100}
CGContextMoveToPoint(context, 10., 100.);
CGContextAddLineToPoint(context, 200., 100.);
CGContextStrokePath(context);

// Draw 3pt horizontal line from {10,105.5} to {200,105.5}
CGContextMoveToPoint(context, 10., 105.5);
CGContextAddLineToPoint(context, 200., 105.5);
CGContextStrokePath(context);
```

Figure 8-3 shows the output of this program on a non-Retina display, scaled to make the differences more obvious.

Figure 8-3 Comparison of line from {10,100} and line from {10,105.5}

The line from {10, 100} to {200, 100} is much more blurry than the line from {10, 105.5} to {200, 105.5}. The reason is because of how iOS interprets coordinates.

When you construct a CGPath, you work in so-called *geometric coordinates*. These are the same kind of coordinates that mathematicians use, representing the zero-dimensional point at the intersection of two grid lines. It's impossible to draw a geometric point or a geometric line, because they're infinitely small and thin. When iOS draws, it has to translate these geometric objects into *pixel coordinates*. These are two-dimensional boxes that can be set to a specific color. A pixel is the smallest unit of display area that the device can control.

Figure 8-4 shows the geometric line from {10, 100} to {200, 100}.

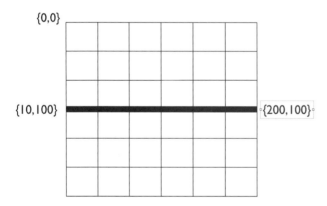

Figure 8-4 Geometric line from {10, 100} to {200, 100}

When you call CGContextStrokePath, iOS centers the line along the path. Ideally, the line would be three pixels wide, from $y = 98.5$ to $y = 101.5$, as shown in Figure 8-5.

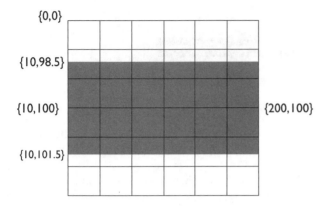

Figure 8-5 Ideal three-pixel line

This line is impossible to draw, however. Each pixel must be a single color, and the pixels at the top and bottom of the line include two colors. Half is the stroke color, and half is the background color. iOS solves this problem by averaging the two. This is the same technique used in anti-aliasing. This is shown in Figure 8-6.

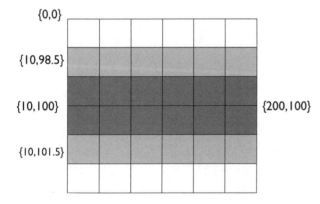

Figure 8-6 Anti-aliased three-pixel line

On the screen, this line will look slightly blurry. The solution to this problem is to move horizontal and vertical lines to the half-point so that when iOS centers the line, the edges fall along pixel boundaries, or to make your line an even width.

You can also encounter this problem with nonintegral line widths, or if your coordinates aren't integers or half-integers. Any situation that forces iOS to draw fractional pixels will cause blurriness.

Fill is not the same as stroke. A stroke line is centered on the path, but fill colors all the pixels up to the path. If you fill the rectangle from {10,100} to {200,103}, each pixel is filled correctly, as shown in Figure 8-7.

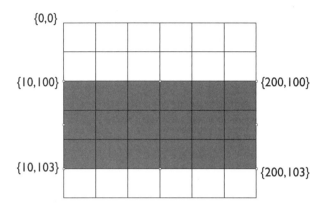

Figure 8-7 Filling the rectangle from {10,100} to {200,103}

The discussion so far has equated points with pixels. On a Retina display, these are not equivalent. The iPhone 4 has four pixels per point and a scale factor of two. That subtly changes things, but generally for the better. Because all the coordinates used in Core Graphics and UIKit are expressed in points, all integral line widths are effectively an even number of pixels. For example, if you request a 1-point stroke width, this is the same as a 2-pixel stroke width. To draw that line, iOS needs to fill one pixel on each side of the path. That's an integral number of pixels, so there's no anti-aliasing. You can still encounter blurriness if you use coordinates that are neither integers nor half-integers.

Offsetting by a half-point is unnecessary on a Retina display, but it doesn't hurt. As long as you intend to support iPhone 3GS or iPad 2, you need to apply a half-point offset for drawing horizontal and vertical lines.

All of this applies only to horizontal and vertical lines. Sloping or curved lines should be anti-aliased so that they're not jagged, so there's generally no reason to offset them.

Resizing and contentMode

Returning to `FlowerView` found in the earlier section, "Paths," if you rotate the device as shown in Figure 8-8, you'll see that the view is distorted, even though you have code that adjusts for the size of the view.

Figure 8-8 Rotated FlowerView

iOS optimizes drawing by taking a snapshot of the view and adjusting it for the new frame. The `drawRect:` method isn't called. The property `contentMode` determines how the view is adjusted. The default, `UIViewContentModeScaleToFill`, scales the image to fill the new view size, changing the aspect ratio if needed. That's why the shape is distorted.

There are a lot of ways to automatically adjust the view. You can move it around without resizing it, or you can scale it in various ways that preserve or modify the aspect ratio. The key is to make sure that any mode you use exactly matches the results of your `drawRect:` in the new orientation. Otherwise, your view will "jump" the next time you redraw. This usually works as long as your `drawRect:` doesn't consider its `bounds` during drawing. In `FlowerView`, you use the `bounds` to determine the size of your shape, so it's hard to get automatic adjustments to work correctly.

Use the automatic modes if you can because they can improve performance. When you can't, ask the system to call `drawRect:` when the frame changes by using `UIViewContentModeRedraw`, as shown in the following code.

```
- (void)awakeFromNib {
    self.contentMode = UIViewContentModeRedraw;
}
```

Transforms

iOS platforms have access to a very nice GPU that can do matrix operations very quickly. If you can convert your drawing calculations into matrix operations, you can leverage the GPU and get excellent performance. Transforms are just such a matrix operation.

iOS has two kinds of transforms: affine and 3D. `UIView` handles only affine transforms, so that's all we discuss right now. An affine transform is a way of expressing rotation, scaling, shear, and translation (shifting) as a matrix. These transforms are shown in Figure 8-9.

A single transform combines any number of these operations into a 3×3 matrix. iOS has functions to support rotation, scaling, and translation. If you want shear, you'll have to write the matrix yourself. (You can also use `CGAffineTransformMakeShear` from Jeff LaMarche; see "Further Reading" at the end of the chapter.)

Transforms can dramatically simplify and speed up your code. Often it's much easier and faster to draw in a simple coordinate space around the origin and then to scale, rotate, and translate your drawing to where you want it. For instance, `FlowerView` includes a lot of code like this:

```
CGPointMake(radius * 2 + xOffset, radius + yOffset)
```

That's a lot of typing, a lot of math, and a lot of things to keep straight in your head. What if, instead, you just draw it in a 4×4 box as shown in Figure 8-10?

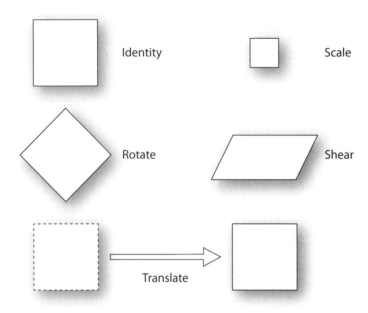

Figure 8-9 Affine transforms

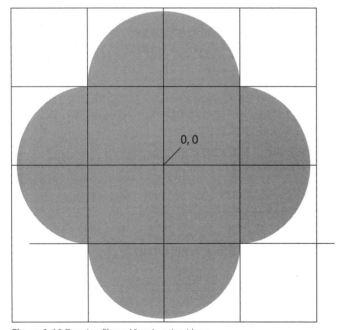

Figure 8-10 Drawing FlowerView in a 4 × 4 box

Now all the interesting points fall on nice, easy coordinates like {0,1} and {1,0}. The following code shows how to draw using this transform. Compare the highlighted sections with the `FlowerView` code earlier in this chapter.

FlowerTransformView.m (Transforms)

```
static inline CGAffineTransform
CGAffineTransformMakeScaleTranslate(CGFloat sx, CGFloat sy,
                                    CGFloat dx, CGFloat dy) {
  return CGAffineTransformMake(sx, 0.f, 0.f, sy, dx, dy);
}

- (void)drawRect:(CGRect)rect {
  CGSize size = self.bounds.size;
  CGFloat margin = 10;

  [[UIColor redColor] set];
  UIBezierPath *path = [UIBezierPath bezierPath];
  [path addArcWithCenter:CGPointMake(0, -1)
                  radius:1
              startAngle:-M_PI
                endAngle:0
               clockwise:YES];
  [path addArcWithCenter:CGPointMake(1, 0)
                  radius:1
              startAngle:-M_PI_2
                endAngle:M_PI_2
               clockwise:YES];
  [path addArcWithCenter:CGPointMake(0, 1)
                  radius:1
              startAngle:0
                endAngle:M_PI
               clockwise:YES];
  [path addArcWithCenter:CGPointMake(-1, 0)
                  radius:1
              startAngle:M_PI_2
                endAngle:-M_PI_2
               clockwise:YES];
  [path closePath];

  CGFloat scale = floor((MIN(size.height, size.width)
                         - margin) / 4);

  CGAffineTransform transform;
  transform = CGAffineTransformMakeScaleTranslate(scale,
                                                  scale,
                                                  size.width/2,
                                                  size.height/2);
  [path applyTransform:transform];
  [path fill];
}
```

When you're done constructing your path, you compute a transform to move it into your view's coordinate space. You scale it by the size you want divided by the size it currently is (4), and you translate it to the center of the view. The utility function `CGAffineTransformMakeScaleTranslate` isn't just for speed (although it is faster). It's easier to get the transform correct this way. If you try to build up the transform one step at a time, each step affects later steps. Scaling and then translating is not the same as translating and then scaling. If you build the matrix all at once, you don't have to worry about that.

This technique can be used to draw complicated shapes at unusual angles. For instance, to draw an arrow pointing to the upper right, it's generally easier to draw it pointing to the right and then rotate it.

You have a choice between transforming the path using `applyTransform:` and transforming the whole view by setting the `transform` property. Which is best depends on the situation, but I usually prefer to transform the path rather than the view when practical. Modifying the view's `transform` makes the results of `frame` and `bounds` more difficult to interpret, so I avoid it when I can. As you see in the following section, you can also transform the current context, which sometimes is the best approach.

Drawing with Core Graphics

Core Graphics, sometimes called Quartz 2D or just Quartz, is the main drawing system in iOS. It provides destination-independent drawing, so you can use the same commands to draw to the screen, layer, bitmap, PDF, or printer. Anything starting with `CG` is part of Core Graphics. Figure 8-11 and the following code provide an example of a simple scrolling graph.

Figure 8-11 Simple scrolling graph

GraphView.h (Graph)

```objc
@interface GraphView : UIView
@property (nonatomic, readonly, strong) NSMutableArray *values;
@end
```

GraphView.m (Graph)

```objc
@implementation GraphView {
  dispatch_source_t _timer;
}
const CGFloat kXScale = 5.0;
const CGFloat kYScale = 100.0;

static inline CGAffineTransform
CGAffineTransformMakeScaleTranslate(CGFloat sx, CGFloat sy,
    CGFloat dx, CGFloat dy) {
  return CGAffineTransformMake(sx, 0.f, 0.f, sy, dx, dy);
}

- (void)awakeFromNib {
  [self setContentMode:UIViewContentModeRight];
  _values = [NSMutableArray array];

  __weak id weakSelf = self;
  double delayInSeconds = 0.25;
  _timer =
      dispatch_source_create(DISPATCH_SOURCE_TYPE_TIMER, 0, 0,
          dispatch_get_main_queue());
  dispatch_source_set_timer(
      _timer, dispatch_walltime(NULL, 0),
      (unsigned)(delayInSeconds * NSEC_PER_SEC), 0);
  dispatch_source_set_event_handler(_timer, ^{
    [weakSelf updateValues];
  });
  dispatch_resume(_timer);
}

- (void)updateValues {
  double nextValue = sin(CFAbsoluteTimeGetCurrent())
      + ((double)rand()/(double)RAND_MAX);
  [self.values addObject:
      [NSNumber numberWithDouble:nextValue]];
  CGSize size = self.bounds.size;
  CGFloat maxDimension = MAX(size.height, size.width);
  NSUInteger maxValues =
      (NSUInteger)floorl(maxDimension / kXScale);

  if ([self.values count] > maxValues) {
    [self.values removeObjectsInRange:
        NSMakeRange(0, [self.values count] - maxValues)];
```

```
    }

    [self setNeedsDisplay];
}

- (void)dealloc {
  dispatch_source_cancel(_timer);
}

- (void)drawRect:(CGRect)rect {
  if ([self.values count] == 0) {
    return;
  }

  CGContextRef ctx = UIGraphicsGetCurrentContext();
  CGContextSetStrokeColorWithColor(ctx,
                                   [[UIColor redColor] CGColor]);
  CGContextSetLineJoin(ctx, kCGLineJoinRound);
  CGContextSetLineWidth(ctx, 5);

  CGMutablePathRef path = CGPathCreateMutable();

  CGFloat yOffset = self.bounds.size.height / 2;
  CGAffineTransform transform =
      CGAffineTransformMakeScaleTranslate(kXScale, kYScale,
                                          0, yOffset);

  CGFloat y = [[self.values objectAtIndex:0] floatValue];
  CGPathMoveToPoint(path, &transform, 0, y);

  for (NSUInteger x = 1; x < [self.values count]; ++x) {
    y = [[self.values objectAtIndex:x] floatValue];
    CGPathAddLineToPoint(path, &transform, x, y);
  }

  CGContextAddPath(ctx, path);
  CGPathRelease(path);
  CGContextStrokePath(ctx);
}
@end
```

Every quarter second, this code adds a new number to the end of the data and removes an old number from the beginning. Then it marks the view as dirty with `setNeedsDisplay`. The drawing code sets various advanced line drawing options not available with `UIBezierPath`, and creates a `CGPath` with all the lines. It then transforms the path to fit into the view, adds the path to the context, and strokes it.

Core Graphics uses the Core Foundation memory management rules. Core Foundation objects require manual retain and release, even under ARC. Note the use of `CGPathRelease`. For full details, see **Chapter 27.**

You may be tempted to cache the `CGPath` here so that you don't have to compute it every time. That's a good instinct, but in this case, it wouldn't help. iOS already avoids calling `drawRect:` except when the view is dirty, which happens only when the data changes. When the data changes, you need to calculate a new path. Caching the old path in this case would just complicate the code and waste memory.

Mixing UIKit and Core Graphics

Within `drawRect:`, UIKit and Core Graphics can generally intermix without issue, but outside of `drawRect:`, you may find that things drawn with Core Graphics appear upside down. UIKit uses an upper-left origin (ULO) coordinate system, whereas Core Graphics uses a lower-left origin (LLO) system by default. As long as you use the context returned by `UIGraphicsGetCurrentContext` inside of `drawRect:`, everything is fine because this context is already flipped. But if you create your own context using functions like `CGBitmapContextCreate`, it'll be LLO. You can either do your math backward or you can flip the context:

```
CGContextTranslateCTM(ctx, 0.0f, height);
CGContextScaleCTM(ctx, 1.0f, -1.0f);
```

This moves (translates) the height of the context and then flips it using a negative scale. When going from UIKit to Core Graphics, the transform is reversed:

```
CGContextScaleCTM(ctx, 1.0f, -1.0f);
CGContextTranslateCTM(ctx, 0.0f, -height);
```

First flip it, and then translate it.

Managing Graphics Contexts

Before calling `drawRect:`, the drawing system creates a graphics context (`CGContext`). A context includes a lot of information such as a pen color, text color, current font, transform, and more. Sometimes you may want to modify the context and then put it back the way you found it. For instance, you may have a function to draw a specific shape with a specific color. There is only one stroke pen, so when you change the color, this would change things for your caller. To avoid side effects, you can push and pop the context using `CGContextSaveGState` and `CGContextRestoreGState`.

Do not confuse this with the similar-sounding `UIGraphicsPushContext` and `UIGraphicsPopContext`. They do not do the same thing. `CGContextSaveGState` remembers the current state of a context. `UIGraphicsPushContext` changes the current context. Here's an example of `CGContextSaveGState`.

```
[[UIColor redColor] setStroke];
CGContextSaveGState(UIGraphicsGetCurrentContext());
[[UIColor blackColor] setStroke];
CGContextRestoreGState(UIGraphicsGetCurrentContext());
UIRectFill(CGRectMake(10, 10, 100, 100)); // Red
```

This code sets the stroke pen color to red and saves off the context. It then changes the pen color to black and restores the context. When you draw, the pen is red again.

The following code illustrates a common error.

```
[[UIColor redColor] setStroke];
// Next line is nonsense
UIGraphicsPushContext(UIGraphicsGetCurrentContext());
[[UIColor blackColor] setStroke];
UIGraphicsPopContext();
UIRectFill(CGRectMake(10, 10, 100, 100)); // Black
```

In this case, you set the pen color to red and then switch context to the current context, which does nothing useful. You then change the pen color to black, and pop the context back to the original (which effectively does nothing). You now will draw a black rectangle, which is almost certainly not what was meant.

The purpose of `UIGraphicsPushContext` is not to save the current *state* of the context (pen color, line width, and so on), but to switch contexts entirely. Say you are in the middle of drawing something into the current view context, and now want to draw something completely different into a bitmap context. If you want to use UIKit to do any of your drawing, you'd want to save off the current UIKit context, including all the drawing that had been done, and switch to a completely new drawing context. That's what `UIGraphicsPushContext` does. When you finish creating your bitmap, you pop the stack and get your old context back. That's what `UIGraphicsPopContext` does . This only matters in cases where you want to draw into the new bitmap context with UIKit. As long as you use Core Graphics functions, you don't need to push or pop contexts because Core Graphics functions take the context as a parameter.

This is a pretty useful and common operation. It's so common that Apple has made a shortcut for it called `UIGraphicsBeginImageContext`. It takes care of pushing the old context, allocating memory for a new context, creating the new context, flipping the coordinate system, and making it the current context. Most of the time, that's just what you want.

Here's an example of how to create an image and return it using `UIGraphicsBeginImageContext`. The result is shown in Figure 8-12.

MYView.m (Drawing)

```
- (UIImage *)reverseImageForText:(NSString *)text {
  const size_t kImageWidth = 200;
  const size_t kImageHeight = 200;
  CGImageRef textImage = NULL;
  UIFont *font = [UIFont boldSystemFontOfSize:17.0];

  UIGraphicsBeginImageContext(CGSizeMake(kImageWidth,
                                         kImageHeight));

  [[UIColor redColor] set];
  [text drawInRect:CGRectMake(0, 0,
                              kImageWidth, kImageHeight)
        withFont:font];

  textImage =
```

(continued)

```
        UIGraphicsGetImageFromCurrentImageContext().CGImage;

    UIGraphicsEndImageContext();

    return [UIImage imageWithCGImage:textImage
                              scale:1.0
                        orientation:UIImageOrientationUpMirrored];

}
```

Figure 8-12 Text drawn with reverseImageForText:

Optimizing UIView Drawing

UIView and its subclasses are highly optimized, and when possible, use them rather than custom drawing. For instance, UIImageView is faster and uses less memory than anything you're likely to put together in an afternoon with Core Graphics. The following sections cover a few things to keep in mind when using UIView to keep it drawing as well as it can.

Avoid Drawing

The fastest drawing is the drawing you never do. iOS goes to great lengths to avoid calling `drawRect:`. It caches an image of your view and moves, rotates, and scales it without any intervention from you. Using an appropriate `contentMode` lets the system adjust your view during rotation or resizing without calling `drawRect:`. The most common cause for `drawRect:` running is when you call `setNeedsDisplay`. Avoid calling `setNeedsDisplay` unnecessarily. Remember, though, `setNeedsDisplay` just schedules the view to be redrawn. Calling `setNeedsDisplay` many times in a single event loop is no more expensive, practically, than calling it once, so don't coalesce your calls. iOS is already doing that for you.

Those familiar with Mac development may be familiar with partial view drawing using `setNeedsDisplayInRect:`. iOS does not perform partial view drawing, and `setNeedsDisplayInRect:` is the same as `setNeedsDisplay`. The entire view will be redrawn. If you want to partially redraw a view, you should use `CALayer` (discussed in Chapter 9) or use subviews.

Caching and Background Drawing

If you need to do a lot of calculations during your drawing, cache the results when you can. At the lowest level, you can cache the raw data you need rather than asking for it from your delegate every time. Beyond that, you can cache static elements like `CGFont` or `CGGradient` objects so that you generate them only once. Fonts and gradients are useful to cache this way because they're often reused. Finally, you can cache the entire result of a complex drawing operation. Often the best place to cache such a result is in a `CGLayer`, which is discussed later in the section "CGLayer." Alternatively, you can cache the result in a bitmap, generally using `UIGraphicsBeginImageContext` as discussed in "Managing Graphics Context," earlier in this chapter.

Much of this caching or precalculation can be done in the background. You may have heard that you must always draw on the main thread, but this isn't completely true. There are several UIKit functions that must be called only on the main thread, such as `UIGraphicsBeginImageContext`, but you are free to create a `CGBitmapContext` object on any thread using `CGBitmapCreateContext` and draw into it. Since iOS 4, you can use UIKit drawing methods like `drawAtPoint:` on background threads as long as you draw into your own `CGContext` and not the main view graphics context (the one returned by `UIGraphicsGetCurrentContext`). You should only access a given `CGContext` on a single thread, however.

Custom Drawing Versus Prerendering

There are two major approaches to managing complex drawing. You can draw everything programmatically with `CGPath` and `CGGradient`, or you can prerender everything in a graphics program like Adobe Photoshop and display it as an image. If you have an art department and plan to have extremely complex visual elements, Photoshop is often the only way to go.

There are a lot of disadvantages to prerendering, however. First, it introduces resolution dependence. You may need to manage 1-scale and 2-scale versions of your images and possibly different images for iPad and iPhone. This complicates workflow and bloats your product. It can make minor changes difficult and lock you into precise element sizes and colors if every change requires a round trip to the artist. Many artists are still unfamiliar with how to draw stretchable images and how to best provide images to be composited for iOS.

Apple originally encouraged developers to prerender because early iPhones couldn't compute gradients fast enough. Since the iPhone 3GS, this has been less of an issue, and each generation makes custom drawing more attractive.

Today, I recommend custom drawing when you can do it in a reasonable amount of code. This is usually the case for small elements like buttons. When you do use prerendered artwork, I suggest that you keep the art files fairly "flat" and composite in code. For instance, you may use an image for a button's background, but handle the rounding and shadows in code. That way, as you want to make minor tweaks, you don't have to rerender the background.

A middle ground in this is automatic Core Graphics code generation with tools like PaintCode and Opacity. These are not panaceas. Typically, the code generated is not ideal, and you may have to modify it, complicating the workflow if you want to regenerate the code. That said, we recommend investigating these tools if you are doing a lot of UI design. See "Further Reading" at the end of this chapter for links to sites with information on these tools.

Pixel Alignment and Blurry Text

One of the most common causes of subtle drawing problems is pixel misalignment. If you ask Core Graphics to draw at a point that is not aligned with a pixel, it performs anti-aliasing as discussed in "Understanding Coordinates" earlier in this chapter. This means it draws part of the information on one pixel and part on another, giving the illusion that the line is between the two. This illusion makes things smoother but also makes them fuzzy. Anti-aliasing also takes processing time, so it slows down drawing. When possible, you want to make sure that your drawing is pixel-aligned to avoid this.

Prior to the Retina display, pixel-aligned meant integer coordinates. As of iOS 4, coordinates are in points, not pixels. There are two pixels to the point on the current Retina display, so half-points (1.5, 2.5) are also pixel-aligned. In the future, there might be four or more pixels to the point, and it could be different from device to device. Even so, unless you need pixel accuracy, it is easiest to just make sure you use integer coordinates for your frames.

Generally, it's the frame origin that matters for pixel alignment. This causes an unfortunate problem for the `center` property. If you set the center to an integral coordinate, your origin may be misaligned. This is particularly noticeable with text, especially with `UILabel`. Figure 8-13 demonstrates this problem. It is subtle and somewhat difficult to see in print, so you can also demonstrate it with the program `BlurryText` available with the online files for this chapter.

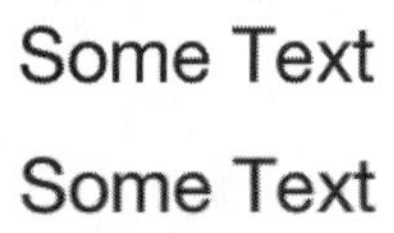

Some Text

Some Text

Figure 8-13 Text that is pixel-aligned (top) and unaligned (bottom)

There are two solutions. First, odd font sizes (13 rather than 12, for instance) will typically align correctly. If you make a habit of using odd font sizes, you can often avoid the problem. To be certain you avoid the problem, you need to make sure that the frame is integral either by using `setFrame:` instead of `setCenter:` or by using a `UIView` category like `setAlignedCenter::`

```
- (void)setAlignedCenter:(CGPoint)center {
  self.center = center;
  self.frame = CGRectIntegral(self.frame);
}
```

Because `setAlignedCenter:` effectively sets the frame twice, it's not the fastest solution, but it is very easy and fast enough for most problems. `CGRectIntegral()` returns the smallest integral rectangle that encloses the given rectangle.

As pre-Retina displays phase out, blurry text will be less of an issue as long as you set `center` to integer coordinates. For now, though, it is still a concern.

Alpha, Opaque, Hidden

Views have three properties that appear related but that are actually orthogonal: `alpha`, `opaque`, and `hidden`.

The `alpha` property determines how much information a view contributes to the pixels within its frame. So an `alpha` of 1 means that all of the view's information is used to color the pixel. An `alpha` of 0 means that none of the view's information is used to color the pixel. Remember, nothing is really transparent on an iPhone screen. If you set the entire screen to transparent pixels, the user isn't going to see the circuit board or the ground. In the end, it's just a matter of what color to draw the pixel. So, as you raise and lower the `alpha`, you're changing how much this view contributes to the pixel versus views "below" it.

Marking a view `opaque` or not doesn't actually make its content more or less transparent. Opaque is a promise that the drawing system can use for optimization. When you mark a view as `opaque`, you're promising the drawing system that you will draw every pixel in your rectangle with fully opaque colors. That allows the drawing system to ignore views below yours and that can improve performance, particularly when applying transforms. You should mark your views `opaque` whenever possible, especially views that scroll like `UITableViewCell`. However, if any partially transparent pixels are in your view, or if you don't draw every pixel in your rectangle, setting `opaque` can have unpredictable results. Setting a nontransparent `backgroundColor` ensures that all pixels are drawn.

Closely related to `opaque` is `clearsContextBeforeDrawing`. This is `YES` by default, and sets the context to transparent black before calling `drawRect:`. This avoids any garbage data in the view. It's a pretty fast operation, but if you're going to draw every pixel anyway, you can get a small benefit by setting it to `NO`.

Finally, `hidden` indicates that the view should not be drawn at all and is generally equivalent to an `alpha` of 0. The `hidden` property cannot be animated, so it's common to hide views by animating `alpha` to 0.

Hidden and transparent views don't receive touch events. The meaning of transparent is not well defined in the documentation, but through experimentation, I've found that it's an `alpha` less than 0.1. Do not rely on this particular value, but the point is that "nearly transparent" is generally treated as transparent. You cannot create a "transparent overlay" to catch touch events by setting the `alpha` very low.

You can make a view transparent and still receive touch events by setting its `alpha` to 1, `opaque` to `NO`, and `backgroundColor` to `nil` or `[UIColor clearColor]`. A view with a transparent background is still considered visible for the purposes of hit detection.

CGLayer

`CGLayer` is a very effective way to cache things you draw often. This should not be confused with `CALayer`, which is a more powerful and complicated layer object from Core Animation. `CGLayer` is a Core Graphics layer that is optimized, often hardware-optimized, for drawing into `CGContext`.

There are several kinds of `CGContext`. The most common is a view graphics context, designed to draw to the screen, which is returned by `UIGraphicsCurrentContext`. Contexts are also used for bitmaps and printing, however. Each of these has different attributes, including maximum resolution, color details, and available hardware acceleration.

At its simplest, a `CGLayer` is similar to a `CGBitmapContext`. You can draw into it, save it off, and use it to draw the result into a `CGContext` later. The difference is that you can optimize `CGLayer` for use with a particular kind of graphics context. If a `CGLayer` is destined for a view graphics context, it can cache its data directly on the GPU, which can significantly improve performance. `CGBitmapContext` can't do this because it doesn't know that you plan to draw it on the screen.

The following example demonstrates caching a `CGLayer`. In this case, it's cached in a static variable the first time the view is drawn. You can then "stamp" the `CGLayer` repeatedly while rotating the context. You use `UIGraphicsPushContext` so that you can use UIKit to draw the text into the layer context, and `UIGraphicsPopContext` to return to the normal context. This could be done with `CGContextShowTextAtPoint` instead, but UIKit makes it very easy to draw an `NSString`. Figure 8-14 shows the output.

LayerView.m (Layer)

```
- (void)drawRect:(CGRect)rect {
  static CGLayerRef sTextLayer = NULL;

  CGContextRef ctx = UIGraphicsGetCurrentContext();

  if (sTextLayer == NULL) {
    CGRect textBounds = CGRectMake(0, 0, 200, 100);
    sTextLayer = CGLayerCreateWithContext(ctx,
                                          textBounds.size,
                                          NULL);

    CGContextRef textCtx = CGLayerGetContext(sTextLayer);
    CGContextSetRGBFillColor (textCtx, 1.0, 0.0, 0.0, 1);
    UIGraphicsPushContext(textCtx);
    UIFont *font = [UIFont systemFontOfSize:13.0];
    [@"Pushing The Limits" drawInRect:textBounds
```

```
                              withFont:font];
    UIGraphicsPopContext();
  }

  CGContextTranslateCTM(ctx, self.bounds.size.width / 2,
                        self.bounds.size.height / 2);

  for (NSUInteger i = 0; i < 10; ++i) {
    CGContextRotateCTM(ctx, 2 * M_PI / 10);
    CGContextDrawLayerAtPoint(ctx,
                              CGPointZero,
                              sTextLayer);
  }
}
```

Figure 8-14 Output of LayerView

Summary

iOS has a rich collection of drawing tools. This chapter focused on Core Graphics and its Objective-C descendant, UIKit. By now, you should have a good understanding of how systems interact and how to optimize your iOS drawing.

Chapter 9 discusses Core Animation, which puts your interface in motion. Also covered is `CALayer`, a powerful addition to `UIView` and `CGLayer` and an important tool for your drawing toolbox even if you're not animating.

iOS 5 added Core Image to iOS for tweaking pictures. iOS also has ever-growing support for OpenGL ES for drawing advanced 3D graphics and textures. OpenGL ES is a book-length subject of its own, so it isn't tackled here, but you can get a good introduction in Apple's "OpenGL ES Programming Guide for iOS" (see the "Further Reading" section).

Further Reading

Apple Documentation

The following documents are available in the iOS Developer Library at `developer.apple.com` or through the Xcode Documentation and API Reference.

Drawing and Printing Guide for iOS

iOS Human Interface Guidelines

iOS App Programming Guide

OpenGL ES Programming Guide for iOS

Quartz 2D Programming Guide

Technical Q&A QA1708: Improving Image Drawing Performance on iOS

View Programming Guide for iOS

Other Resources

LaMarche, Jeff. "iPhone Development." Jeff has several articles that provide a lot of insight into using `CGAffineTransform`
`iphonedevelopment.blogspot.com/search/label/CGAffineTransform`.

PaintCode. Fairly simple vector editor that exports Core Graphics code. Particularly well-suited to common UI elements.
`www.paintcodeapp.com`

Opacity. More powerful vector editor that exports Core Graphics code, and can be used to generate more general vector drawings.
`likethought.com/opacity`

Chapter 9
Layers Like an Onion: Core Animation

The iPhone has made animation central to the mobile experience. Views slide in and out, applications zoom into place, pages fly into the bookmark list. Apple has made animation not just a beautiful part of the experience, but a better way to let the user know what's happening and what to expect. When views slide into place from right to left, it's natural to press the left-pointing button to go back to where you were. When you create a bookmark and it flies to the toolbar, it's obvious where you should look to get back to that bookmark. These subtle cues are a critical part of making your user interface intuitive as well as engaging. To facilitate all this animation, iOS devices include a powerful GPU and frameworks that let you harness that GPU easily.

In this chapter, you discover the two main animation systems of iOS: view animations and the Core Animation framework. You find out how to draw with Core Animation layers and how to move layers around in two and three dimensions. Common decorations like rounded corners, colored borders, and shadows are trivial with `CALayer`, and you discover how to apply them quickly and easily. You learn how to create custom automatic animations, including animating your own properties. Finally, Core Animation is all about performance, so you find out how to manage layers in multithreaded applications.

This chapter focuses on animations for view-based programming. These frameworks are ideal for most iOS applications except games. Game development is outside the scope of this book, and it's usually best served by built-in frameworks like OpenGL ES or third-party frameworks like Cocos2D. For more information on OpenGL ES, see the OpenGL ES for iOS portal at `developer.apple.com`. For more information on Cocos2D, see `cocos2d-iphone.org`.

View Animations

`UIView` provides rich animation functionality that's very easy to use and well optimized. Most common animations can be handled with `+animateWithDuration:animations:` and related methods. You can use `UIView` to animate `frame`, `bounds`, `center`, `transform`, `alpha`, `backgroundColor`, and `contentStretch`. Most of the time, you'll animate `frame`, `center`, `transform`, and `alpha`.

It's likely that you're familiar with basic view animations, so I'll just touch on the high points in this section and then move on to more advanced layer-based drawing and animation.

Let's start with a very simple animation of a ball that falls when you tap the view. `CircleView` just draws a circle in its frame. The following code creates the animation shown in Figure 9-1.

ViewAnimationViewController.m (ViewAnimation)

```
- (void)viewDidLoad {
  [super viewDidLoad];
  self.circleView = [[CircleView alloc] initWithFrame:
                        CGRectMake(0, 0, 20, 20)];
  self.circleView.center = CGPointMake(100, 20);
  [[self view] addSubview:self.circleView];

  UITapGestureRecognizer *g;
  g = [[UITapGestureRecognizer alloc]
        initWithTarget:self
        action:@selector(dropAnimate)];
  [[self view] addGestureRecognizer:g];
}


...


- (void)dropAnimate {
  [UIView animateWithDuration:3 animations:^{
    self.circleView.center = CGPointMake(100, 300);
  }];
}
```

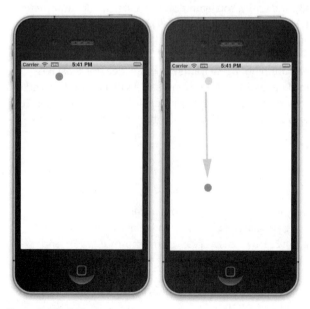

Figure 9-1 CircleView animation

This is the simplest kind of view-based animation, and it can handle most common problems, particularly animating size, location, and opacity. It's also common to animate `transform` to scale, rotate, or translate the view over time. Less commonly, you can animate `backgroundColor` and `contentStretch`. Animating the background color is particularly useful in HUD-style interfaces to move between mostly transparent and mostly opaque backgrounds. This can be more effective than just animating the overall `alpha`.

Chaining animations is also straightforward, as shown in the following code.

```
- (void)dropAnimate {
  [UIView
   animateWithDuration:3 animations:^{
     self.circleView.center = CGPointMake(100, 300);
   }
   completion:^(BOOL finished){
     [UIView animateWithDuration:1 animations:^{
       self.circleView.center = CGPointMake(250, 300);
     }
      ];
   }];
}
```

Now the ball will drop and then move to the right. But there's a subtle problem with this code. If you tap the screen while the animation is in progress, the ball will jump to the lower-left and then animate to the right. That's probably not what you want. The issue is that every time you tap the view, this code runs. If an animation is in progress, then it's canceled and the `completion` block runs with `finished==NO`. You look at how to handle that next.

Managing User Interaction

The problem mentioned in the previous section is caused by a user experience mistake: allowing the user to send new commands while you're animating the last command. Sometimes that's what you want, but in this case, it isn't. Anytime you create an animation in response to user input, you need to consider this issue.

When you animate a view, by default it automatically stops responding to user interaction. So, while the ball is dropping, tapping it won't generate any events. In this example, however, tapping the main view causes the animation. There are two solutions. First, you can change your user interface so that tapping the ball causes the animation:

```
[self.circleView addGestureRecognizer:g];
```

The other solution is to ignore taps while the ball is animating. The following code shows how to disable the `UIGestureRecognizer` in the gesture recognizer callback and then enable it when the animation completes.

```
- (void)dropAnimate:(UIGestureRecognizer *)recognizer {
  [UIView
   animateWithDuration:3 animations:^{
     recognizer.enabled = NO;
     self.circleView.center = CGPointMake(100, 300);
   }
   completion:^(BOOL finished){
     [UIView
      animateWithDuration:1 animations:^{
        self.circleView.center = CGPointMake(250, 300);
      }
      completion:^(BOOL finished){
        recognizer.enabled = YES;
```

(continued)

```
        }];
    }];
}
```

This technique is nice because it minimizes side effects to the rest of the view, but you might want to prevent all user interaction for the view while the animation runs. In that case, you replace `recognizer.enabled` with `self.view.userInteractionEnabled`.

Drawing with Layers

View animations are powerful, so rely on them whenever you can, especially for basic layout animation. They also provide a small number of stock transitions that you can read about in the Animations section of the *View Programming Guide for iOS* available at `developer.apple.com`. If you have basic needs, these are great tools.

But you're here to go beyond the basic needs, and view animations have a lot of limitations. Their basic unit of animation is `UIView`, which is a pretty heavyweight object, so you need to be careful about how many of them you use. `UIView` also doesn't support three-dimensional layout, except for basic z-ordering, so it can't create anything like Cover Flow. To move your UI to the next level, you need to use Core Animation.

Core Animation provides a variety of tools, several of which are useful even if you don't intend to animate anything. The most basic and important part of Core Animation is `CALayer`. This section explains how to draw with `CALayer` without animations. You explore animating later in the chapter.

In many ways, `CALayer` is very much like `UIView`. It has a location, size, transform, and content. You can override a draw method to draw custom content, usually with Core Graphics. There is a layer hierarchy exactly like the view hierarchy. You might ask, why even have separate objects?

The most important answer is that `UIView` is a fairly heavyweight object that manages drawing and event handling, particularly touch events. `CALayer` is all about drawing. In fact, `UIView` relies on a `CALayer` to manage its drawing, which allows the two to work very well together.

Every `UIView` has a `CALayer` to do its drawing. And every `CALayer` can have sublayers, just like every `UIView` can have subviews. Figure 9-2 shows the hierarchy.

A layer draws whatever is in its `contents` property, which is a `CGImage` (see the note at the end of this section). It's your job to set this somehow, and there are various ways of doing so. The simplest approach is to assign it directly, as shown here (and discussed more fully in "Setting Contents Directly" later in this section).

```
UIImage *image = ...;
CALayer *layer = ...;
layer.contents = (__bridge id)[image CGImage];
```

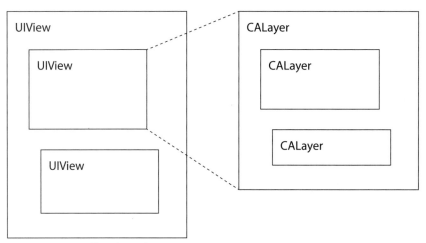

Figure 9-2 View and layer hierarchies

If you do not set the `contents` property directly, then Core Animation will go through the following `CALayer` and delegate methods in the order presented in the following list to create it.

1. [CALayer setNeedsDisplay]—Your code needs to call this. It marks the layer as dirty, requesting that `contents` be updated using the following steps in this list.

   Unless `setNeedsDisplay` is called, the `contents` property is never updated, even if it's `nil`.

2. [CALayer displayIfNeeded]—The drawing system automatically calls this as needed. If the layer has been marked dirty by a call to `setNeedsDisplay`, then the drawing system will continue with the next steps.

3. [CALayer display]—This is called by `displayIfNeeded` when appropriate. You shouldn't call it directly. The default implementation calls the delegate method `displayLayer:` if the delegate implements it. If not, `display` calls `drawInContext:`. You can override `display` in a subclass to set `contents` directly.

4. [delegate displayLayer:]—The default [CALayer display] calls this if the delegate implements it. Its job is to set `contents`. If this method is implemented, even if it does nothing, then no further custom drawing code will be run.

5. [CALayer drawInContext:]—The default `display` method creates a view graphics context and passes it to `drawInContext:`. This is similar to [UIView drawRect:], but no UIKit context is set up for you automatically. To draw with UIKit, you need to call `UIGraphicsPushContext()` to make the passed context the current context. Otherwise, just use the passed context to draw with Core Graphics. The default `display` method takes the resulting context, creates a CGImage (see note below) and assigns it to `contents`. The default [CALayer drawInContext:] calls [delegate drawLayer:inContext:] if it's implemented. Otherwise, it does nothing. Note that you may call this directly. See the section "Drawing in Your Own Context," later in this section, for information on why you would call this directly.

6. `[delegate drawLayer:inContext:]`—If implemented, the default `drawInContext:` calls this to update the context so that `display` can create a `CGImage`.

As you can see, there are several ways to set the contents of a layer. You can set it directly with `setContent:`, you can implement `display` or `displayLayer:`, or you can implement `drawInContext:` or `drawLayer:inContext:`. In the rest of this section, we discuss each approach.

The drawing system almost never automatically updates `contents` in the way that `UIView` is often automatically refreshed. For instance, `UIView` draws itself the first time it's put onscreen. `CALayer` does not. Marking a `UIView` as dirty with `setNeedsDisplay` automatically redraws all the subviews as well. Marking a `CALayer` as dirty with `setNeedsDisplay` doesn't impact sublayers. The thing to remember is that the default behavior of a `UIView` is to draw when it thinks you need it. The default behavior of a `CALayer` is to never draw unless you explicitly ask for it. `CALayer` is a much lower-level object, and it's optimized to not waste time doing anything that isn't explicitly asked for.

> The `contents` property is usually a `CGImage`, but this is not always the case. If you use custom drawing, Core Animation will use a private class, `CABackingStorage`, for `contents`. You can set `contents` to either a `CGImage` or the `contents` of another layer.

Setting Contents Directly

Providing a content image (shown in the following code) is the easiest solution if you already have an image handy.

LayersViewController.m (Layers)

```
#import <QuartzCore/QuartzCore.h>
. . .
  UIImage *image = [UIImage imageNamed:@"pushing.png"];
  self.view.layer.contents = (__bridge id)[image CGImage];
```

> You must always import `QuartzCore.h` and link with `QuartzCore.framework` to use Core Animation. This is an easy thing to forget.

The cast to `__bridge id` is needed because `contents` is defined as an `id`, but actually expects a `CGImageRef`. To make this work with ARC, a cast is required. A common error is to pass a `UIImage` here instead of a `CGImageRef`. You won't get a compiler error or runtime warning. Your view will just be blank.

By default, the contents are scaled to fill the view, even if that distorts the image. As with `contentMode` and `contentStretch` in `UIView`, you can configure `CALayer` to scale its image in different ways using `contentsCenter` and `contentsGravity`.

Implementing Display

The job of `display` or `displayLayer:` is to set `contents` to a correct `CGImage`. You can do this any way you'd like. The default implementation creates a `CGContext`, passes it to `drawInContext:`, turns the result into a `CGImage`, and assigns it to `contents`. The most common reason to override this is if your layer has several states and you have an image for each. Buttons often work this way. You can create those images by loading them from your bundle, drawing them with Core Graphics, or any other way you'd like.

Whether to subclass `CALayer` or use a delegate is really a matter of taste and convenience. `UIView` has a layer, and it must be that layer's delegate. In our experience, it's dangerous to make a `UIView` the delegate for any of the sublayers. Doing so can create infinite recursion when the `UIView` tries to copy its sublayers in certain operations such as transitions. So you can implement `displayLayer:` in `UIView` to manage its layer, or you can have some other object be the delegate for sublayers.

Having `UIView` implement `displayLayer:` seldom makes sense in our opinion. If your view content is basically several images, it's usually a better idea to use a `UIImageView` or a `UIButton` rather than a custom `UIView` with hand-loaded layer content. `UIImageView` is highly optimized for displaying images. `UIButton` is very good at switching images based on state, and includes a lot of good user interface mechanics that are a pain to reproduce. Don't try to reinvent UIKit in Core Animation. UIKit likely does it better than you will.

What can make more sense is to make your `UIViewController` the delegate for the layers, particularly if you aren't subclassing `UIView`. This avoids extra objects and subclasses if your needs are pretty simple. Just don't let your `UIViewController` become overcomplicated.

Custom Drawing

As with `UIView`, you can provide completely custom drawing with `CALayer`. Typically, you'll draw with Core Graphics, but using `UIGraphicsPushContext`, you can also draw with UIKit.

See Chapter 8 for information on how to draw with Core Graphics and UIKit.

Using `drawInContext:` is just another way of setting `contents`. It's called by `display`, which is called only when the layer is explicitly marked dirty with `setNeedsDisplay`. The advantage of this over setting `contents` directly is that `display` automatically creates a `CGContext` appropriate for the layer. In particular, the coordinate system is flipped for you. (See Chapter 8 for a discussion of Core Graphics and flipped coordinate systems.) The following code shows how to implement the delegate method `drawLayer:inContext:` to draw the string "Pushing The Limits" at the top of the layer using UIKit. Because Core Animation does not set a UIKit graphics context, you need to call `UIGraphicsPushContext` before calling UIKit methods, and `UIGraphicsPopContext` before returning.

DelegateView.m (Layers)

```
- (id)initWithFrame:(CGRect)frame {
    self = [super initWithFrame:frame];
```

(continued)

```
    if (self) {
      [self.layer setNeedsDisplay];
    }
    return self;
}

- (void)drawLayer:(CALayer *)layer inContext:(CGContextRef)ctx {
    UIGraphicsPushContext(ctx);
    [[UIColor whiteColor] set];
    UIRectFill(layer.bounds);
    [[UIColor blackColor] set];
    UIFont *font = [UIFont systemFontOfSize:48.0];
    [@"Pushing The Limits" drawInRect:[layer bounds]
                            withFont:font
                       lineBreakMode:NSLineBreakByWordWrapping
                           alignment:NSTextAlignmentCenter];
    UIGraphicsPopContext();
}
```

Note the call to `setNeedsDisplay` in `initWithFrame:`. As discussed earlier, layers don't automatically draw themselves when put onscreen. You need to mark them as dirty with `setNeedsDisplay`.

You may also notice the hand-drawing of the background rather than using the `backgroundColor` property. This is intentional. Once you engage in custom drawing with `drawLayer:inContext:`, most automatic layer settings like `backgroundColor` and `cornerRadius` are ignored. Your job in `drawLayer:inContext:` is to draw everything needed for the layer. There isn't helpful compositing going on for you like in `UIView`. If you want layer effects like rounded corners together with custom drawing, then put the custom drawing onto a sublayer and round the corners on the superlayer.

Drawing in Your Own Context

Unlike `[UIView drawRect:]`, it's completely legal to call `[CALayer drawInContext:]` yourself. You just need to generate a context and pass it in, which is nice for capturing the contents of a layer onto a bitmap or PDF so you can save it or print it. Calling `drawInContext:` this way is mostly useful if you want to composite this layer with something else, because, if all you want is a bitmap, you can just use `contents`.

`drawInContext:` only draws the current layer, not any of its sublayers. If you want to draw the layer and its sublayers, use `renderInContext:`, which also captures the current state of the layer if it's animating. `renderInContext:` uses the current state of the render tree that Core Animation maintains internally, so it doesn't call `drawInContext:`.

Moving Things Around

Now that you can draw in a layer, take a look into how to use those layers to create powerful animations.

Layers naturally animate. In fact, you need to do a small amount of work to prevent them from animating. Consider the following example.

LayerAnimationViewController.m (LayerAnimation)

```objc
- (void)viewDidLoad {
  [super viewDidLoad];
  CALayer *squareLayer = [CALayer layer];
  squareLayer.backgroundColor = [[UIColor redColor] CGColor];
  squareLayer.frame = CGRectMake(100, 100, 20, 20);
  [self.view.layer addSublayer:squareLayer];

  UIView *squareView = [UIView new];
  squareView.backgroundColor = [UIColor blueColor];
  squareView.frame = CGRectMake(200, 100, 20, 20);
  [self.view addSubview:squareView];

  [self.view addGestureRecognizer:
    [[UITapGestureRecognizer alloc]
      initWithTarget:self
      action:@selector(drop:)]];
}

- (void)drop:(UIGestureRecognizer *)recognizer {
  NSArray *layers = self.view.layer.sublayers;
  CALayer *layer = [layers objectAtIndex:0];
  [layer setPosition:CGPointMake(200, 250)];

  NSArray *views = self.view.subviews;
  UIView *view = [views objectAtIndex:0];
  [view setCenter:CGPointMake(100, 250)];
}
```

This draws a small red sublayer and a small blue subview. When the view is tapped, both are moved. The view jumps immediately to the new location. The layer animates over a quarter-second. It's fast, but it's not instantaneous like the view.

CALayer implicitly animates all properties that support animation. You can prevent this by disabling actions:

```objc
[CATransaction setDisableActions:YES];
```

We discuss actions further in the "Auto-Animate with Actions" section, later in this chapter.

disableActions **is a poorly named method. Because it begins with a verb, you expect it to have a side effect (disabling actions) rather than returning the current value of the property. It should be** actionsDisabled **(or** actionsEnabled **to be parallel with** userInteractionEnabled**). Apple may remedy this eventually, as it has with other misnamed properties. In the meantime, make sure to call** setDisableActions: **when you mean to change its value. You won't get a warning or error if you call** [CATransaction disableActions] **in a void context.**

Implicit Animations

You now know all the basics of animation. Just set layer properties, and your layers animate in the default way. But what if you don't like the defaults? For instance, you may want to change the duration of the animation. First, you need to understand transactions.

Most of the time when you change several layer properties, you want them all to animate together. You also don't want the renderer to waste the renderer's time calculating animations for one property change if the next property change affects it. For instance, `opacity` and `backgroundColor` are interrelated properties. Both affect the final displayed pixel color, so the renderer needs to know about both animations when working out the intermediate values.

Core Animation bundles property changes into atomic transactions (`CATransaction`). An implicit `CATransaction` is created for you the first time you modify a layer on a thread that includes a run loop. (If that last sentence piqued your interest, see the "Core Animation and Threads" section, later in this chapter.) During the run loop, all layer changes are collected, and when the run loop completes, all the changes are committed to the layer tree.

To modify the animation properties, you need to make changes to the current transaction. The following changes the duration of the current transaction to 2 seconds rather than the default quarter-second.

```
[CATransaction setAnimationDuration:2.0];
```

You can also set a completion block to run after the current transaction finishes animating using `[CATransaction setCompletionBlock:]`. You can use this to chain animations together, among other things.

Although the run loop creates a transaction for you automatically, you can also create your own explicit transactions using `[CATransaction begin]` and `[CATransaction commit]`. These allow you to assign different durations to different parts of the animation or to disable animations for only a part of the event loop.

> See the "Auto-Animate with Actions" section of this chapter for more information on how implicit animations are implemented and how you can extend them.

Explicit Animations

Implicit animations are powerful and convenient, but sometimes you want more control. That's where `CAAnimation` comes in. With `CAAnimation`, you can manage repeating animations, precisely control timing and pacing, and employ layer transitions. Implicit animations are implemented using `CAAnimation`, so everything you can do with an implicit animation can be done explicitly as well.

The most basic animation is a `CABasicAnimation`. It interpolates a property over a range using a timing function, as shown in the following code:

```
CABasicAnimation *anim = [CABasicAnimation
                            animationWithKeyPath:@"opacity"];
anim.fromValue = [NSNumber numberWithDouble:1.0];
anim.toValue = [NSNumber numberWithDouble:0.0];
anim.autoreverses = YES;
anim.repeatCount = INFINITY;
anim.duration = 2.0;
[layer addAnimation:anim forKey:@"anim"];
```

This pulses the layer forever, animating the opacity from one to zero and back over 2 seconds. When you want to stop the animation, remove it:

```
[layer removeAnimationForKey:@"anim"];
```

An animation has a `key`, `fromValue`, `toValue`, `timingFunction`, `duration`, and some other configuration options. The way it works is to make several copies of the layer, send `setValue:forKey:` messages to the copies and then `display`. It captures the generated `contents` and displays them.

If you have custom properties in your layer, you may notice that they're not set correctly during animation. This is because the layer is copied. You must override `initWithLayer:` to copy your custom properties if you want them to be available during an animation. We discuss this later in the "Animating Custom Properties" section of this chapter.

`CABasicAnimations` are basic, as the name implies. They're easy to set up and use, but they're not very flexible. If you want more control over the animation, you can move to `CAKeyframeAnimation`. The major difference is that instead of giving a `fromValue` and `toValue`, you now can give a path or a sequence of points to animate through, along with individual timing for each segment. The *Animation Types and Timing Programming Guide* at `developer.apple.com` provides excellent examples. They're not technically difficult to set up. Most of the work is on the creative side to find just the right path and timing.

Model and Presentation

A common problem in animations is the dreaded "jump back." The mistake looks like this:

```
CABasicAnimation *fade;
fade = [CABasicAnimation animationWithKeyPath:@"opacity"];
fade.duration = 1;
fade.fromValue = [NSNumber numberWithDouble:1.0];
fade.toValue = [NSNumber numberWithDouble:0.0];
[circleLayer addAnimation:fade forKey:@"fade"];
```

This fades the circle out over 1 second, just as expected, and then suddenly the circle reappears. To understand why this happens, you need to be aware of the difference between the model layer and the presentation layer.

The *model layer* is defined by the properties of the "real" `CALayer` object. Nothing in the preceding code modifies any property of `circleLayer` itself. Instead, `CAAnimation` makes copies of `circleLayer` and modifies those. These become the *presentation layer*. They represent roughly what is shown on the screen. There is technically another layer called the *render layer* that really represents what's on the screen, but it's internal to Core Animation, and you very seldom encounter it.

So what happens in the preceding code? `CAAnimation` modifies the presentation layer, which is drawn to the screen, and when it completes, all its changes are thrown away and the model layer is used to determine the new state. The model layer hasn't changed, so you snap back to where you started. The solution to this is to set the model layer, as shown here:

```
circleLayer.opacity = 0;
CABasicAnimation *fade;
fade = [CABasicAnimation animationWithKeyPath:@"opacity"];
...
[circleLayer addAnimation:fade forKey:@"fade"];
```

Sometimes this works fine, but sometimes the implicit animation in `setOpacity:` fights with the explicit animation from `animationWithKeyPath:`. The best solution to that is to turn off implicit animations if you're doing explicit animations:

```
[CATransaction begin];
[CATransaction setDisableActions:YES];
circleLayer.opacity = 0;
CABasicAnimation *fade;
fade = [CABasicAnimation animationWithKeyPath:@"opacity"];
...
[circleLayer addAnimation:fade forKey:@"fade"];
[CATransaction commit];
```

Sometimes you see people recommend setting `removedOnCompletion` to `NO` and `fillMode` to `kCAFillModeBoth`. **This is not a good solution. It essentially makes the animation go on forever, which means the model layer is never updated. If you ask for the property's value, you continue to see the model value, not what you see on the screen. If you try to implicitly animate the property afterward, it won't work correctly because the** `CAAnimation` **is still running. If you ever remove the animation by replacing it with another with the same name, calling** `removeAnimationForKey:` **or** `removeAllAnimations`, **the old value snaps back. On top of all of that, it wastes memory.**

All of this becomes a bit of a pain, so you may like the following category on `CALayer` that wraps it all together and lets you set the duration and delay. Most of the time, I still prefer implicit animation, but this can make explicit animation a bit simpler.

CALayer+RNAnimation.m (LayerAnimation)

```
@implementation CALayer (RNAnimations)
- (void)setValue:(id)value
      forKeyPath:(NSString *)keyPath
        duration:(CFTimeInterval)duration
           delay:(CFTimeInterval)delay
{
  [CATransaction begin];
  [CATransaction setDisableActions:YES];
```

```
    [self setValue:value forKeyPath:keyPath];

    CABasicAnimation *anim;
    anim = [CABasicAnimation animationWithKeyPath:keyPath];
    anim.duration = duration;
    anim.beginTime = CACurrentMediaTime() + delay;
    anim.fillMode = kCAFillModeBoth;
    anim.fromValue = [[self presentationLayer] valueForKey:keyPath];
    anim.toValue = value;
    [self addAnimation:anim forKey:keyPath];

    [CATransaction commit];
}
@end
```

A Few Words on Timings

As in the universe at large, in Core Animation, time is relative. A second does not always have to be a second. Just like coordinates, time can be scaled.

CAAnimation conforms to the CAMediaTiming protocol, and you can set the speed property to scale its timing. Because of this, when considering timings between layers, you need to convert them just like you need to convert points that occur in different views or layers.

```
    localPoint = [self convertPoint:remotePoint fromLayer:otherLayer];
    localTime = [self convertTime:remotetime fromLayer:otherLayer];
```

This isn't very common, but it comes up when you're trying to coordinate animations. You might ask another layer for a particular animation and when that animation will end so that you can start your animation.

```
    CAAnimation *otherAnim = [layer animationForKey:@"anim"];
    CFTimeInterval finish = otherAnim.beginTime + otherAnim.duration;
    myAnim.beginTime = [self convertTime:finish fromLayer:layer];
```

Setting beginTime like this is a nice way to chain animations, even if you hard-code the time rather than ask the other layer. To reference "now," just use CACurrentMediaTime().

This raises another issue, however. What value should your property have between now and when the animation begins? You would assume that it would be the fromValue, but that isn't how it works. It's the current model value because the animation hasn't begun. Typically, this is the toValue. Consider the following animation:

```
    [CATransaction begin];
    anim = [CABasicAnimation animationWithKeyPath:@"opacity"];
    anim.fromValue = [NSNumber numberWithDouble:1.0];
    anim.toValue = [NSNumber numberWithDouble:0.5];
    anim.duration = 5.0;
    anim.beginTime = CACurrentMediaTime() + 3.0;
    [layer addAnimation:anim forKey:@"fade"];
```

(continued)

```
layer.opacity = 0.5;
[CATransaction commit];
```

This animation does nothing for 3 seconds. During that time, the default property animation is used to fade `opacity` from 1.0 to 0.5. Then the animation begins, setting the opacity to its `fromValue` and interpolating to its `toValue`. So the layer begins with `opacity` of 1.0, fades to 0.5 over a quarter-second, then 3 seconds later, and jumps back to 1.0 and fades again to 0.5 over 5 seconds. This almost certainly isn't what you want.

You can resolve this problem using `fillMode`. The default is `kCAFillModeRemoved`, which means that the animation has no influence on the values before or after its execution. This can be changed to "clamp" values before or after the animation by setting the fill mode to `kCAFillModeBackwards`, `kCAFillModeForwards`, or `kCAFillModeBoth`. Figure 9-3 illustrates this.

In most cases, you want to set `fillMode` to `kCAFillModeBackwards` or `kCAFillModeBoth`.

Into the Third Dimension

Chapter 8 discussed how to use `CAAffineTransform` to make `UIView` drawing much more efficient. This technique limits you to two-dimensional transformations: translate, rotate, scale, and skew. With layers, however, you can apply three-dimensional transformations by adding perspective. This is often called 2.5D rather than 3D because it doesn't make layers into truly three-dimensional objects in the way that OpenGL ES does. But it does allow you to give the illusion of three-dimensional movement.

You rotate layers around an anchor point. By default, the anchor point is in the center of the layer, designated {0.5, 0.5}. It can be moved anywhere within the layer, making it convenient to rotate around an edge or corner. The anchor point is described in terms of a unit square rather than in points. So the lower-right corner is {1.0, 1.0}, no matter how large or small the layer is.

Here's a simple example of a three-dimensional box.

BoxViewController.h (Box)

```
@interface BoxViewController : UIViewController
@property (nonatomic, readwrite, strong) CALayer *topLayer;
@property (nonatomic, readwrite, strong) CALayer *bottomLayer;
@property (nonatomic, readwrite, strong) CALayer *leftLayer;
@property (nonatomic, readwrite, strong) CALayer *rightLayer;
@property (nonatomic, readwrite, strong) CALayer *frontLayer;
@property (nonatomic, readwrite, strong) CALayer *backLayer;
@end
```

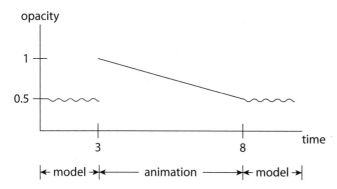

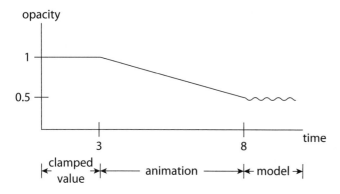

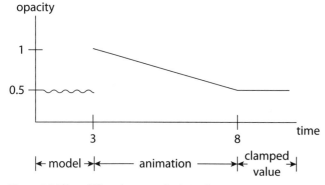

Figure 9-3 Effect of fill modes on media timing functions

BoxViewController.m (Box)

```
@implementation BoxViewController

const CGFloat kSize = 100.;
const CGFloat kPanScale = 1./100.;

- (CALayer *)layerWithColor:(UIColor *)color
                   transform:(CATransform3D)transform {
  CALayer *layer = [CALayer layer];
  layer.backgroundColor = [color CGColor];
  layer.bounds = CGRectMake(0, 0, kSize, kSize);
  layer.position = self.view.center;
  layer.transform = transform;
  [self.view.layer addSublayer:layer];
  return layer;
}

static CATransform3D MakePerspetiveTransform() {
  CATransform3D perspective = CATransform3DIdentity;
  perspective.m34 = -1./2000.;
  return perspective;
}

- (void)viewDidLoad {
  [super viewDidLoad];

  CATransform3D transform;
  transform = CATransform3DMakeTranslation(0, -kSize/2, 0);
  transform = CATransform3DRotate(transform, M_PI_2, 1.0, 0, 0);
  self.topLayer = [self layerWithColor:[UIColor redColor]
                             transform:transform];

  transform = CATransform3DMakeTranslation(0, kSize/2, 0);
  transform = CATransform3DRotate(transform, M_PI_2, 1.0, 0, 0);
  self.bottomLayer = [self layerWithColor:[UIColor greenColor]
                                transform:transform];

  transform = CATransform3DMakeTranslation(kSize/2, 0, 0);
  transform = CATransform3DRotate(transform, M_PI_2, 0, 1, 0);
  self.rightLayer = [self layerWithColor:[UIColor blueColor]
                               transform:transform];

  transform = CATransform3DMakeTranslation(-kSize/2, 0, 0);
  transform = CATransform3DRotate(transform, M_PI_2, 0, 1, 0);
  self.leftLayer = [self layerWithColor:[UIColor cyanColor]
                              transform:transform];

  transform = CATransform3DMakeTranslation(0, 0, -kSize/2);
  transform = CATransform3DRotate(transform, M_PI_2, 0, 0, 0);
  self.backLayer = [self layerWithColor:[UIColor yellowColor]
```

```
                                     transform:transform];

        transform = CATransform3DMakeTranslation(0, 0, kSize/2);
        transform = CATransform3DRotate(transform, M_PI_2, 0, 0, 0);
        self.frontLayer = [self layerWithColor:[UIColor magentaColor]
                                     transform:transform];

        self.view.layer.sublayerTransform = MakePerspetiveTransform();

        UIGestureRecognizer *g = [[UIPanGestureRecognizer alloc]
                                  initWithTarget:self
                                  action:@selector(pan:)];
        [self.view addGestureRecognizer:g];
}

-  (void)pan:(UIPanGestureRecognizer *)recognizer {
     CGPoint translation = [recognizer translationInView:self.view];
     CATransform3D transform = MakePerspetiveTransform();
     transform = CATransform3DRotate(transform,
                                     kPanScale * translation.x,
                                     0, 1, 0);
     transform = CATransform3DRotate(transform,
                                     -kPanScale * translation.y,
                                     1, 0, 0);
     self.view.layer.sublayerTransform = transform;
}
@end
```

`BoxViewController` shows how to build a simple box and rotate it based on panning. All the layers are created with `layerWithColor:transform:`. Notice that all the layers have the same `position`. They only appear to be in the shape of a box through transforms that translate and rotate them.

You apply a perspective `sublayerTransform` (a transform applied to all sublayers, but not the layer itself). We won't go into the math here, but the `m34` position of the 3D transform matrix should be set to $-1/EYE\_DISTANCE$. For most cases, 2000 units works well, but you can adjust this to "zoom the camera."

You could also build this box by setting `position` and `zPosition` rather than translating, as shown in the following code. This may be more intuitive for some developers.

BoxTransformViewController.m (BoxTransform)

```
- (CALayer *)layerAtX:(CGFloat)x y:(CGFloat)y z:(CGFloat)z
               color:(UIColor *)color
          transform:(CATransform3D)transform {
     CALayer *layer = [CALayer layer];
     layer.backgroundColor = [color CGColor];
     layer.bounds = CGRectMake(0, 0, kSize, kSize);
     layer.position = CGPointMake(x, y);
     layer.zPosition = z;
```

(continued)

```
    layer.transform = transform;
    [self.contentLayer addSublayer:layer];
    return layer;
}

- (void)viewDidLoad {
    [super viewDidLoad];
    CATransformLayer *contentLayer = [CATransformLayer layer];
    contentLayer.frame = self.view.layer.bounds;
    CGSize size = contentLayer.bounds.size;
    contentLayer.transform =
        CATransform3DMakeTranslation(size.width/2, size.height/2, 0);
    [self.view.layer addSublayer:contentLayer];

    self.contentLayer = contentLayer;

    self.topLayer = [self layerAtX:0 y:-kSize/2 z:0
                             color:[UIColor redColor]
                         transform:MakeSideRotation(1, 0, 0)];
    ...
}

- (void)pan:(UIPanGestureRecognizer *)recognizer {
    CGPoint translation = [recognizer translationInView:self.view];
    CATransform3D transform = CATransform3DIdentity;
    transform = CATransform3DRotate(transform,
                                    kPanScale * translation.x,
                                    0, 1, 0);
    transform = CATransform3DRotate(transform,
                                    -kPanScale * translation.y,
                                    1, 0, 0);
    self.view.layer.sublayerTransform = transform;
}
```

You now need to insert a CATransformLayer to work with. If you just use a CALayer, then zPosition is used only for calculating layer order. It's not used to determine location in space. This makes the box look completely flat. CATransformLayer supports zPosition without requiring you to apply a perspective transform.

Decorating Your Layers

A major advantage of CALayer over UIView, even if you're working only in 2D, is the automatic border effects that CALayer provides. For instance, CALayer can automatically give you rounded corners, a colored border, and a drop shadow. All of these can be animated, which can provide some nice visual effects. For instance, you can adjust the position and shadow to give the illusion of clicking as the user presses and releases a layer. The following code will create the layer shown in Figure 9-4.

DecorationViewController.m (Decoration)

```
CALayer *layer = [CALayer layer];
layer.frame = CGRectMake(100, 100, 100, 100);
layer.cornerRadius = 10;
layer.backgroundColor = [[UIColor redColor] CGColor];
layer.borderColor = [[UIColor blueColor] CGColor];
layer.borderWidth = 5;
layer.shadowOpacity = 0.5;
layer.shadowOffset = CGSizeMake(3.0, 3.0);
[self.view.layer addSublayer:layer];
```

Figure 9-4 Layer with colored, rounded border and shadow

Auto-Animate with Actions

Most of the time, implicit animations do what you want, but there are times you'd like to configure them. You can turn off all implicit animations using `CATransaction`, but that applies only to the current transaction (generally the current run loop). To modify how an implicit animation behaves, and especially if you want it to always behave that way for this layer, you need to configure the layer's actions. This allows you to configure your animations when you create the layer rather than apply an explicit animation every time you change a property.

Layer actions are fired in response to various changes on the layer, such as adding or removing the layer from the hierarchy or modifying a property. When you modify the `position` property, for instance, the default action is to animate it over a quarter-second. In the following examples, `CircleLayer` is a layer that draws a red circle in its center with the given `radius`.

ActionsViewController.m (Actions)

```
CircleLayer *circleLayer = [CircleLayer new];
circleLayer.radius = 20;
circleLayer.frame = self.view.bounds;
[self.view.layer addSublayer:circleLayer];
...
[circleLayer setPosition:CGPointMake(100, 100)];
```

Let's modify this so that changes in position always animate over 2 seconds:

```
CircleLayer *circleLayer = [CircleLayer new];
circleLayer.radius = 20;
circleLayer.frame = self.view.bounds;
[self.view.layer addSublayer:circleLayer];

CABasicAnimation *anim =
  [CABasicAnimation animationWithKeyPath:@"position"];
anim.duration = 2;
NSMutableDictionary *actions =
  [NSMutableDictionary dictionaryWithDictionary:
                                [circleLayer actions]];
[actions setObject:anim forKey:@"position"];
circleLayer.actions = actions;
...
[circleLayer setPosition:CGPointMake(100, 100)];
```

Setting the action to `[NSNull null]` disables implicit animations for that property. A dictionary cannot hold `nil`, so you need to use the `NSNull` class.

There are some special actions for when the layer is added to the layer tree (`kCAOnOrderIn`) and when it's removed (`kCAOnOrderOut`). For example, you can make a group animation of growing and fade-in like this:

```
CABasicAnimation *fadeAnim = [CABasicAnimation
                         animationWithKeyPath:@"opacity"];
fadeAnim.fromValue = [NSNumber numberWithDouble:0.4];
```

```
fadeAnim.toValue = [NSNumber numberWithDouble:1.0];

CABasicAnimation *growAnim = [CABasicAnimation
                                animationWithKeyPath:
                                @"transform.scale"];
growAnim.fromValue = [NSNumber numberWithDouble:0.8];
growAnim.toValue = [NSNumber numberWithDouble:1.0];

CAAnimationGroup *groupAnim = [CAAnimationGroup animation];
groupAnim.animations = [NSArray arrayWithObjects:fadeAnim,
                        growAnim, nil];

[actions setObject:groupAnim forKey:kCAOnOrderIn];
```

Actions are also important when dealing with transitions (kCATransition) when one layer is replaced with another. This is commonly used with a CATransition (a special type of CAAnimation). You can apply a CATransition as the action for the contents property to create special effects like slide show whenever the contents change. By default, the fade transition is used.

Animating Custom Properties

Core Animation implicitly animates several layer properties, but what about custom properties on CALayer subclasses? For instance, in the CircleLayer, you have a radius property. By default, radius is not animated, but contents is (using a fade CATransition). So changing the radius causes your current circle to cross-fade with your new circle. This probably isn't what you want. You want radius to animate just like position. There are a few steps to make this work correctly, as shown in the following example.

CircleLayer.m (Actions)

```
@implementation CircleLayer
@dynamic radius;

- (id)init {
    self = [super init];
    if (self) {
      [self setNeedsDisplay];
    }

    return self;
}

- (id)initWithLayer:(id)layer {
  self = [super initWithLayer:layer];
  [self setRadius:[layer radius]];
  return self;
}

- (void)drawInContext:(CGContextRef)ctx {
  CGContextSetFillColorWithColor(ctx,
```

(continued)

```
                                       [[UIColor redColor] CGColor]);
    CGFloat radius = self.radius;
    CGRect rect;
    rect.size = CGSizeMake(radius, radius);
    rect.origin.x = (self.bounds.size.width - radius) / 2;
    rect.origin.y = (self.bounds.size.height - radius) / 2;
    CGContextAddEllipseInRect(ctx, rect);
    CGContextFillPath(ctx);
}

+ (BOOL)needsDisplayForKey:(NSString *)key {
    if ([key isEqualToString:@"radius"]) {
      return YES;
    }
    return [super needsDisplayForKey:key];
}

- (id < CAAction >)actionForKey:(NSString *)key {
    if ([self presentationLayer] != nil) {
      if ([key isEqualToString:@"radius"]) {
        CABasicAnimation *anim = [CABasicAnimation
                                  animationWithKeyPath:@"radius"];
        anim.fromValue = [[self presentationLayer]
                          valueForKey:@"radius"];
        return anim;
      }
    }

    return [super actionForKey:key];
}
@end
```

We'll start with a reminder of the basics. You call `setNeedsDisplay` in init so that the layer's `drawInContext:` is called the first time it's added to the layer tree. You override `needsDisplayForKey:` so that whenever `radius` is modified, you automatically redraw.

Core Animation makes several copies of the layer in order to animate. It uses `initWithLayer:` to perform the copy, so you need to implement that method to copy your custom properties.

Now you come to your actions. You implement `actionForKey:` to return an animation with a `fromValue` of the currently displayed (`presentationLayer`) radius. This means that you'll animate smoothly if the animation is changed midflight.

It's critical to note that you implemented the `radius` property using @dynamic here, not @ `synthesize`. CALayer automatically generates accessors for its properties at runtime, and those accessors have important logic. It's vital that you not override these CALayer accessors by either implementing your own accessors or using @synthesize to do so.

Core Animation and Threads

It's worth noting that Core Animation is very tolerant of threading. You can generally modify CALayer properties on any thread, unlike UIView properties. drawInContext: may be called from any thread (although a given CGContext should be modified on only one thread at a time). Changes to CALayer properties are batched into transactions using CATransaction. This happens automatically if you have a run loop. If you don't have a run loop, you need to call [CATransaction flush] periodically. If at all possible, though, you should perform Core Animation actions on a thread with a run loop to improve performance.

Summary

Core Animation is one of the most important frameworks in iOS. It puts a fairly easy-to-use API in front of an incredibly powerful engine. There are still a few rough edges to it, however, and sometimes things need to be "just so" to make it work correctly (for example, implementing your properties with @dynamic rather than @synthesize). When it doesn't work correctly, it can be challenging to debug, so having a good understanding of how it works is crucial. Hopefully, this chapter has made you confident enough with the architecture and the documentation to dive in and make some really beautiful apps.

Further Reading

Apple Documentation

The following documents are available in the iOS Developer Library at developer.apple.com or through the Xcode Documentation and API Reference.

Animation Types and Timing Programming Guide

Core Animation Programming Guide

Other Resources

millen.me. (Milen Dzhumerov), "CA's 3D Model" An excellent overview of the math behind the perspective transform, including the magic -1/2000.
http://milen.me/technical/core-animation-3d-model/

Cocoa with Love, (Matt Gallagher), "Parametric acceleration curves in Core Animation," Explains how to implement timing curves that cannot be implemented with CAMediaTimingFunction, such as damped ringing and exponential decay.
cocoawithlove.com/2008/09/parametric-acceleration-curves-in-core.html

Chapter 10

Tackling Those Pesky Errors

Error management can be one of the most frustrating parts of development. It's hard enough getting everything to work when things go well, but to build really great apps, you need to manage things gracefully when they go wrong. Cocoa provides some tools to make the job easier.

In this chapter, you find the major patterns that Cocoa uses to handle errors that you should use in your own projects. You also discover the major error-handling tools, including assertions, exceptions, and NSError objects. Because your program may crash in the field, you learn how to get those crash reports from your users, and how to log effectively and efficiently.

Error-Handling Patterns

There are several useful approaches to handling errors. The first and most obvious is to crash. This isn't a great solution, but don't discount it too quickly. I've seen a lot of very elaborate code around handling extremely unlikely errors, or errors you won't be able to recover from anyway. The most common of these is failure to allocate memory. Consider the following code:

```
NSString *string = [NSString stringWithFormat:@"%d", 1];
NSArray *array = [NSArray arrayWithObject:string];
```

It's conceivable (not really, but let's pretend) that stringWithFormat: might fail because Foundation isn't able to allocate memory. In that case, it returns nil, and the call to arrayWithObject: throws an exception for trying to insert nil into an array, and your app probably crashes. You could (and in C you often would) include a check here to make sure that doesn't happen. Don't do that. Doing so needlessly complicates the code, and there's nothing you're going to be able to do anyway. If you can't allocate small amounts of memory, the OS is very likely about to shut you down anyway. Besides, it's almost impossible to write error-handling code in Objective-C that does not itself allocate memory. Accept that in this impossible case you may crash, and keep the code simple.

The next closely related error-handling approach is NSAssert. This raises an NSInternalInconsistencyException, which by default crashes your program. Particularly during development, this is a very good thing. It "fails fast," which means the failure tends to happen close to the bug. One of the worst things I see in code is something like this:

```
// Do not do this
- (void)doSomething:(NSUInteger)index {
  if (index > self.maxIndex) {
    return;
```

(continued)

```
    }
    ...
}
```

Passing an out-of-range index is a programming error. This code swallows that error, turning it into a no-op. That is incredibly difficult to debug. Note how NSArray handles this situation. If you pass an index out of range, it raises an exception very similar to NSAssert. It's the caller's job to pass good values. The worst thing NSArray could do is to silently ignore bad values. It's better to crash. I'll discuss assertions more in the following two sections, "Assertions" and "Exceptions," including how to manage development and release builds, and how to make these a bit more graceful.

The lesson here is that crashing is not the worst-possible outcome. Data corruption is generally the worst-possible outcome, and if getting into a deeply unknown state could corrupt user data, it's definitely better to crash.

Expected errors should be handled gracefully and should never crash. The common pattern for managing expected errors is to return an NSError object by reference. I discuss this in the section "Errors and NSError" later in this chapter.

There is a major difference between expected and unexpected errors. In iOS, failure to allocate small amounts of memory is an unexpected error. It should never happen in normal operation. You should have received a memory warning and been terminated long before you got to that state. You can generally ignore truly unexpected errors and let them crash you. On the other hand, running out of disk space is a rare but expected error. It can easily happen if the user has requested that iTunes fill the device with music. You need to recover gracefully when you cannot write a file.

In the middle are programming errors. You generally handle these errors with assertions.

Assertions

Assertions are an important defense against programming errors. An assertion requires that something must be true at a certain point in the program. If it is not true, then the program is in an undefined state and should not proceed. Consider the following example of NSAssert:

```
    NSAssert(x == 4, @"x must be four");
```

NSAssert tests a condition, and if it returns NO, raises an exception, which is processed by the current exception handler, which by default calls abort and crashes the program. If you're familiar with Mac development, you may be used to exceptions terminating only the current run loop, but iOS calls abort by default, which terminates the program no matter what thread it runs on.

> Technically, abort sends the process a SIGABRT, which can be caught by a signal handler. Generally, we don't recommend catching SIGABRT except as part of a crash reporter. See "Catching and Reporting Crashes," later in this chapter, for information about how to handle crashes.

You can disable `NSAssert` by setting `NS_BLOCK_ASSERTIONS` in the build setting "Preprocessor Macros" (`GCC_PREPROCESSOR_DEFINITIONS`). Opinions differ on whether `NSAssert` should be disabled in release code. It really comes down to this: When your program is in an illegal state, would you rather it stop running, or would you prefer that it run in a possibly random way? Different people come to different conclusions here. My opinion is that it's generally better to disable assertions in release code. I've seen too many cases where the programming error would have caused only a minor problem, but the assertion causes a crash. Xcode 4 templates automatically disable assertions when you build for the Release configuration.

That said, although I like removing assertions in the Release configuration, I don't like ignoring them. They're exactly the kind of "this should never happen" error condition that you'd want to find in your logs. Setting `NS_BLOCK_ASSERTIONS` completely eliminates them from the code. My solution is to wrap assertions so that they log in all cases. The following code assumes you have an `RNLogBug` function that logs to your log file. It's mapped to `NSLog` as an example. Generally I don't like to use `#define`, but it's necessary here because `__FILE__` and `__LINE__` need to be evaluated at the point of the original caller.

This also defines `RNCAssert` as a wrapper around `NSCAssert` and a helper function called `RNAbstract`. `NSCAssert` is required when using assertions within C functions, rather than Objective-C methods.

RNAssert.h

```
#import <Foundation/Foundation.h>

#define RNLogBug NSLog // Use DDLogError if you're using Lumberjack

// RNAssert and RNCAssert work exactly like NSAssert and NSCAssert
// except they log, even in release mode

#define RNAssert(condition, desc, ...) \
  if (!(condition)) { \
    RNLogBug((desc), ## __VA_ARGS__); \
    NSAssert((condition), (desc), ## __VA_ARGS__); \
  }

#define RNCAssert(condition, desc) \
  if (!(condition)) { \
    RNLogBug((desc), ## __VA_ARGS__); \
    NSCAssert((condition), (desc), ## __VA_ARGS__); \
  }
```

Assertions often precede code that would crash if the assertion were not valid. For example (assuming you're using `RNAssert` to log even in the Release configuration):

```
RNAssert(foo != nil, @"foo must not be nil");
[array addObject:foo];
```

The problem is that this still crashes, even with assertions turned off. What was the point of turning off assertions if you're going to crash anyway in many cases? That leads to code like this:

```
RNAssert(foo != nil, @"foo must not be nil");
if (foo != nil) {
   [array addObject:foo];
}
```

That's a little better, using `RNAssert` so that you log, but you've got duplicated code. This raises more opportunities for bugs if the assertion and conditional don't match. Instead, I recommend this pattern when you want an assertion:

```
if (foo != nil) {
   [array addObject:foo];
}
else {
   RNAssert(NO, @"foo must not be nil");
}
```

This pattern ensures that the assertion always matches the conditional. Sometimes assertions are overkill, but this is a good pattern in cases where you want one. I almost always recommend an assertion as the default case of a `switch` statement, however.

```
switch (foo) {
   case kFooOptionOne:
      ...
      break;
   case kFooOptionTwo:
      ...
      break;
   default:
      RNAssert(NO, @"Unexpected value for foo: %d", foo):
      break;
}
```

This way, if you add a new enumeration item, it will help you catch any `switch` blocks that you failed to update.

Exceptions

Exceptions are not a normal way of handling errors in Objective-C. From *Exception Programming Topics* (`developer.apple.com`):

> *The Cocoa frameworks are generally not exception-safe. The general pattern is that exceptions are reserved for programmer error only, and the program catching such an exception should quit soon afterwards.*

In short, exceptions are not for handling recoverable errors in Objective-C. Exceptions are for handling those things that should never happen and which should terminate the program. This is similar to `NSAssert`, and in fact, `NSAssert` is implemented as an exception.

Objective-C has language-level support for exceptions using directives such as `@throw` and `@ catch`, but you generally should not use these. There is seldom a good reason to catch exceptions except at the top level of your program, which is done for you with the global exception handler. If you want to raise an exception to indicate a programming error, it's best to use `NSAssert` to raise an `NSInternalInconsistencyException`, or create and raise your own `NSException` object. You can build these by hand, but we recommend `+raise:format:` for simplicity.

```
[NSException raise:NSRangeException
            format:@"Index (%d) out of range (%d...%d)",
               index, min, max];
```

There seldom is much reason to do this. In almost all cases, it would be just as clear and useful to use `NSAssert`. Because you generally shouldn't catch exceptions directly, the difference between `NSInternalInconsistencyException` and `NSRangeException` is rarely useful.

Automatic Reference Counting is not exception-safe by default in Objective-C. You should expect significant memory leaks from exceptions. In principle, ARC is exception-safe in Objective-C++, but `@autoreleasepool` blocks are still not released, which can lead to leaks on background threads. Making ARC exception-safe incurs performance penalties, which is one of many reasons to avoid significant use of Objective-C++. The Clang flag `-fobjc-arc-exceptions` controls whether ARC is exception-safe.

Catching and Reporting Crashes

iTunes Connect is supposed to provide crash reports, but it has a lot of limitations. Apple makes a single blanket request to the user for permission to upload crash reports. Many users decline. Reports are updated only once a day. iTunes Connect only supports applications deployed on the App Store, so you need a different system during development and internal betas. In short, if iTunes Connect works for you, great, but often it doesn't.

The best replacement I've found is Quincy Kit (`quincykit.net`), which is integrated with HockeyApp (`hockeyapp.net`). It's easy to integrate into an existing project, and it uploads reports to your own web server or the HockeyApp server after asking user permission. Currently, it doesn't handle uploading logs to go along with the crash report.

Quincy Kit is built on top of PLCrashReporter from Plausible Labs. PLCrashReporter handles the complex problem of capturing crash information. Quincy Kit provides a friendly front end for uploading that information. If you need more flexibility, you might consider writing your own version of Quincy Kit. It's handy and nice, but not all that complicated. You probably should not try to rewrite PLCrashReporter. While a program is in the middle of crashing, it can be in a bizarre and unknown state. Properly handling all of the subtle issues that go with that is not simple, and Landon Fuller has been working on PLCrashReporter for years. Even something as simple as allocating or freeing memory can deadlock the system and rapidly drain the battery. That's why Quincy Kit uploads the crash files when the program restarts rather than during the crash. You should do as little work as possible during the crash event.

When you get your crash reports, depending on how your image was built, they may have symbols or they may not. Xcode generally does a good job of automatically symbolicating the reports (replacing addresses with method names) in Organizer as long as you keep the `.dSYM` file for every binary you ship. Xcode uses Spotlight to find these files, so make sure they're available in a place that Spotlight can search. You can also upload your symbol files to HockeyApp if you're using it for managing crash reports.

Errors and NSError

There's a major difference between a user or environment error and a programming error. Programming errors need to be handled with exceptions in debug mode and with logging in release mode. If data corruption is possible, programming errors should also raise exceptions (crash) in release mode. Failure to allocate small amounts of memory needs to be treated as a programming error in iOS because it shouldn't be possible and almost certainly indicates a programming error.

User errors or environment errors (network failures, disk full, and so on) should never raise exceptions. They should return errors, generally using an `NSError` object. `NSFileManager` is a good example of an object that uses `NSError` extensively.

```
- (BOOL)copyItemAtPath:(NSString *)srcPath
               toPath:(NSString *)dstPath
               error:(NSError **)error
```

This method copies a file or directory from one location to another. Obviously that might fail for a variety of reasons. If it does, the method returns `NO` and updates an `NSError` object that the caller passes by reference (pointer to a pointer), as shown in this example.

```
NSError *error;
if (! [fileManager copyItemAtPath:srcPath
                          toPath:toPath
                          error:&error]) {
   [self handleError:error];
}
```

This pattern is convenient because the return value is consistent with the success of the operation. If the method were, instead, to return an `NSError`, then `nil` would indicate success. This would be confusing and error prone.

Internally, the method might look something like this:

```
- (BOOL)copyItemAtPath:(NSString *)srcPath
               toPath:(NSString *)dstPath
               error:(NSError **)error {

  BOOL success = ...;
  if (! success) {
    if (error != NULL) {
      *error = [NSError errorWithDomain:...];
    }
  }
  return success;
}
```

Note how this checks that `error` (a pointer to a pointer) is non-`NULL` before dereferencing it. This allows callers to pass `NULL` to indicate that they don't care about the error details. They might still check the return value to determine the overall success or failure of the operation.

`NSError` encapsulates information about an error in a consistent package that is easy to pass around. Since it conforms to `NSCoding`, it's also easy to write `NSError` objects to disk or over a network.

Errors are primarily defined by their domain and a code. The code is an integer, and the domain is a string that allows you to identify the meaning of that integer. For instance, in `NSPOSIXErrorDomain`, the error code 4 indicates that a system call was interrupted (`EINTR`), but in `NSCocoaErrorDomain`, the error code 4 indicates that a file was not found (`NSFileNoSuchFileError`). Without a domain, the caller would have to guess how to interpret the error code. You're encouraged to create your own domains for your own errors. You should generally use a Uniform Type Indicator (UTI) for this, such as `com.example.MyApp.ErrorDomain`.

`NSError` includes a user info dictionary that can contain any information you like. There are several predefined keys for this dictionary, such as `NSLocalizedDescriptionKey`, `NSUnderlyingErrorKey`, and `NSRecoveryAttempterErrorKey`. You're free to create new keys to provide domain-specific information. Several domains already do this, such as `NSStringEncodingErrorKey` for passing the relevant string encoding or `NSURLErrorKey` for passing an URL.

Error Localization

Where to localize errors is always a tricky subject. Low-level frameworks tend to present errors in very user-unfriendly ways. Errors like "Interrupted system call (4)" are generally not useful to the user. Translating such an error message into French and Spanish doesn't help anything. It just wastes money and confuses users in more languages. Localizing these kinds of error messages makes things more difficult to debug because logs may be sent to you with errors in a language you can't read.

> This last point bears emphasizing. Never localize a string that you don't intend to display to a user.

Because errors often need to be logged in the developer's language, I recommend against using `NSLocalizedDescriptionKey` and its relatives in most cases for `NSError`. Instead, localize at the point of displaying the error. You can keep localized strings for various error codes using a localized string table with the same name as your error domain with `.strings` appended. For instance, for the error domain `com.example.MyApp.ErrorDomain`, you have a localized strings file named `com.example.MyApp.ErrorDomain.strings`. In that file, just map the error code to the localized value:

```
"1" = "File not found."
```

Then, to read the file, just use `NSBundle`:

```
NSString *key = [NSString stringWithFormat:@"%d", [error code]];
NSString *localizedMessage = [[NSBundle mainBundle]
                          localizedStringForKey:key
                                          value:nil
                                          table:[error domain]];
```

Error Handler Blocks

Blocks provide a very flexible way to handle errors. Passing error-handling blocks is particularly useful for asynchronous operations. They're also useful as a more flexible version of error recovery attempters.

> See the Error Handling Programming Guide's (`developer.apple.com/`) section on "Error Responders and Error Recovery" for more information on error responders and error recovery attempters. Blocks are almost always a better error recovery solution on iOS because iOS offers no UI integration with error responders.

Often error handling can be easily combined with completion blocks. For example, the `<NSFilePresenter>` protocol includes this method:

```
- (void)savePresentedItemChangesWithCompletionHandler:
                (void (^)(NSError *errorOrNil))completionHandler
```

This is an excellent example of using blocks to manage error handling. Because this method may perform time-intensive operations, it's not possible to immediately inform the caller of success or failure. You might implement this method as follows:

```
- (void)savePresentedItemChangesWithCompletionHandler:
                (void (^)(NSError *errorOrNil))completionHandler
{
  dispatch_queue_t queue = ...;

  // Dispatch to the background queue and immediately return
  dispatch_async(queue, ^{
    //
    // ... Perform some operations ...
    //
    if (completionHandler) {
      NSError *error = nil;
      if (anErrorOccurred) {
        error = [NSError errorWithDomain:...];
      }
      // Run the completion handler on the main thread
      dispatch_sync(dispatch_get_main_queue(), ^{
          completionHandler(error);
      });
    }
  });
}
```

This pattern is often more convenient for the caller than the typical error-handling delegate callbacks. Rather than defining a delegate method such as `filePresenter:didFailWithError:`, the caller can keep the error-handling code close to the calling code. The preceding method would be used like this:

```
[presenter savePresentedItemChangesWithCompletionHandler:^(NSError *e) {
  if (e) {
    ... respond to error ...
  }
  else {
    ... cleanup after success ...
  }];
```

Note that the previous code doesn't require that `completionHandler` be stored as an ivar. This approach avoids the problem of retain loops. There may be a short-lived retain loop until the operation completes, but this is generally a good thing. As soon as the operation completes and the completion handler fires, the retain loop (if any) is automatically cleared. For more information on blocks and their patterns, see Chapter 23.

Logs

Logging is a critical part of debugging. It's also very hard to get right. You want to log the right things, and you want to log in the right way. Let's start with logging in the right way.

Foundation provides a single logging call: `NSLog`. The only advantage `NSLog` has is that it's convenient. It's inflexible and incredibly slow. Worst of all, it logs to the console, which is never appropriate in released code. `NSLog` should never appear in production code.

Some people deal with this issue simply:

```
#ifdef DEBUG
#define MYLog NSLog
#else
#define MYLog
#end
```

That's fine for pulling out `NSLog`, but now you have no logs at all, which is not ideal. What you need is a logging engine that adapts to both development and release. Here are some of the things to consider in your logging engine:

- It should log to console in debug mode and to a file in release mode. If you don't log to console in debug mode, you won't see logging output in Xcode. Ideally, it should be able to log to both at the same time.

- It should include logging levels (error, warning, info, verbose).

- It should make sure that logging to disabled logging levels is cheap.

- It should not block the calling thread while it writes to a file or the console.

- It must support log aging to avoid filling the disk.

- It should be very easy to call, generally using a C syntax with varargs rather than an Objective-C syntax. The `NSLog` interface is very easy to use, and you want something that looks basically like that. You definitely don't want simple logging statements to require multiple lines of code.

My current recommendation for iOS logging is Lumberjack from Robbie Hanson of Deusty Designs. See "Further Reading" at the end of this chapter for the link. In general, it requires only a few extra lines of code to configure, and a simple substitution of NSLog to DDLog.

This still leaves the question of what to log. If you log too little, you won't have the information you need to debug issues. If you log too much, you'll overwhelm even the best system, hurt performance, and age your logs so quickly that you probably still won't have the information you need. Middle ground is very application-specific, but there are some general rules.

When adding a logging statement, ask yourself what you would ever do with it. Are you just relogging something that's already covered by another log statement? This is particularly important if you're logging data rather than just "I'm in this method now."

Avoid calculating complex data if you might not log it. Consider the following code:

```
NSString *expensiveValue = [self expensiveCall];
DDLogVerbose(@"expensiveValue=%@", expensiveValue);
```

If you never use expensiveValue in the upcoming code and verbose logging isn't turned on, you've wasted time calculating it. Lumberjack is written in such a way that this kind of logging is efficient:

```
DDLogVerbose(@"expensiveValue=%@", [self expensiveCall]);
```

This translates to

```
do {
  if(ddLogLevel && LOG_FLAG_VERBOSE)
    [DDLog log:...
          format:@"expensiveValue=%@", [self expensiveCall]];
} while(0);
```

In this case, expensiveCall is not executed unless needed. The log level is checked twice (once in the macro and once in [DDLog log:...]), but this is a very fast operation compared to expensiveCall. If you build your own logging engine, this is a good technique to emulate.

A similar logging trick is to make sure you need to log before entering a loop. In Lumberjack, it's done this way:

```
if (LOG_VERBOSE) {
  for (id object in list) {
    DDLogVerbose(@"object=%@", object);
  }
}
```

The point is to avoid repeatedly calculating whether to log and to avoid calculating the log string. That's even more important if complex work needs to be done to generate the log.

Most of the time, verbose logging is turned off, so even if DDLogVerbose checks the level again, the preceding code is cheaper in most cases. It also avoids creating a string for object. When verbose logging is turned on, the extra LOG_VERBOSE check is trivial compared to the rest of the loop.

Logging Sensitive Information

Logging opens up serious privacy concerns. Many applications process information that should never go into a log. Obviously, you should not log passwords or credit card numbers, but this is sometimes trickier than it sounds. What if sensitive information is sent over a network and you log the packets? You may need to filter your logs before writing them to avoid this situation.

Don't ask your customers to "just trust you" with their sensitive information. Not only does that put the customer at risk, but the more of their information in your possession, the more legal issues you have to consider. Few things eat up profits as quickly as consulting lawyers.

Regularly audit your logs to make sure you're not logging sensitive information. After running your program at the maximum logging level, search the logs for your password and any other sensitive information. If you have automated tests, this generally can be added fairly easily.

Encrypting your logs does nothing to help this situation. The problem is that the users send their logs to you, and you have the decryption key. If you feel you need to encrypt your logs, you're probably logging something you shouldn't be.

During development, it's occasionally important to see the real data in the logs. I spent quite some time tracking down a bug where we were dropping the last character of the password. Had we logged the password, this bug would have been much easier to discover. If you need this kind of functionality, just make sure it doesn't stay in place in production code.

Getting Your Logs

Logs aren't very useful if you can't get to them. Don't forget to include some way to get the logs from the user. If you have a network protocol, you could upload them. Otherwise, you can use `MFMailComposeViewController` to send them as an attachment. Keep in mind the potential size of your logs. You often will want to compress them first. I've had good luck using Minizip for this (see "Further Reading"). There are some wrappers for Minizip such as Objective-Zip and ZipArchive, but I'm not particularly impressed with them.

TestFlight (`testflightapp.com`) supports uploading logs to their servers by simply using `TFLog()` instead of `NSLog()`. HockeyApp (`hockeyapp.net`) currently has no support for this.

Be sure to ask permission before sending logs. Not only are there privacy concerns, but sending logs can use a lot of bandwidth and battery. Generally, you only need to send logs in response to a problem report.

Summary

Error handling is one of the trickiest parts of any environment. It's much easier to manage things when they go right than when they go wrong. In this chapter, you've seen how to best handle things when they go wrong. Nothing will make this an easy process, but you need to have the tools to make the process a manageable one.

Further Reading

Apple Documentation

The following documents are available in the iOS Developer Library at `developer.apple.com` or through the Xcode Documentation and API Reference.

Exception Programming Topics

Error Handling Programming Guide

TN2151: Understanding and Analyzing iPhone OS Application Crash Reports

Other Resources

Clang documentation, "Automatic Reference Counting." This is the official documentation on how ARC and exceptions interact. `clang.llvm.org/docs/AutomaticReferenceCounting.html#misc.exceptions`

Lumberjack. Mac and iOS logger.
`github.com/robbiehanson/CocoaLumberjack`

Olsson, Fredrik. "Exceptions and Errors on iOS," *Jayway Team Blog*. A good discussion of programmer versus user errors and how to deal with exceptions versus other kinds of errors.
`www.blog.jayway.com/2010/10/13/exceptions-and-errors-on-ios`

Quincy Kit. A nice crash-catcher for iOS. This is integrated into HockeyApp.
`quincykit.net`

Volant, Gilles. *zLib and Minizip*. The standard for ZIP file handling. Don't let the "win" and "Dll" fool you. This is highly portable.
`http://winimage.com/zLibDll/minizip.html`

## Chapter 11

Location Services: Know Where You Are

The ability to know the exact location of the device/user is arguably one of the main differentiating factors that set mobile programming apart from conventional web or native Mac development. That information about where your user is opens up a wealth of ideas and opportunities, and location-based services are arguably one of the hottest categories in the App Store.

The earliest location-based apps showed points of interest within your vicinity. Later, with the gyroscope hardware, some of these apps implemented augmented reality to show the same information. Another category of apps used location information to build a complete business around it—for example, foursquare and Facebook Places. The latest and hottest categories of location-based apps are the turn-by-turn navigation services and location-based reminders.

Apple came up with a couple of location-based apps, Find My Friends and Find My iPhone. Find My Friends helps you locate friends in your vicinity, and Find My iPhone can help you track a lost iDevice.

In this chapter, I briefly go through the basics of using location-based services in your app. Then, in the "Background Location" section I explain in detail the most important aspect of location services: getting a user's location while your app is running in the background on his or her device. You also learn about managing the life of the device's battery effectively.

Core Location Framework

Any iOS app that requires a user's location needs to link with the Core Location framework (available as a part of the SDK). You can use classes and methods in this framework to get the location information on a user's device.

> **If your application cannot function without location information (such as turn-by-turn navigation), you need to use the** `UIRequiredDeviceCapabilities` **key and add** `location-services` **or** `gps`**. This setting will ensure that the app cannot be installed on devices without those capabilities.** `UIRequiredDeviceCapabilities` **is discussed more in detail in Chapter 16.**

If you were writing code directly on the hardware, getting the location of a device wouldn't be an easy job. So far (at least as of this writing), no iOS device has shipped with a standalone/autonomous GPS chip. The GPS chip inside your iPhone or the 3G iPad is an assisted GPS (aGPS) chip that requires network data to calculate the location information of the device. (This is why you cannot locate yourself while roaming overseas with Data Roam disabled.) iOS uses a wealth of information including values from the assisted GPS chip, Wi-Fi and cell tower triangulation to calculate the location of the device. On devices without Wi-Fi (some iPhones sold in China have Wi-Fi disabled) or cellular network connectivity (all iPod touches and Wi-Fi-only iPads), iOS automatically uses other available information to calculate the location. However, location data obtained this way isn't as precise.

The best part of Apple's Core Location framework is that it abstracts you from all these complexities. Core Location framework provides a clean and elegant class that just reports the device's location and accuracy when you ask for the location. You, as a developer, don't have to deal with MAC addresses, cell tower signal strength, and so on.

Tapping into the User's Location

Getting the user's location with the Core Location framework is straightforward. You create an instance of `CLLocationManager` and call the `startUpdatingLocation` method.

Starting the Location Manager

```
locationManager = [[CLLocationManager alloc] init];
locationManager.delegate = self;
[locationManager startUpdatingLocation];
```

Optionally, you can set preferences in the `CLLocationManager` instance for the desired accuracy and the distance filter. You guessed it, a large distance filter and a coarse accuracy result in faster responses, whereas a narrow (or no) distance filter and a higher accuracy setting result in slower responses. iOS supports accuracy ranges from ten meters to three kilometers. Use them according to your application's needs. For example, if you were writing a weather app, you probably wouldn't need a very high accuracy. Weather information isn't going to be different within a three-kilometer radius, which means you should consider using `kCLLocationAccuracyThreeKilometers`.

Getting the User's Heading with the Built-In Compass

Similar to location, you can get the `heading` in which a user is traveling (or the direction that the device is pointed) using the Core Location framework. The built-in maps app rotates the map according to the direction users are facing so as to provide a better idea of exactly where they are. You can use `heading` in your own app in similar innovative ways. You can get heading information using the instance method `startUpdatingHeading` of `CLLocationManager`.

Getting Heading Updates

```
locationManager = [[CLLocationManager alloc] init];
locationManager.delegate = self;
[locationManager startUpdatingHeading];
```

You then conform to the `CLLocationManagerDelegate` protocol by implementing the methods below.

Location and Heading Delegates

```
- (void)locationManager:(CLLocationManager *)manager
     didUpdateToLocation:(CLLocation *)newLocation
          fromLocation:(CLLocation *)oldLocation {
  NSTimeInterval interval= [[newLocation timestamp] timeIntervalSinceNow];
  if(interval > -30 && interval < 0) {
    NSLog(@"%.4f, %.4f", newLocation.coordinate.latitude,
    newLocation.coordinate.longitude);
    [self.locationManager stopUpdatingLocation];
  }
}

- (void)locationManager:(CLLocationManager *)manager
       didFailWithError:(NSError *)error {
  [self.locationManager stopUpdatingLocation];
  if([error code] != kCLErrorDenied) {
    //your code goes here
  }
  else {
    //Ask user to enable location
  }
}
```

Heading information is available only on devices with a built-in compass. Before using heading information, be sure to check for heading availability using [CLLocationManager headingAvailable]. Devices older than iPhone 3GS don't have compass hardware to provide heading information.

The most important piece in this code to check is whether the location information you receive is new. iOS caches the last-known location and sends you that information first, along with the timestamp telling when it was cached. This cached location is shared across apps. For example, if the user is currently using a navigation app and opens your app, iOS already knows the location of the device and sends this information to your application; this information is sent to you almost immediately. But since this information might be stale, you should wait for more accurate and recent data. If you're fine with the results, call stopUpdatingLocation.

Location Services and Privacy

For those writing software over the past decade, privacy has been a monumental concern. Web programmers have always been aware of the privacy issues associated with placing tracking cookies on a user's computer, and now mobile programmers have their own list of similar concerns. In protecting users' privacy, it's important to let them know you want to access their location and then wait for their consent before accessing it. iOS shows a privacy alert to the users on your behalf and you have no control over it. So you might want to explain to users why you need access to their location.

Setting the Location Usage Description Text

The SDK allows developers to provide a "purpose" text stating why the app needs location information. iOS displays this information to users when asking for their consent, as illustrated in Figures 11-1 and 11-2.

Figure 11-1 Location Services dialog shown without a location usage description text

You set the purpose text in the `Info.plist` file for the key `NSLocationUsageDescription`. The `CLLocationManager` instance's purpose text is deprecated in iOS 6 and you shouldn't use that anymore. When you use location services, you need to consider setting this key in your `Info.plist`. The location usage description key helps users feel more comfortable in sharing their location with you.

Working Without Location Information

When asking a user for consent to use his location, it's important to be on the watch for his denial. Your app must be able to handle this denial of consent and to work without location information. Failing to do so will result in your app being rejected by the app review team. The following code shows how to handle the denial of consent:

```
if([error code] != kCLErrorDenied) {
}
```

Figure 11-2 Location Services dialog showing the location usage description text

Other particular cases are when users have disabled location services altogether (for all apps) or when location services are disabled as a part of parental control restrictions. In either case, you need to check for location services availability and disable necessary UI elements if parental control restrictions disable use of location services. You can do that using the following code snippet:

```
if([CLLocationManager locationServicesEnabled]) {

}
```

Alternatively, if your application cannot work without location information, you should require it in the `UIRequiredDeviceCapablities` key.

Background Location

Prior to iOS 4, no third-party application could use location services when the app is not in the foreground. In fact, iOS 3 didn't even support multitasking. Beginning with iOS 4, however, you have multiple ways to get location information when you're running in the background.

You can get location updates continuously while running in the background, you can let the operating system notify you when a significant location change occurs, or you can monitor whether the device/user enters or exits a particular region. Which service you use is entirely application-dependent. However, as a developer, you should also be aware of how location updates can affect battery life. I explain about battery issues in the section, "Keeping an Eye on the Battery," later in this chapter.

Getting Continuous Location Updates in the Background

Some apps, such as fitness trackers or apps that track your running and other workouts, need location information continuously in the background. If your application's use case is similar, request iOS to let you access GPS information in the background by adding `location` to the `UIBackgroundModes` in your application's `Info.plist`. Applications that require continuous location information are GPS-based fitness trackers and turn-by-turn navigation.

The GPS chip is easily one of the most power-hungry chips on the iPhone. Hence, using continuous location updates in background drains the battery quickly. If your app alerts the user about something when he "nears" a place or when he "leaves" a place, you should *not* use continuous background location. Apple provides another method to get updates when the user's location changes significantly.

Subscribing to Significant Change Notification

Subscribing to significant location change is very similar to starting the location manager service you learnt in the previous section. You used `startUpdatingLocation` to get the device's location. Instead of this method, you get significant location change notifications by calling the `startMonitoring SignificantLocationChanges`. For all other methods like the `CLLocationManagereDelegate`, setting the optional properties in `CLLocationManager` instance remains the same.

> Significant location change monitoring will also bring your app to the foreground if it's in the background. You can process the location information as though your app is in the foreground.

The significant location change API brings up some interesting app ideas, such as those in the following sections.

WeatherLah

Singapore is known for spontaneous showers throughout the year. Wouldn't it be great if an app reminds you of an impending shower just before you move into a new location? Turns out, "There is an app for that." WeatherLah is one among the list of the top 100 must-have apps in Singapore. What makes it unique and makes it stand out from other weather apps (including Apple's weather app) is its innovative use of background location and crowd-sourced data. The screenshots in Figures 11-3 and 11-4 show two of the user interface's important screens.

The iOS device monitors for significant location change and notifies the server of the latest location. When this location (or any location that you're interested in) gets "enough" input from other users of the app that it's raining, a push notification will be sent to you that it's raining in the neighboring district.

foursquare Radar

foursquare introduced a feature called foursquare Radar that alerts you if you're near some of the important locations that you are following. For example, you can "follow" a curated list such as "10 must-see places in Singapore," and foursquare Radar will alert you when you are near one. You start the significant location service on iOS, send the location to the server whenever a significant change occurs, and send a push notification if the new location is near a place of interest.

Figure 11-3 Setting locations in WeatherLah

Figure 11-4 Gathering crowd-sourced data

Facebook Places

Facebook introduced a feature called Facebook Places that notifies you when a friend checks into a location near you. This feature, again, uses significant location changes. The iOS app registers for significant location changes and updates the server of the new location. The server sends a notification whenever a friend comes near your location.

From the examples in the preceding sections, you can see that just by using the significant location change API, you can build a product and a complete business using it. Next, I'll show you yet another powerful feature called *region monitoring* provided by the Core Location framework.

Region Monitoring (Geo-Fencing)

In some cases, rather than being notified for every significant location change, your app might just require a notification when the user enters or exits a *region*. The Core Location framework provides APIs to monitor a region. Region monitoring is slightly trickier than the previous two techniques I described.

Gist

First, you create a region or an area of interest. This region could be one that's specified by users or automatically calculated by your app based on the users' input.

You create a region using the CLRegion class. The CLRegion class takes a center coordinate and a radius. As of iOS6, you can monitor only circular regions.

Creating a CLRegion and Monitoring It

```
CLRegion* region = [[CLRegion alloc] initCircularRegionWithCenter:overlay.
coordinate
                    radius:radius identifier:identifier];
[self.locationManager startMonitoringForRegion:region
desiredAccuracy:kCLLocationAccuracyHundredMeters];
```

You then register this region and start region monitoring by implementing the delegate callbacks as follows:

Region Monitoring delegates

```
- (void)locationManager:(CLLocationManager *)manager
didEnterRegion:(CLRegion
     *)region;
- (void)locationManager:(CLLocationManager *)manager
didExitRegion:(CLRegion
     *)region;
```

These delegate methods will be called even when your app is in the background. In fact, your app will be brought to the foreground before the system calls this delegate.

Apple's Reminders App

You can create reminders using Siri from iOS 5 using your location. That is, you can ask Siri to remind you to drop books at the library when you "*leave the office.*" What actually happens behind the scenes is that Siri picks up your office location from your "Me" card on the Contacts app and uses the coordinates to create a region. The Reminders app will now start monitoring that region for entry and exit events and will display a local notification when an exit event (`didExitRegion:`) is received.

iOS handles region monitoring and significant location changes internally using cell tower triangulation. To aggressively preserve battery life, iOS doesn't turn on the GPS chip. Instead, region monitoring and significant location changes notifications are triggered when the tower that your device is connected to changes. Because cell towers are denser in cities and sparser in the outskirts, you might get more accurate notifications in a densely populated city than when you're driving on a freeway. In my testing, I got a region-exited message only after I moved more than 100 kilometers away when I was driving across cities in Malaysia, though within Singapore, it's much more accurate (in fact, it's less than a kilometer). The same would be true for most countries.

Location Simulation

While I recommend using a device to test your location-based application, some apps (like a turn-by-turn navigation app) require the location to change while you test the app. The iOS Simulator allows you to simulate this navigation. You can turn this on from the menu option Debug⇨Location. You can also set Xcode to simulate a location using a scheme option. Choose the Edit Scheme⇨Run Action⇨Options tab to choose a location.

Additionally you can add a GPX file by creating one from File⇨New⇨File and choosing the GPX template from the resource section. When you have one or more GPX files, you can set Xcode to launch with a particular GPX file selected.

Keeping an Eye on the Battery

As a developer, you need to be aware of the battery life of your customers' devices. If for some reason installing your app drains the battery, your users will promptly delete your app. To be a good citizen in the iOS land, use location services wisely. As I mentioned before, iOS is intelligent enough not to turn on the GPS chip unless it's needed. Avoid setting the `location` string in `UIBackgroundModes` in your `Info.plist`. In most cases, when in the background, your app probably won't need continuous location updates (as long as it isn't a GPS-based fitness tracker or a turn-by-turn navigation app). Using a GPS fitness tracker on an iPhone 4S drains approximately 25 percent of the battery for a single run (that takes about 45 minutes), which is clearly not acceptable on any other app. Use significant location change or region monitoring as much as possible. This technique of getting location updates in the background drains little or no battery.

Secondly, when you register for region monitoring or significant location changes, avoid expensive network operations when your app is brought to the foreground. Network operations turn on the 3G chip, which further drains the battery.

Summary

In this chapter, you read about the ins and outs of location services. I explained the different methods that you can use to get location updates from a device. I also showed you how some of your commonly used apps use the background location feature to build useful features. Later, you discovered the pros and cons of every method and found out how to optimally use the battery on a user's iDevice.

Further Reading

Apple Documentation

The following document is available in the iOS Developer Library at `developer.apple.com` or through the Xcode Documentation and API Reference.

Location Awareness Programming Guide

Part III

The Right Tool for the Job

## Chapter 12

Common UI Paradigms Using Table Views

In Chapter 6, I introduced table views and showed you how to get a good scrolling performance when you use table views. Apart from showing a list of data, table views are also used as a substitute for creating complex structured scrollable views, and in fact most of the time, table views are used as cheap substitutes for creating vertically scrollable views even if the content they display is not a list of data. For example, in the built-in contacts app, the contacts list is a `UITableView`, and so is the view for adding a new contact. Additionally, third-party application developers have introduced new interaction patterns, and they have been commonly used on other apps as well.

This chapter focuses on the advanced aspects of table views and shows you how to create complex (yet common) UIs such as Pull-To-Refresh and infinite scrolling lists. It also briefly explains how to use table view row animations to create accordion and options drawer (a UI that shows available toolbar elements just below the table view cell that is acted upon) and several other interesting UI paradigms.

iOS has been around for more than five years, so this chapter is based on the assumption that you're well versed in concepts like `UITableViewDelegate` and `UITableViewDataSource`. In addition, though you don't have to read this book sequentially, this chapter is also based on the assumption that you have read Chapter 6, which as I just mentioned, introduces you to table views.

> If you're not familiar with `UITableViewDelegates` and `UITableViewDataSource`, I suggest you read Chapter 8 in *Beginning iPhone Development: Exploring the iPhone SDK* by Dave Mark and Jeff Lamarche (Apress 2009, ISBN 978-1430216261) before finishing this chapter.

Pull-To-Refresh

Pull-to-Refresh is an interesting UI paradigm that was invented (and patented) by Loren Brichter. Prior to iOS 6, several third-party developers such as Dr. Touch (Cocoanetics) and enormego released code that allowed you to implement Pull-to-Refresh in your own applications. With iOS 6, Apple added it as a built-in feature of `UITableViewController`, and you can use the all-new `UIRefreshControl` to add Pull-to-Refresh support to your table view controllers.

In this section, I show you how to write a `PullToRefreshTableView` class based on enormego's excellent open source implementation (see the "Further Reading" section at the end of this chapter). This class isolates most of the Pull-To-Refresh code into a superclass, and later on, when you need to add a Pull-To-Refresh feature to your table view, all you need to do is inherit your view controller from

`PullToRefreshTableViewController` instead of `UIViewController` and override methods to perform the actual refresh. Later, I'll show you how to implement Pull-to-Refresh using the built-in `UIRefreshControl` in iOS 6. After explaining both the techniques, I'll show you how to selectively use `UIRefreshControl` on iOS 6 and fall back gracefully to the previous implementation on devices running the older operating system.

Now, it's time delve into the code. First, download the files for this chapter from the book's website.

Create a project using the Master/Detail template and add these files:

EGORefreshTableHeaderView.h

EGORefreshTableHeaderView.m

PullToRefreshViewController.h

PullToRefreshViewController.m

RefreshArrow.png

RefreshArrow@2x.png

The `PullToRefreshTableViewController` is a subclass of `UITableViewController` that abstracts the mechanics behind Pull-To-Refresh. It handles the `UIScrollView` delegates and adds the `EGORefreshHeaderView` to the top of your `UITableView` when the table view is pulled beyond a certain threshold. It also remembers the last refreshed state. By default, the last refreshed state is stored in a key that uses your subclass name and a suffix string. In case this isn't enough and you have multiple instances of the same class displaying different data, you can customize the key in which the last refreshed date is remembered. The key is stored in a property called `keyNameForDataStore`.

To implement Pull-To-Refresh in your code, inherit the MasterViewController (the controller that is originally inheriting from `UITableViewController`) from `PullToRefreshViewController` and override the `doRefresh` method to perform the actual refresh. Once the refresh is done, set the loading state to NO. It's as simple as that. The `PullToRefreshViewController` also needs you to link your target with `QuartzCore.Framework`.

When you inherit your view controller from `PullToRefreshViewController`, you will see a `view` (from the superclass `UITableViewController`) in the IBOutlet list in IB. Connect this `view` to the table in your nib file.

Now, in the controller, override the `doRefresh` method and perform your network call (or any time-consuming refresh operation). When the refresh operation is complete, set the loading state to NO.

Following is the sample code snippet for your view controller.

Sample doRefresh Implementation

```
-(void) doRefresh {

// Do your time-consuming operation here.
// The dispatch_after is just for your illustration
   double delayInSeconds = 2.0;
```

```
    dispatch_time_t popTime = dispatch_time(DISPATCH_TIME_NOW,
      delayInSeconds * NSEC_PER_SEC);
    dispatch_after(popTime, dispatch_get_main_queue(), ^(void){
      self.loading = NO;
// the loading property is exposed by
PullToRefreshViewController. When you set this to NO,
it restores the tableview back to its normal position.
    });
}
```

iOS Pull-to-Refresh

With iOS 6, Apple introduced a new control called `UIRefreshControl` that can be attached to any `UITableViewController`. Implementing the iOS 6 Pull-To-Refresh couldn't be easier. Apple has abstracted away nearly everything so that you only need to instantiate a `UIRefreshControl` and set it to the `refreshControl` property of your `UITableViewController`. Handle the events sent by the refresh control to do the actual refresh. The following lines of code illustrate the use of `UIRefreshControl` for implementing Pull-to-Refresh.

Pull-To-Refresh Using UIRefreshControl

```
self.refreshControl = [[UIRefreshControl alloc] init];
    [self.refreshControl addTarget:self
      action:@selector(refreshedByPullingTable:)
      forControlEvents:UIControlEventValueChanged];

// handler that gets called when the refresh control is pulled
-(void) refreshedByPullingTable:(id) sender {

  [self.refreshControl beginRefreshing];
  double delayInSeconds = 2.0;
  dispatch_time_t popTime = dispatch_time(DISPATCH_TIME_NOW,
    delayInSeconds * NSEC_PER_SEC);
  dispatch_after(popTime, dispatch_get_main_queue(), ^(void){
    [self.refreshControl endRefreshing];
  });
}
```

Note that `dispatch_after` is used here for simulating a delay. You need to replace it with your time-consuming network operation. The most important lines are shown in **bold**. That was easy, wasn't it? The (slightly) hard part is to support iOS 5 and fall back to enormego's mechanism while showing the cool iOS 6-style refresh control on supported devices.

You do this by selectively instantiating a `UIRefreshControl` or enormego's `EGORefreshHeaderView` in the `viewDidLoad` method of the `PullToRefreshViewController`, as shown here:

```
if(NSClassFromString(@"UIRefreshControl")) {

    self.refreshControl = [[UIRefreshControl alloc] init];
    [self.refreshControl addTarget:self
```

(continued)

```
      action:@selector(refreshedByPullingTable:)
      forControlEvents:UIControlEventValueChanged];
  } else {

    self.refreshHeaderView = [[EGORefreshTableHeaderView alloc]
    initWithFrame:CGRectMake(0.0f,
    0.0f - self.tableView.bounds.size.height,
    self.tableView.bounds.size.width,
    self.tableView.bounds.size.height)];

    self.refreshHeaderView.keyNameForDataStore = [NSString
    stringWithFormat:@"%@_LastRefresh", [self class]];
    self.tableView.showsVerticalScrollIndicator = YES;
    [self.tableView addSubview:self.refreshHeaderView];
  }
```

Well, that wasn't hard. Right? The complete code is available in the code download for this chapter on the book's website. When you use this code, you don't have to take care of iOS 5/iOS 6 switching. If you've been using the previous version of the code (from *iOS 5 Programming Pushing the Limits*), replacing the files with the newer version will automatically handle the switching. The previous section explains how this is done internally in `PullToRefreshViewController`.

enormego did an excellent job of writing the mechanics behind Pull-to-Refresh. This takes it to the next level by abstracting the logic out and providing a super-easy way to implement it in any of your view controllers with under five lines of code.

Now, it's time to find out about another cool UI paradigm: *infinite scrolling.*

Infinite Scrolling

Infinite scrolling is normally used in Twitter clients or any app that displays chronologically ordered data. You generally use infinite scrolling when you don't know the number of items in the dataset or when the number of items is immensely large (unlike a contacts list).

You will now implement infinite scrolling support to the `PullToRefreshViewController` written in the previous section. The superclass, `PullToRefreshViewController`, adds a section to the end of your table view that shows a single loading cell. You implement this by adding a couple of properties called `numberOfSections` and `endReached` to the class `PullToRefreshTableView`:

```
@property (nonatomic) NSInteger numberOfSections;

@property (nonatomic) BOOL endReached;
```

You then add a method, `loadMore`, that is called when the user reaches the end of the current page in the table view. The superclass implementation for the `loadMore` method will be empty, and you will leave that for the subclasses to implement. Essentially, the super class adds a new section to your tableview and adds a loading cell and calls the `loadMore` method. The complete code can be found in the code download for this chapter, in the `Chapter 11/InfiniteScrollingExample` folder.

Implementing the infinite scrolling code in your app is easy. Override the method `loadMore` (defined in the superclass) and provide implementation for loading more content. Do *not* implement the `numberOfSectionsInTableView:` in your subclass. The superclass (`PullToRefreshViewController`) does this automatically for you. Instead, set the number of sections using the superclass property `numberOfSections`. The superclass adds an additional section to the end of your table to show the loading cell. When your server returns no content, you can set the `endReached` property to `YES`. This prevents the loading cell from being shown again. The following sample code snippet explains this.

Sample loadMore Implementation

```
-(void) loadMore {

  double delayInSeconds = 2.0;
  dispatch_time_t popTime = dispatch_time(DISPATCH_TIME_NOW,
   delayInSeconds * NSEC_PER_SEC);
  dispatch_after(popTime, dispatch_get_main_queue(), ^(void){
    self.pageCount ++;
    if(self.pageCount == 5) self.endReached = YES;
    [self.tableView reloadData];
  });
}
```

The superclass adds a loading section as the last section of the table view, and your table view data source methods will be called for sections you're not aware of. You need to forward these calls to the superclass of the `tableView:numberOfRowsInSection:` method and `tableView:cellForRowAtIndexPath:` for sections greater than your section count. In other words, let the superclass handle sections greater than the `numberOfSections` for you. This implementation shows the `loadingCell`.

The following code snippet explains this.

Sample TableView Data Source

```
- (NSInteger)tableView:(UITableView *)tableView
   numberOfRowsInSection:(NSInteger)section {
   // Return the number of rows in the section.
   if(section == self.numberOfSections)   {
       return [super tableView:tableView numberOfRowsInSection:section];
   }
   return 20 * self.pageCount; // we are assuming 20 rows per page
}

- (UITableViewCell *)tableView:(UITableView *)tableView
  cellForRowAtIndexPath:(NSIndexPath *)indexPath {

    if(indexPath.section == self.numberOfSections)   {
        return [super tableView:tableView
  cellForRowAtIndexPath:indexPath];
    }
```

(continued)

```
        static NSString *CellIdentifier = @"Cell";

        UITableViewCell *cell =
[tableView dequeueReusableCellWithIdentifier:CellIdentifier];

    cell.textLabel.text = [NSString stringWithFormat:
@"Row %d", indexPath.row];
        cell.detailTextLabel.text = [NSString stringWithFormat:
@"Row %d", indexPath.row];

        cell.imageView.image = [UIImage imageNamed:@"ios6"];
        cell.selectionStyle = UITableViewCellSelectionStyleNone;
        return cell;
    }
```

That completes it. With very few changes, you've added infinite scrolling support to the Pull-to-Refresh example code. Implementing infinite scrolling in your apps couldn't be easier.

Inline Editing and Keyboard

Form-filling is a common UI pattern found on both web and mobile environments. On iOS, forms are usually implemented using `UITableView`, with each cell representing one data entry field. For reasons stated in Chapter 6, a table view is preferred rather than a scroll view in these scenarios.

The most important point to remember here is to show data entry fields above the keyboard when the keyboard is shown. To do so, you need to adjust your table views dynamically.

If your table view contains data entry fields such as a `UITextField` or a `UITextView` and the table view is long enough to cover the screen, you'll have a problem accessing data entry fields that are hidden by the keyboard. The easiest—and recommended—way to overcome this problem is to use a `UITableViewController` for your form. Otherwise, if you use a `UIViewController` and a `UITableView` as its subview, you must explicitly code for scrolling your UI so that elements that might get hidden by the keyboard stay visible.

You can scroll your UI's frame by observing the `UIKeyBoardDidShowNotification` and `UIKeyBoardDidHideNotification`. The notification posts an object (`NSDictionary`) containing information pertaining to the size of the keyboard, the animation duration, and the animation curve.

The following code snippet illustrates this in practice.

In `viewDidLoad` or one of the `init` methods, add these observer keyboard notifications and remove them in the `didReceiveMemoryWarning` method:

```
    [[NSNotificationCenter defaultCenter]
     addObserver:self selector:@selector(keyboardWillShow:)
     name:UIKeyboardWillShowNotification object:nil];

    [[NSNotificationCenter defaultCenter]
```

```
    addObserver:self selector:@selector(keyboardWillHide:)
    name:UIKeyboardWillHideNotification object:nil];
```

Then adjust the table view's height accordingly in the selectors. You can get the size of the keyboard, the animation curve, and the animation duration from the notification object, as illustrated in the following code:

```
-(void) keyboardWillShow:(NSNotification*) notification {

  CGRect keyboardFrame;
  [[notification.userInfo valueForKey:UIKeyboardFrameEndUserInfoKey]
    getValue: &keyboardFrame];
  keyboardFrame = [self.view convertRect:keyboardFrame toView:nil];
  double keyboardHeight = keyboardFrame.size.height;

  double animationDuration;
  UIViewAnimationCurve animationCurve;

  [[notification.userInfo valueForKey:UIKeyboardAnimationDurationUserInfo
Key]
    getValue: &animationDuration];
  [[notification.userInfo valueForKey:UIKeyboardAnimationCurveUserInfoKey]
    getValue: &animationCurve];

  [UIView animateWithDuration:animationDuration
                  delay:0.0f
                options:animationCurve
              animations:^{

                  CGRect frame = self.tableView.frame;
                  frame.size.height -= keyboardHeight;
                  self.tableView.frame = frame;
              } completion:^(BOOL finished) {

              }];
}

-(void) keyboardWillHide:(NSNotification*) notification {

  CGRect keyboardFrame;
  [[notification.userInfo valueForKey:UIKeyboardFrameEndUserInfoKey]
    getValue: &keyboardFrame];
  keyboardFrame = [self.view convertRect:keyboardFrame toView:nil];
  double keyboardHeight = keyboardFrame.size.height;

  double animationDuration;
  UIViewAnimationCurve animationCurve;

  [[notification.userInfo valueForKey:UIKeyboardAnimationDurationUserInfo
Key]
    getValue: &animationDuration];
  [[notification.userInfo valueForKey:UIKeyboardAnimationCurveUserInfoKey]
```

(continued)

```
            getValue: &animationCurve];

    [UIView animateWithDuration:animationDuration
                          delay:0.0f
                        options:animationCurve
                     animations:^{

                    CGRect frame = self.tableView.frame;
                    frame.size.height += keyboardHeight;
                    self.tableView.frame = frame;
                } completion:^(BOOL finished) {

                }];
    }
```

Also note that, as the table view is scrolled, your cells are recycled and reused. Any data entered is lost when the cell is recycled, so you need to copy the entered data from the UI to your model classes (or `NSString`) immediately after the entry is made, which is quite easy to do. One way is to set the delegate of your `UITextField` to the table view controller and handle `textFieldDidEndEditing`. But a good design practice is to let the table cell handle the delegate and notify its superclass. (You find out more about such best practices throughout this chapter.) The superclass saves the data to the corresponding model object and prepopulates the table view cell with values from the model when it's created or dequeued in `cellForRowAtIndexPath`.

The following code segment shows you how to save data entered in a table view cell.

Saving and Restoring Data from UITextField Inside a Custom UITableViewCell

```
cell.inputText.text = [self.data objectAtIndex:indexPath.row];
    cell.onTextEntered = ^(NSString* enteredString) {

        [self.data insertObject:enteredString
   atIndex:indexPath.row];
    };
```

This code assumes that the cell handles `IBAction` and `textFieldDidEndEditing` and passes the `enteredString` value to the table view controller using a block. The data entered is stored in a member variable (`data`) and is restored in the first line. Two lines and that's it. You can use delegates as well, but blocks are cleaner and result in much less code. (You discover more about blocks in Chapter 23.)

Animating a UITableView

You've now seen some practical implementations of `UITableView`. Now, you go to the next level by finding out how to make the best use of the animations a table view provides. The `UIKit` framework provides some default animation styles for animating rows in a `UITableView`. These animations play a very important role in giving subtle hints to the user about what's happening behind the scenes.

A good example is the phone app on your iPhone. When you toggle between all calls and missed calls, the complete list animates to show or hide the relevant data. Similarly, on the settings app, when you turn on

Airplane mode, the Carrier row hides because it's no longer relevant. But if these actions happen without animations, users will be confused about what's happening. One thing that sets iOS apart from its competitor is that creating a compelling user experience that blends well with the OS is easy. In this case, implementing these animations is very easy with `UITableView`. Using methods in `UITableView`, you can animate a row insertion, row deletion, or row updates with fewer than ten lines of code.

The most important thing to remember here is that prior to your table updates, you need to update your models. Failure to do so will result in an `NSInteralInconsistencyException` (crash). In other words, if you're displaying a list of items, you need to insert a new row in the table view after the model is updated.

Remember, if you have to perform a batch of animated updates on `UITableView`, you can sandwich them between calls to `beginUpdates` and `endUpdates`. iOS automatically computes the changes and performs the correct animation sequence for you. The following are the commonly used methods for performing animated updates to a `UITableView`:

```
insertRowsAtIndexPaths:withRowAnimation:

deleteRowsAtIndexPaths:withRowAnimation:

reloadRowsAtIndexPaths:withRowAnimation:
```

In the following list, the first parameter is the array of index paths you need to add, and the second is the animation style that should be used. The animation style can be one of the following values

```
UITableViewRowAnimationFade, UITableViewRowAnimationNone

UITableViewRowAnimationRight, UITableViewRowAnimationLeft

UITableViewRowAnimationTop, UITableViewRowAnimationBottom

UITableViewRowAnimationMiddle

UITableViewRowAnimationAutomatic
```

> From iOS 5, you can use a new style, `UITableViewRowAnimationAutomatic`, and the system automatically chooses the correct animation for you. iOS 5 also introduces two new methods to move a complete section from one location to another. This is helpful in case you want to visually show movement of a complete section.

The following are methods for moving rows and/or sections in a `UITableView`:

```
moveSection: toSection:

moveRowAtIndexPath: toIndexPath:
```

Partially Reloading Tables

You can use the `reloadRowsAtIndexPaths:withRowAnimation:` method to partially reload a table view. For example, if you get a push notification that data currently displayed on the table view should be updated, you can reload just that single row in a `UITableView`. I recommend using the `UITableViewRowAnimationFade` or `UITableViewRowAnimationNone` style on iOS 4 and earlier and `UITableViewRowAnimationAutomatic` on iOS 5 for the animation style.

Practical Implementations of Table View Animations

With the built-in `UITableView` animations, you can easily implement custom controls such as accordion or show and hide drawers that expose additional controls. The next sections provide some ideas for implementing them. Custom controls like these can be implemented in multiple ways. So, instead of focusing on code, you find out about the process behind building them.

Implementing an Accordion List

Accordion is a control often found on content-rich websites to categorize navigational links. It contains a list of sections and subitems under each section. Sections can be opened to reveal the items within and can be closed or collapsed to hide them. On iOS, accordions are often used to model a single-level hierarchical navigation menu. An example is the *USA Today* app's Pictures tab. I'll dissect the view and analyze how a control like that can be created.

From the UI, it appears that the section headers are tappable and that every section has either one row or zero rows based on whether it is in an expanded state or not. This means you need a custom section view that tells the parent controller (your table view controller) that it was tapped.

For this example, design a custom `UIView` that has one big tappable button. You will use this view as the custom section view for your table. Override the `tableView:viewForHeaderInSection:` method, create your `UIView`, and return it. These views will notify (via a delegate or handler) the table view of the button-tapped event back to the table view. On this handler, the table view controller will do two things. First, it updates the models and then the table view. For updating models, you can save the tapped section's index as the currently expanded index. After this is done, you can refresh the table view in one of two ways: either by firing `reloadData` to the table view or by calculating the changes and calling the necessary `insertRowsAtIndexPaths:withRowAnimation:` and `deleteRowsAtIndexPaths:withRowAnimation:` methods. The reload data method refreshes the entire table, and users will not know what happened behind the scenes. For getting the accordion-like UI effect, you call `deleteRowsAtIndexPaths:withRowAnimation:` for the old section (currently expanded row) and `addRowsAtIndexPaths:withRowAnimation:` for the tapped section. Because you're doing two operations on the table view and you don't want the table to update itself for every operation, you sandwich them between the methods `beginUpdates` and `endUpdates`.

The most complicated part here is matching the changes to the model and the UI synchronously. If your model doesn't exactly reflect the UI, your code will crash with an `NSInternalInconsitencyException`.

In a nutshell, the following code explains how to "expand" your accordion table view. The complete code, `Ch11/MKAccordion`, is available on the book's website.

Opening an accordion

```objc
-(void) openAccordionAtIndex:(int) index {

  NSMutableArray *indexPaths = [NSMutableArray array];

  int sectionCount = [[self.objects objectForKey:[[self.objects
    allKeys] objectAtIndex:self.currentlyExpandedIndex]] count];

  for(int i = 0; i < sectionCount; i ++) {

    NSIndexPath *indexPath = [NSIndexPath indexPathForRow:i
      inSection:self.currentlyExpandedIndex];
    [indexPaths addObject:indexPath];
  }

  self.currentlyExpandedIndex = -1;

  [self.tableView deleteRowsAtIndexPaths:indexPaths
                    withRowAnimation:UITableViewRowAnimationTop];

  self.currentlyExpandedIndex = index;

  sectionCount = [[self.objects objectForKey:[[self.objects
    allKeys] objectAtIndex:self.currentlyExpandedIndex]] count];

  [indexPaths removeAllObjects];
  for(int i = 0; i < sectionCount; i ++) {

    NSIndexPath *indexPath = [NSIndexPath indexPathForRow:i
      inSection:self.currentlyExpandedIndex];
    [indexPaths addObject:indexPath];
  }

  double delayInSeconds = 0.35;
  dispatch_time_t popTime = dispatch_time(DISPATCH_TIME_NOW,
    delayInSeconds * NSEC_PER_SEC);
  dispatch_after(popTime, dispatch_get_main_queue(), ^(void){
    [self.tableView insertRowsAtIndexPaths:indexPaths
                      withRowAnimation:UITableViewRowAnimationFade];
  });

  [self.tableView endUpdates];
}
```

As stated previously, table views are powerful enough to create highly customized user interfaces. The good news is that the coding part is simple. In the next section, I'll show you how to implement a drawer control.

Implementing a Drawer

Implementing a drawer-like UI is quite similar to the accordion you implemented in the previous section. A drawer is a unique row in the table view that, instead of showing data, shows tools to manipulate the data. The Twitter client Tweetbot (and many other apps) uses this to show context-sensitive menu options for a table view row.

Implementing a drawer is slightly easier (programmatically) than implementing an accordion. Create a custom `UITableViewCell` for your drawer. Next, maintain an `NSIndexPath` pointer that will store the currently tapped row. The following code explains this process.

Drawer Implementation

```
@property (strong, nonatomic) MKControlsCell *controlsCell;
@property (strong, nonatomic) NSIndexPath *selectedIndexPath;
```

Now, in `tableView:numberOfRowsInSection:`, return an additional row if your drawer needs to be displayed.

```
- (NSInteger)tableView:(UITableView *)tableView
  numberOfRowsInSection:(NSInteger)section
{
  if(self.selectedIndexPath)
    return self.objects.count + 1;
  else
    return self.objects.count;
}
```

`tableView:cellForRowAtIndexPath:` returns the control cell for the selected row and returns a data cell for all other rows, as explained in the following code.

tableView:cellForRowAtIndexPath: Method

```
if(self.selectedIndexPath.row == indexPath.row && self.selectedIndexPath.
    section == indexPath.section && self.selectedIndexPath) {

    return self.controlsCell;
  } else {

    UITableViewCell *cell = [tableView
    dequeueReusableCellWithIdentifier:@"Cell"];
    if(self.selectedIndexPath && self.selectedIndexPath.row < indexPath.
row)
      cell.textLabel.text = self.objects[indexPath.row-1];
    else
      cell.textLabel.text = self.objects[indexPath.row];
    return cell;
  }
```

Update the `selectedIndexPath` when a row is selected (`tableView:didSelectRowAtIndexPath:`). Insert a new row below the selected row and remove the previously added drawer (if any) using `insertRowsAtIndexPaths:withRowAnimation:` methods.

Now comes the tricky part. Use `dispatch_after` to time the animation so that it's "just nice."

```
double delayInSeconds = 0.0;
  BOOL shouldAdjustInsertedRow = YES;
  if(self.selectedIndexPath) {
    NSArray *indexesToDelete = @[self.selectedIndexPath];
    if(self.selectedIndexPath.row <= indexPath.row)
      shouldAdjustInsertedRow = NO;
    self.selectedIndexPath = nil;
    delayInSeconds = 0.2;
    [tableView deleteRowsAtIndexPaths:indexesToDelete
      withRowAnimation:UITableViewRowAnimationAutomatic];
  }

  dispatch_time_t popTime = dispatch_time(DISPATCH_TIME_NOW,
   delayInSeconds * NSEC_PER_SEC);
  dispatch_after(popTime, dispatch_get_main_queue(), ^(void){
    if(shouldAdjustInsertedRow)
      self.selectedIndexPath = [NSIndexPath indexPathForRow:indexPath.row
+ 1 inSection:indexPath.section];
    else
      self.selectedIndexPath = indexPath;
    [tableView insertRowsAtIndexPaths:@[self.
     selectedIndexPath] withRowAnimation:UITableViewRowAnimationAutomat
ic];
  });
```

That's it. The complete code is also available on the book's website (`Chapter 11/MKDrawerDemo`).

Using Gesture Recognizers in Table View Cells

Swipe gestures like the swipe-to-delete or swipe-to-reveal options on Twitter for iPhone are another interesting type of interaction pattern. With gesture recognizers introduced in iOS 3.2, you can attach a swipe gesture recognizer (`UISwipeGestureRecognizer`) to your table cells' `contentView`. Attaching a long press gesture recognizer (`UILongPressGestureRecognizer`) can help in showing a context-sensitive menu (using a `UIActionSheet`) for a given table view cell element.

Summary

This chapter discussed some of the advanced table view concepts and how to create custom controls like accordions and drawers using table views. Table views can be used for purposes other than just displaying a vanilla list of data. The examples in this chapter unleashed the power of table views and explained how to create interesting controls and interaction paradigms. Customizing a table view is fairly easy, and you can invent your own interaction patterns and a wealth of UI elements similar to the two illustrated in this chapter.

Further Reading

Apple Documentation

The following documents are available in the iOS Developer Library at `developer.apple.com` or through the Xcode Documentation and API Reference.

What's New in iOS 6

TableView Programming Guide: "iOS Developer Documentation"

TableViewSuite

UIViewController Programming Guide: "iOS Developer Documentation (Storyboards)"

Chapter 13
Controlling Multitasking

There are two broad meanings of *multitasking* in iOS. First, it refers to running multiple applications at the same time by allowing one or more applications to run "in the background." Second, it refers to when a single application runs multiple operations simultaneously. Both are important parts of many iOS applications, and this chapter discusses both.

You discover the best practices for multitasking and the major iOS frameworks for multitasking: run loops, threads, operations, and Grand Central Dispatch (GCD). If you're familiar with thread-based multitasking from other platforms, you find out how to reduce your reliance on explicit threads and make the best use of iOS's frameworks that avoid threads or handle threading automatically. Perhaps most importantly, you learn how to give your users the illusion of multitasking without wasting system resources.

In this chapter, I assume that you understand the basics of running tasks in the background and that you're familiar with `beginBackgroundTaskWithExpirationHandler:`, registering an app as location aware, and similar background issues. If you need information about the fundamental technologies, see "Executing Code in the Background" in the *iOS Application Programming Guide* (`developer.apple.com`).

Similarly, this chapter assumes that you have at least a passing familiarity with operation queues and Grand Central Dispatch, though you may never have used them in real code. If you've never heard of them, I suggest you skim the *Concurrency Programming Guide* before continuing.

You can find the sample code for this chapter in the projects `SimpleGCD`, `SimpleOperation`, and `SimpleThread`.

Best Practices for Backgrounding: With Great Power Comes Great Responsibility

In iPhoneOS 3, only one third-party application could run at a time. When the user left your application, it was terminated. This ensured that third-party background applications couldn't waste resources like memory or battery. Apple wanted to make sure that the iPhone didn't suffer the same performance and stability problems of earlier mobile platforms, most pointedly Windows Mobile.

Starting with iOS 4, Apple began to permit third-party applications to run in the background, but only in limited ways. This continued Apple's focus on not allowing third-party applications to destabilize the platform or waste resources. It can be very frustrating, but the policy has generally met its goal. iOS remains focused on the user, not the developer.

Your application should give the illusion that it's always running, even though it isn't. Although your application may be terminated without warning any time it's suspended, it should give the impression that nothing has changed when it launches again. This means that you need to avoid displaying a splash screen during loading and that you need to save sufficient state when you enter the background to resume seamlessly if terminated. `NSUserDefaults` is a good place to stash small amounts of data during `applicationWillResignActive:`. Larger data structures need to be written to files, usually in `~/Library/Caches`. See the section "When We Left Our Heroes: State Restoration," later in this chapter, to find out how to do much of this work automatically.

Reducing your app's memory footprint is important when going into the background, and so is minimizing the time required to resume. If throwing away your cached information makes resuming from the background as expensive as launching from scratch, there isn't any point to suspending. Be thoughtful about what you throw away and how long it will take you to re-create it. Everything you do drains the battery, so always look to avoid wasteful processing, even if it doesn't visibly delay your app.

When your application is suspended, it doesn't receive memory warnings. If its memory footprint is very large, your application is likely to be terminated when there is memory pressure, and you won't have an opportunity to do anything about it. `NSCache` and `NSPurgeableData` are invaluable in addressing this issue. `NSPurgeableData` is an `NSData` object that you can flag as currently in use or currently purgeable. If you store it in an `NSCache` object and mark it as purgeable with `endContentAccess`, the OS saves it until there is memory pressure. At that point, it discards that data, even if your app is suspended at the time. This saves you the cost of throwing away this object and re-creating it every time the user leaves your app briefly, while ensuring that it can be thrown away if needed.

A lot of framework data is automatically managed for you when your app goes into the background. The data for images loaded with `imageNamed:` are discarded automatically and reread from disk when needed again. Views automatically throw away their backing stores (their bitmap cache). You should expect your `drawRect:` methods to be called when you resume. There is a major exception to this rule: `UIImageView` doesn't discard its data, and this can be quite a lot of memory. If you have a large image shown in a `UIImageView`, you should generally remove it before going into the background. However, decompressing images can be very expensive, so you don't throw them away too often. There's no one right answer. You need to profile your application.

In Instruments, the VM Tracker is useful for determining how much memory you're using while the application is in the background. It's part of the Allocations template. First, create a "memory pressure" app that displays a massive image. Then run your program with the VM Tracker. Note the amount of memory you're using. Press the Home button and note the amount of memory you're using now. This is what you're releasing when you go into the background. Now, launch the memory pressure app. Note how much memory you release. Ideally, your background usage should be less than your normal usage, without being so low that you delay resuming. Your usage under memory pressure needs to be as low as possible.

In Instruments, you see two kinds of memory: dirty memory and resident memory. *Dirty memory* is the memory that iOS can't automatically reclaim when you go into the background. *Resident memory* is your total current memory utilization. Both are important for different reasons. Minimizing dirty memory reduces the likelihood that you'll be terminated in the background. Reducing it should be your primary focus. Your application needs to consume as few resources as possible when it's not the foreground application. `NSCache` and `NSPurgeableData` are excellent tools for reducing dirty memory. Resident memory is your entire memory footprint. Minimizing this helps prevent low memory warnings while you're in the foreground.

> In the Instruments VM Tracker, you may see references to *Memory Tag 70*. That refers to memory for decompressed images and is primarily caused by `UIImage`.

Memory is important, but it's not the only resource. Avoid excessive network activity, disk access, or anything else that will waste battery life. Generally, you want to complete any user-initiated download using `beginBackgroundTaskWithExpirationHandler:`. Users don't want to have to sit and stare at your application while their data downloads; they want to go play a game. However, you want to abort or pause any downloads that weren't requested by the user, provided that you can resume them later.

Some actions are forbidden while in the background. The most significant ones are OpenGL calls. You must stop updating OpenGL views when you go into the background. A subtle issue here is application termination. The application is allowed to run for a brief time after `applicationWillTerminate:` is called. During that time, the application is "in the background" and must not make OpenGL calls. If it does, it's killed immediately, which could prevent it from finishing other application-termination logic.

> Be sure to shut down your OpenGL updates when the application is terminating as well as when going into the background. `GLKViewController` **automatically handles all of this for you. The OpenGL Game template in Xcode uses** `GLKViewController`**.**

Running in the background creates new challenges for developers, but users expect this key feature. Just be sure to keep the user as your top priority, test heavily, and watch your resource utilization. Your application should delight, even when it's not onscreen.

When We Left Our Heroes: State Restoration

As discussed in the previous section, your application needs to give the illusion that it is always running, even when it's terminated in the background. If the user is in the middle of typing a message and flips over to another application, that message should be waiting when the user returns. Before iOS 6, this often meant storing a lot of values in `NSUserDefaults`. For example, you need to keep track of which tab is visible, the entire navigation controller stack, the position of the scrollbar, and the partially entered text. The work isn't that difficult, but it is very tedious and somewhat error prone.

In iOS 6, UIKit will handle much of this work for you if you request it to do so. The system is called *state restoration*. It's an opt-in, so you have to tell UIKit that you want it. In this example, you start with a simple, existing app called "FavSpots," available in the sample code. The final version is in the project "FavSpotsRestore" in the same directory. This application lets you bookmark places on a map and add a name and some notes, which are saved in Core Data. To implement state restoration, do the following:

1. Opt in by requesting state saving and restoring in the application delegate.

2. Modify the application delegate's startup handling if needed.

3. Assign restoration identifiers to all the view controllers.

4. Add state encoders and decoders to view controllers with state.

5. Implement `UIDataSourceModelAssociation` for the table view controller.

This example represents a fairly common set of problems, but it doesn't cover every situation that state restoration can handle. For full details, see "State Preservation and Restoration" in the *iOS App Programming Guide*, and the other references in the "Further Reading" section at the end of the chapter.

While implementing state preservation, always remember that state preservation is a way to store only *state*. You never use it to store *data*. The information stored in state preservation may be thrown away at any time (and the OS is somewhat aggressive in throwing state preservation away if it suspects a problem). Do not encode actual data objects in the state preservation system. You need to encode identifiers that will allow you to later fetch those data objects. I explain this topic more in the section "State Encoders and Decoders," later in this chapter.

Note that all the code in this section assumes that you're exclusively targeting iOS 6. If you're targeting earlier versions, you'll need to check that certain methods exist before calling them, particularly `setRestorationIdentifier:`.

Testing State Restoration

Before adding state restoration, you need to understand how programs behave without it, and how to test it to make sure it's working. When the user presses the Home button, your app may not immediately terminate. It typically moves to the background state, and later may be moved to the suspended state. In either case, however, your program is still in memory. When the user returns to your application, it typically resumes exactly where it left off. The same screen is showing, any text the user has entered is still in the text fields, and so on. The goal of state restoration is to achieve this same user experience even though the program terminates while in the background. A common error while implementing state restoration is mistaking this default behavior for your restoration code.

In order to properly test state restoration, you need to make sure that the application actually terminates, but you also need to make sure that it terminates in a way that allows state restoration to happen. In order for state restoration to run, three things must be true:

- **State preservation must have successfully completed.** This means that the application needs to have entered the background before being terminated.

- **The application must not have been forced to quit.** If you double-tap the Home button and delete the app from the "recent apps" list, the state preservation files are deleted. The assumption is that if the user force-quit the app, there may have been a serious bug and state restoration might create an endless loop.

- **The application must not have failed launching since the last state preservation.** This means that if your application terminates during startup, the state preservation information will be deleted. Otherwise a small bug in your state restoration code could lead to an unlaunchable app.

With these things in mind, here is the correct way to test state restoration:

1. Run the app in Xcode and navigate to whatever situation you want to test.

2. Press the Home button in the simulator or on the device.

3. Stop the app in Xcode.

4. Relaunch the app.

I also recommend adding some logging to `application:willEncodeRestorableStateWithCoder:` and `application:didDecodeRestorableStateWithCoder:` in the application delegate just so you can see when state preservation and restoration are running.

Opting In

State preservation doesn't automatically apply to old programs. Requesting it is trivial. Just add the following to your application delegate.

AppDelegate.m (FavSpotsRestore)

```
- (BOOL)application:(UIApplication *)application
shouldSaveApplicationState:(NSCoder *)coder {
   return YES;
}

- (BOOL)application:(UIApplication *)application
shouldRestoreApplicationState:(NSCoder *)coder {
   return YES;
}
```

It's possible to add conditional logic here, but this should be rare.

Startup Changes

There are some subtle startup changes due to state restoration. You may need to perform some initialization work prior to state restoration and some after state restoration. There is a new method, `application:willFinishLaunchingWithOptions:`, that is called prior to state restoration. After state restoration is complete, Cocoa will call `application:didFinishLaunchingWithOptions:`. You need to think carefully about which logic should go in each.

You can save application-level state in `application:willEncodeRestorableStateWithCoder:`. View controllers manage the majority of state preservation, but you may need to store some global state. You can then restore this state in `application:didDecodeRestorableStateWithCoder:`. I will explain encoding and decoding state in the section "State Encoders and Decoders."

This change doesn't apply to the FavSpots project.

Restoration Identifiers

Every subclass of `UIView` and `UIViewController` can have a restoration identifier. These identifiers do not have to be unique, but the full restoration path of every restorable object must be unique. The *restoration path* is a sequence of restoration identifiers, starting at the root view controller and separated by slashes like an URL. The restoration path gives the system a way to glue the UI back together, encoding things like the entire navigation controller stack. For example, you might encounter the path `RootTab/FirstNav/Detail/Edit` indicating a tab bar controller with a navigation controller in one of its tabs. The navigation controller has a root view controller of Detail, and Edit has been pushed on top of that.

For custom, unique view controller classes, I recommend using the name of the custom class. For standard classes like `UITabBarController`, I recommend using some more descriptive name to make them unique.

If you use storyboards, managing restoration identifiers is very simple. For each view controller you want to save, set the name as the storyboard ID in the Identity pane. Then select "Use Storyboard ID" for the Restoration ID. See Figure 13-1 for an example. Although you can have different storyboard and restoration identifiers, doing so is confusing and should be avoided.

Figure 13-1 Setting the Restoration Identifier

You need to assign a restoration identifier to every view controller and view that you want to save. This will typically be most of your view controllers. If you have a particular view controller that wouldn't make sense to come back to, you can leave its identifier blank, and the user will be returned to the parent.

Most views should be reconfigured by their view controller, but sometimes it's useful to save the view's state directly. Several UIKit views support state restoration, specifically `UICollectionView`, `UIImageView`, `UIScrollView`, `UITableView`, `UITextField`, `UITextView`, and `UIWebView`. These will save their state, but not their contents. So a text field will save the selection range but will not save the text. See the documentation for each view for information on exactly what state it saves.

In FavSpots, add restoration identifiers to all of the view controllers, the table view, the text field, and both text views.

State Encoders and Decoders

The heart of state restoration is the encoding and decoding state. Much like `NSKeyedArchiver`, the state restoration system will pass an `NSCoder` to each object via `encodeRestorableStateWithCoder:` and `decodeRestorableStateWithCoder:`.

Unlike `NSKeyedArchiver`, the goal of the restoration system is not to store the entire object. You want to store only the information required to reconstruct your state. For example, if you have a `PersonViewController` that displays a `Person` object, don't encode the `Person` object. You only encode a unique identifier that will allow you to look up that object in the future. If you're using iCloud syncing or a network data source of any kind, it's possible that the record you stored will have been modified, moved, or deleted by the time you restore. Your restoration system should deal gracefully with a missing object.

If you're using Core Data, the best way to save references to your object is `[[obj objectID] URIRepresentation]`. Once an object has been saved, this identifier is guaranteed to be unique and persistent. However, if the object hasn't been saved yet, the `objectID` may be temporary. You need to check `[objectID isTemporary]` prior to relying on it. Generally, you save prior to state preservation. Luckily, `applicationDidEnterBackground:` is called prior to state preservation, so if you save your context there, all your identifiers will be permanent.

It bears repeating that you need to save user-provided data in a persistent store, not in state preservation. If your user has started writing a long review comment, don't rely on the state preservation system to save that draft. State preservation data can be deleted at any time. If you can't put the data into the persistent store yet, at least write it to a file in `Library/Application Support` so you don't lose it.

Here are the changes for FavSpots in `MapViewController`. It needs to save the region being viewed and the tracking mode. The various RN methods on `NSCoder` are from a category to simplify encoding `MKCoordinateRegion`. Notice the use of `containsValueForKey:` to distinguish between a missing value and 0. You may be restoring state from a previous version of your program, so be tolerant of missing keys.

MapViewController.m (FavSpotsRestore)

```
- (void)encodeRestorableStateWithCoder:(NSCoder *)coder {
  [super encodeRestorableStateWithCoder:coder];

  [coder RN_encodeMKCoordinateRegion:self.mapView.region
                             forKey:kRegionKey];
  [coder encodeInteger:self.mapView.userTrackingMode
             forKey:kUserTrackingKey];
}

- (void)decodeRestorableStateWithCoder:(NSCoder *)coder {
  [super decodeRestorableStateWithCoder:coder];

  if ([coder containsValueForKey:kRegionKey]) {
    self.mapView.region =
    [coder RN_decodeMKCoordinateRegionForKey:kRegionKey];
  }

  self.mapView.userTrackingMode =
  [coder decodeIntegerForKey:kUserTrackingKey];
}
```

`TextEditViewController` just needs a `Spot` object. In order to protect the user's notes (which could be lengthy), `TextEditViewController` automatically saves them when the application resigns active. This happens prior to state preservation, which makes state preservation very simple. It just stores the URI for the object ID. Notice that in `decodeRestoreableStateWithCoder:`, you don't use an accessor. This method is very similar to `init` in that you need to be very careful of side effects. As a general rule, add this method to the short list of places that direct ivar access is appropriate. Here are the additions to `TextEditViewController` and the helper methods to encode and decode `Spot`.

TextEditViewController.m (FavSpotsRestore)

```objc
- (void)encodeRestorableStateWithCoder:(NSCoder *)coder {
    [super encodeRestorableStateWithCoder:coder];
    [coder RN_encodeSpot:self.spot forKey:kSpotKey];
}

- (void)decodeRestorableStateWithCoder:(NSCoder *)coder {
    [super decodeRestorableStateWithCoder:coder];
    _spot = [coder RN_decodeSpotForKey:kSpotKey];
}
```

NSCoder+FavSpots.m (FavSpotsRestore)

```objc
- (void)RN_encodeSpot:(Spot *)spot forKey:(NSString *)key {
    NSManagedObjectID *spotID = spot.objectID;
    NSAssert(! [spotID isTemporaryID],
            @"Spot must not be temporary during state saving. %@",
            spot);

    [self encodeObject:[spotID URIRepresentation] forKey:key];
}

- (Spot *)RN_decodeSpotForKey:(NSString *)key {
    Spot *spot = nil;
    NSURL *spotURI = [self decodeObjectForKey:key];

    NSManagedObjectContext *
    context = [[ModelController sharedController]
                managedObjectContext];
    NSManagedObjectID *
    spotID = [[context persistentStoreCoordinator]
                managedObjectIDForURIRepresentation:spotURI];
    if (spotID) {
        spot = (Spot *)[context objectWithID:spotID];
    }

    return spot;
}
```

`DetailViewController` is more complicated. It includes an editable text field. Usually, I recommend saving any changes to the persistent store as I did in `TextEditViewController`, in which case saving state is simple. The name is very short, however, and losing it would not be a great burden on the user, so I will demonstrate here how you can store this kind of data in the preservation system.

This is complicated because the value of the text field can be set two ways. If you're restoring, you use the restored value; if you're not restoring, you use the value from Core Data. This means you have to keep track of whether you're restoring, as shown here.

DetailViewController.m (FavSpotsRestore)

```
@property (nonatomic, readwrite, assign, getter = isRestoring) BOOL
restoring;
...
- (void)encodeRestorableStateWithCoder:(NSCoder *)coder {
  [super encodeRestorableStateWithCoder:coder];
  [coder RN_encodeSpot:self.spot forKey:kSpotKey];
  [coder RN_encodeMKCoordinateRegion:self.mapView.region
                            forKey:kRegionKey];
  [coder encodeObject:self.nameTextField.text forKey:kNameKey];
}


- (void)decodeRestorableStateWithCoder:(NSCoder *)coder {
  [super decodeRestorableStateWithCoder:coder];
  _spot = [coder RN_decodeSpotForKey:kSpotKey];

  if ([coder containsValueForKey:kRegionKey]) {
    _mapView.region =
    [coder RN_decodeMKCoordinateRegionForKey:kRegionKey];
  }

  _nameTextField.text = [coder decodeObjectForKey:kNameKey];
  _restoring = YES;
}
...
- (void)configureView {
  Spot *spot = self.spot;

  if (! self.isRestoring || self.nameTextField.text.length == 0) {
    self.nameTextField.text = spot.name;
  }

  if (! self.isRestoring ||
      self.mapView.region.span.latitudeDelta == 0 ||
      self.mapView.region.span.longitudeDelta == 0) {
    CLLocationCoordinate2D center =
    CLLocationCoordinate2DMake(spot.latitude, spot.longitude);
    self.mapView.region =
    MKCoordinateRegionMakeWithDistance(center, 500, 500);
  }

  self.locationLabel.text =
  [NSString stringWithFormat:@"(%.3f, %.3f)",
   spot.latitude, spot.longitude];
  self.noteTextView.text = spot.notes;

  [self.mapView removeAnnotations:self.mapView.annotations];
  [self.mapView addAnnotation:
   [[MapViewAnnotation alloc] initWithSpot:spot]];

  self.restoring = NO;
}
```

In general, it's easier and less error-prone to save the data directly to the persistent store, but if you need the ability to cancel edits of very small amounts of data, the preceding approach is useful.

Table Views and Collection Views

`UITableView` and `UICollectionView` have a special problem for state restoration. The user basically wants the view to return to the same state it was left in, but the underlying data may have changed. The specific records being displayed may be in a different order, or records may have been added or removed. In all but the simplest cases, it may be impossible to restore a table view to its exact previous state.

To make restoration predictable, UIKit restores table and collection views such that the first index shown represents the same record. So if a record representing "Bob Jones" was at the top of the screen when your application quit, then that's the record that should be at the top of the screen when you resume. All the other records on the screen may be different, and even the "Bob Jones" record might display a different name, but it should be the same record.

To achieve this, UIKit uses the protocol `UIDataSourceModelAssociation`. This protocol maps an index path to a string identifier. UIKit doesn't care how you perform this mapping. During state preservation, it will ask you for the identifier for an index path. During state restoration, it will ask you for the index path for an identifier. The index paths do not need to match, but the referenced object should. The easiest way to implement this is using Core Data's `objectID` and taking the `URIRepresentation`, which returns a string. Without Core Data, you'll need to devise your own mapping that is guaranteed to be consistent and unique.

MasterViewController.m FavSpots

```objc
- (NSString *)modelIdentifierForElementAtIndexPath:(NSIndexPath *)idx
                                            inView:(UIView *)view {
  if (idx && view) {
    Spot *spot = [self.fetchedResultsController objectAtIndexPath:idx];
    return [[[spot objectID] URIRepresentation] absoluteString];
  }
  else {
    return nil;
  }
}

- (NSIndexPath *)
indexPathForElementWithModelIdentifier:(NSString *)identifier
                                inView:(UIView *)view {
  if (identifier && view) {
    NSUInteger numberOfRows =
    [self tableView:self.tableView numberOfRowsInSection:0];
    for (NSUInteger index = 0; index < numberOfRows; ++index) {
      NSIndexPath *indexPath = [NSIndexPath indexPathForItem:index
                                                   inSection:0];
      Spot *spot = [self.fetchedResultsController
                    objectAtIndexPath:indexPath];
      if ([spot.objectID.URIRepresentation.absoluteString
           isEqualToString:identifier]) {
        return indexPath;
```

```
        }
      }
    }
    return nil;
  }
```

> Although the preceding code matches the documentation, as of iOS 6 Beta 4, I'm unable to get UIKit to restore the table view scroll using it. This may be a bug in the code or may be a bug in iOS. Watch `iosptl.com/code` **for updates.**

Advanced Restoration

In the previous sections, I discussed the common problems facing state restoration. I assumed the use of an application using storyboards and with a typical view controller hierarchy. But state restoration can handle more complicated situations.

Although storyboards are definitely the easiest way to manage state restoration, you also can perform state restoration without them. First, you can set restoration identifiers in XIB files rather than the storyboard. Alternatively, you can set the restoration identifier at runtime by calling `setRestorationIdentifier:`. Setting this to `nil` prevents preserving its state.

You can also change which class handles restoration. By default, the view controller is created using the storyboard, but you may require different initialization for restoration. You can set the `restorationClass` property to the `Class` of whatever object you want to construct your view controller. This is normally the view controller's class (such that view controllers create themselves). If this property is set, the view controller will be re-created by calling `+viewControllerWithRestorationIdentifierPath: coder:` rather than pulling it from the storyboard. You are responsible for creating the view controller in that method and returning it. If you return `nil`, this view controller will be skipped. The system will still call `decodeRestorableStateWithCoder:` on the resulting object, unless the object you return isn't of the same class as the object that was encoded (or a subclass of that object). See "Restoring Your View Controllers at Launch Time" in the *iOS App Programming Guide* for full details.

State preservation and restoration is a powerful system for simplifying this complicated problem. It relies heavily on you using UIKit in "normal" ways. If you use storyboards, Core Data, and don't get overly "clever" in how you use view controllers, it will do much of the work for you.

Introduction to Multitasking and Run Loops

So far in this chapter, I've discussed how your application can multitask with other applications. For the rest of the chapter, I'll discuss how to multitask within your application. As iOS devices add additional CPU cores, multitasking is becoming a key part of harnessing their power.

The most basic form of multitasking is the run loop. This is the form of multitasking that was originally developed for NeXTSTEP.

Every iOS program is driven by a `do`/`while` loop that blocks waiting for an event, dispatches that event to interested listeners, and repeats until something tells it to stop. The object that handles this is called a *run loop* (`NSRunLoop`).

You almost never need to understand the internals of a run loop. There are mach ports and message ports and `CFRunLoopSourceRef` types, and a variety of other arcana. These are incredibly rare in normal programs, even in very complex programs. What's important to understand is that the run loop is just a big `do`/`while` loop, running on one thread, pulling events off of various queues and dispatching them to listeners one at a time on that same thread. This is the heart of an iOS program.

When your `applicationWillResignActive:` method, `IBAction`, or any other entry point to your program is called, it's because an event fired somewhere that traced its way to a delegate call that you implemented. The system is waiting for you to return so it can continue. While your code is running on the main thread, scroll views can't scroll, buttons can't highlight, and timers can't fire. The entire UI is hanging, waiting for you to finish. Keep that in mind when you're writing your code.

This doesn't mean everything is on the main run loop. Each thread has its own run loop. Animations generally run on background threads, as does much of `NSURLConnection` network handling. But the heart of the system runs on a single, shared run loop.

> Although each thread *has* a run loop, this doesn't mean that each thread *processes* its run loop. Run loops only execute their `do`/`while` loop in response to commands like `runUntilDate:`. The call to `UIApplicationMain` in `main.m` of almost every project runs the main run loop.

`NSTimer` relies on the run loop to dispatch messages. When you schedule an `NSTimer`, it asks the current run loop to dispatch a selector at a certain time. Each time the run loop iterates, it checks what time it is and fires any timers that have expired. Delay-action methods like `performSelector:withObject:afterDelay:` are implemented by scheduling a timer.

Most of the time, all of this happens behind the scenes, and you don't need to worry about the run loop very much. `UIApplicationMain` sets up the main thread's run loop for you and keeps it running until the program terminates.

Although it's important to understand the basics of run loops, Apple has developed better approaches to handling common multitasking problems. Most of these are built on top of Grand Central Dispatch. The first step toward these better approaches is breaking your program into operations, as I discuss in the next section.

Developing Operation-Centric Multitasking

Many developers from other platforms are used to thinking in terms of threads. iOS has good objects for dealing directly with threads, particularly `NSThread`, but I recommend avoiding them and moving toward operation-centric multitasking. Operations are a much more powerful abstraction than threads and can lead to substantially better and faster code if used correctly. Threads are expensive to create and sustain, so threaded designs often employ a small to moderate number of long-lived threads. These threads perform heavyweight operations such as "the network" or "the database" or "the calculations." Because they're focused on large areas of responsibility, they need access to lots of inputs and outputs, and that means locking. Locking is very expensive and a significant source of bugs.

Of course, threaded programs don't have to be written this way. It would be better if you spawned short-lived threads that focus on a single input and output, minimizing the need for locking. But threads are expensive to spawn, so you need to manage a thread pool. And small operations often have ordering dependencies, so ideally you create some kind of queue to keep things straight. Now, what if you could take the thread pool and the queue and let the operating system worry about them? That's exactly what `NSOperation` is for. It's the solution you come to by optimizing threaded programs to minimize locking. But operations can be better than that. Instead of worrying about the mechanics of semaphores and mutexes, you worry about the work you want to do.

An *operation* is an encapsulated unit of work, often expressed in the form of an Objective-C block. Blocks are more fully covered in Chapter 23. Operations support priority, dependencies, and cancelation, so they're ideal for situations where you need to schedule some work that might not actually be required. For example, you may want to update images in a scrolling view, favoring images currently onscreen and canceling updates for images that have been scrolled off-screen. In the following example, you'll create a `UICollectionView` that contains random fractals. These are very expensive to calculate, so you'll generate them asynchronously. In order to improve display performance, you'll calculate the fractal at various resolutions, displaying a low-resolution fractal quickly and then improving the resolution if the user stops scrolling.

In this example, you'll use an `NSOperation` subclass to generate Julia sets (a kind of fractal) and display them in a `UICollectionView`. The work being done by the operation is somewhat complicated, and so it's nice to put it in its own file. For smaller operations, you could put the work into an inline block. The sample code is available in the project JuliaOp. Here is the basic structure of the operation's `main` method (I skipped the actual mathematics in the interest of space).

JuliaOperations.m (JuliaOp)

```
- (void)main {
  // ...
  // Configure bits[] to hold bitmap data
  // ...

  for (NSUInteger y = 0; y < height; ++y) {
    for (NSUInteger x = 0; x < width; ++x) {
      if (self.isCancelled) {
        return;
      }
      // ...
      // Calculate Julia values and update bits[]
      // May iterate up to 255 times per pixel.
      // ...
    }
```

(continued)

```
    }

    // ...
    // Create bitmap and store in self.image
    // ...
}
```

Some key features of this operation:

- **All required data is given to the operation before it starts.** The operation doesn't need to interact with other objects during its run, so there's no need for locking.

- **When the operation is complete, it stores its result in a local ivar.** Again, this avoids the need for locking. You could use a delegate method here to update something outside the operation, but as you'll see shortly, completion blocks can be a simpler approach.

- **The operation periodically checks** `isCancelled` **so it can abort if requested.** This is an important point. Calling `cancel` on a running operation does not cause it to stop running. That would leave the system in an unknown state. It's up to you to check `isCancelled` and abort in whatever way is appropriate. It's possible for an operation to return YES for both `isExecuting` and `isCancelled`.

I'll talk more about cancellation shortly. First, I'll discuss how this operation is created and queued in `JuliaCell`. Each cell in the `UICollectionView` creates a separate operation for each resolution. The operation is just responsible for taking a set of input values and a resolution and returning an image. The operation doesn't know anything about how that image will be used. It doesn't even know that other resolutions are being calculated. In the `UICollectionViewController`, you configure each cell by passing the row as a random seed and a queue to work with.

CollectionViewController.m (JuliaOp)

```
- (UICollectionViewCell *)collectionView:(UICollectionView *)
collectionView
                  cellForItemAtIndexPath:(NSIndexPath *)indexPath {
  JuliaCell *
  cell = [self.collectionView
          dequeueReusableCellWithReuseIdentifier:@"Julia"
          forIndexPath:indexPath];
  [cell configureWithSeed:indexPath.row queue:self.queue];
  return cell;
}
```

The queue is just a shared `NSOperationQueue` that you create in the `UICollectionViewController`. I'll explain `maxConcurrentOperationCount` and `countOfCores()` in the section "Setting Maximum Concurrent Operations."

```
    self.queue = [[NSOperationQueue alloc] init];
    self.queue.maxConcurrentOperationCount = countOfCores();
```

In `JuliaCell`, you configure the cell by creating a set of operations and adding them to the queue.

JuliaCell.m (JuliaOp)

```
- (void)configureWithSeed:(NSUInteger)seed
                    queue:(NSOperationQueue *)queue {
  // ...
  JuliaOperation *prevOp = nil;
  for (CGFloat scale = minScale; scale <= maxScale; scale *= 2) {
    JuliaOperation *op = [self operationForScale:scale seed:seed];
    if (prevOp) {
      [op addDependency:prevOp];
    }
    [self.operations addObject:op];
    [queue addOperation:op];
    prevOp = op;
  }
}
```

Notice that each operation is dependent on the previous operation. This ensures that the high-resolution image isn't scheduled before the low-resolution image. Every cell creates several operations, and all the operations from all the cells are put onto the same queue. NSOperationQueue automatically orders the operations to manage dependencies.

In operationForScale:seed:, you configure the actual operation for a given resolution.

JuliaCell.m (JuliaOp)

```
- (JuliaOperation *)operationForScale:(CGFloat)scale
                                 seed:(NSUInteger)seed {
  JuliaOperation *op = [[JuliaOperation alloc] init];
  op.contentScaleFactor = scale;

  CGRect bounds = self.bounds;
  op.width = (unsigned)(CGRectGetWidth(bounds) * scale);
  op.height = (unsigned)(CGRectGetHeight(bounds) * scale);

  srandom(seed);

  op.c = (long double)random()/LONG_MAX + I*(long double)random()/LONG_
MAX;
  op.blowup = random();
  op.rScale = random() % 20;  // Biased, but simple is more important
  op.gScale = random() % 20;
  op.bScale = random() % 20;

  __weak JuliaOperation *weakOp = op;
  op.completionBlock = ^{
    if (! weakOp.isCancelled) {
      [[NSOperationQueue mainQueue] addOperationWithBlock:^{
        JuliaOperation *strongOp = weakOp;
        if (strongOp && [self.operations containsObject:strongOp]) {
          self.imageView.image = strongOp.image;
          self.label.text = strongOp.description;
```

(continued)

```
                [self.operations removeObject:strongOp];
            }
        }];
    }
};

if (scale < 0.5) {
    op.queuePriority = NSOperationQueuePriorityVeryHigh;
}
else if (scale <= 1) {
    op.queuePriority = NSOperationQueuePriorityHigh;
}
else {
    op.queuePriority = NSOperationQueuePriorityNormal;
}

return op;
}
```

The operation has various data associated with it (c, blowup, and the various scales). These are configured once, on the main thread, so there are no race conditions.

The completionBlock is a convenient way to handle the operation's results. The completion block is called whenever the block finishes, even if it's cancelled. That's why you check isCancelled before dealing with the results. Because the UI updates are processed on the main thread, it's possible that this cell will have scrolled off the screen and been reused prior to the UI update code running. That's why you check that the operation still exists and that it's still one of the operations that this cell cares about (containsObject:) before applying the UI changes. Exactly how much of this kind of double-checking code is required depends on your specific problem.

Finally, you set the priority based on the scale. This encourages low-resolution images to be generated prior to high-resolution images.

Because this is a UICollectionView, the cells may be reused at any time. When that happens, you want to cancel all the current operations, which is easily done with cancel:

```
- (void)prepareForReuse {
    [self.operations makeObjectsPerformSelector:@selector(cancel)];
    [self.operations removeAllObjects];
    self.imageView.image = nil;
    self.label.text = @"";
}
```

At the same time, you remove all the operations from the operations ivar, ensuring that any pending operations won't update the UI. Remember that all of the methods in JuliaOp except for the completionBlock run on the main thread (and the completionBlock moves the UI calls to the main thread as well). This minimizes locking concerns, improving reliability and performance.

Setting Maximum Concurrent Operations

Finally, there is the matter of `maxConcurrentOperationCount`. `NSOperationQueue` tries to manage the number of threads it generates, but in many cases, it doesn't do an ideal job. This is likely your fault (and my fault, but read on). The common cause of trouble is flooding the main queue with lots of small operations when you want to update the UI.

Apple suggests in the "Asynchronous Design Patterns" session from WWDC 2012 that you use fewer operations. For instance, they suggest that in JuliaOp, you use a single operation to manage one full screen of images. If the user scrolls, you could cancel that one operation rather than the many I've generated (including their many small main-thread updates). The problem is that then you don't get any parallelism on a multi-core machine, which was the whole point of using operation queues.

In my experience, creating small operations as I demonstrate in JuliaOp, even if you have to cancel them later, makes coding much simpler and the code much more robust and understandable. The problem is that `NSOperationQueue` will schedule too many of them at a time. If you explicitly tell `NSOperationQueue` not to schedule more CPU-bound operations than you have CPUs, simple code runs very fast.

My recommendation may change in the future, but for CPU-bound operations, I recommend setting `maxConcurrentOperationsCount` to the number of cores. You can determine this with the function `countOfCores`:

```
unsigned int countOfCores() {
  unsigned int ncpu;
  size_t len = sizeof(ncpu);
  sysctlbyname("hw.ncpu", &ncpu, &len, NULL, 0);

  return ncpu;
}
```

Multitasking with Grand Central Dispatch

Grand Central Dispatch (GCD) is at the heart of multitasking in iOS. It is used throughout the system layer for nearly everything. `NSOperationQueue` is implemented on top of GCD, and the basic queue concept is similar. Rather than adding an `NSOperation` to `NSOperationQueue`, you add a block to a dispatch queue.

Dispatch queues are more low-level than operation queues, however. There's no way to cancel a block once it is added to a dispatch queue. Dispatch queues are strictly first-in-first-out (FIFO), so there's no way to apply priorities or reorder blocks within a queue. If you need those kinds of features, definitely use `NSOperationQueue` instead of trying to re-create them with GCD.

You can do many things with dispatch queues that you cannot do with operations. For example, GCD offers `dispatch_after`, allowing you to schedule the next operation rather than sleeping. The time is in nanoseconds, which can lead to some confusion because nearly every time interval in iOS is in seconds. Luckily, Xcode automatically provides a conversion snippet if you type `dispatch_after` and press Enter. Using nanoseconds is optimized for the hardware, not the programmer. Passing the time in seconds would require floating point math, which is more expensive and less precise. GCD is a very low-level framework and does not waste many cycles on programmer convenience.

GCD and ARC

In iOS 6, GCD objects are ARC-compatible, which means you no longer use `dispatch_retain` or `dispatch_release` to manage them. Most of the time this is great because you can now have GCD strong properties that correctly auto-synthesize. But sometimes it can be a problem in existing code. In particular, if you've been storing GCD objects like `dispatch_queue` in context pointers or `dispatch_queue_set_specific`, things are more complicated. (See the later section "Queue-Specific Data" for details on `dispatch_queue_set_specific`.) Apple hasn't given guidance on how to convert this code. I'm currently recommending the following approach. This may change as I develop more code in this space.

Consider the case where you have attached a sub-queue as a `set_specific` property (non-ARC):

```
dispatch_queue_t subqueue = dispatch_queue_create(...);
dispatch_queue_set_specific(queue, &kSubqueueKey, (void*)subqueue,
                            (dispatch_function_t)dispatch_release);
```

Under ARC, I currently recommend this approach:

```
dispatch_queue_t subqueue = dispatch_queue_create(...);
void* subqueuePtr = (void*)CFBridgingRetain(subqueue);
dispatch_queue_set_specific(queue, &kSubqueueKey, subqueuePtr,
                            (dispatch_function_t)CFBridgingRelease);
```

I believe this correctly employs the intent of `CFBridgingRetain` and `CFBridgingRelease` to move objects in and out of ARC.

If you have complex GCD code and don't want to rework it for ARC, you can opt out of ARC for GCD objects by passing `-DOS_OBJECT_USE_OBJC=0` to the compiler. See `/usr/include/os/object.h` for more details.

Sources and Timers

A dispatch source provides an efficient way to handle events. You register an event handler, and when the event occurs, you are notified. If the event occurs multiple times before the system is able to notify you, the events are coalesced into a single event. This is highly useful for low-level, high performance I/O, but it's very rare for an iOS app developer to need this functionality. Similarly, dispatch sources can respond to UNIX signals, file system changes, changes in other processes, and mach port events. Many of these are very useful on Mac, but again, don't often apply to iOS developers. There are custom sources, but they have significant restrictions in what information they can pass, and I have yet to find a case where they are useful on iOS.

One dispatch source, however, is very useful on iOS: the timer source. GCD timers are based on dispatch queues rather than run loops like `NSTimer`, which means that they are much easier to use in multithreaded apps. Because they use blocks rather than selectors, GCD timers don't require a separate method to process the timer. They also make it easier to avoid retain cycles with repeating GCD timers.

`RNTimer` (`https://github.com/rnapier/RNTimer`) implements a simple GCD timer that avoids retain loops (as long as the block doesn't capture `self`), and automatically invalidates when deallocated. It creates a timer dispatch source with `dispatch_source_create` and ties it to the main dispatch queue. This means that the timer will always fire on the main thread. You could, of course, use a different queue for this if you like. It then sets the timer and the event handler and calls `dispatch_resume` to start the timer. Because dispatch sources often require configuration, they're created in a suspended state and begin delivering events only when they're resumed. Following is the full code.

RNTimer.m (RNTimer)

```
+ (RNTimer *)repeatingTimerWithTimeInterval:(NSTimeInterval)seconds
                                      block:(dispatch_block_t)block {
  RNTimer *timer = [[self alloc] init];
  timer.block = block;
  timer.source = dispatch_source_create(DISPATCH_SOURCE_TYPE_TIMER,
                              0, 0,
                              dispatch_get_main_queue());
  uint64_t nsec = (uint64_t)(seconds * NSEC_PER_SEC);
  dispatch_source_set_timer(timer.source,
                            dispatch_time(DISPATCH_TIME_NOW, nsec),
                            nsec, 0);
  dispatch_source_set_event_handler(timer.source, block);
  dispatch_resume(timer.source);
  return timer;
}

- (void)invalidate {
  if (self.source) {
    dispatch_source_cancel(self.source);
    dispatch_release(self.source);
    self.source = nil;
  }
  self.block = nil;
}

- (void)dealloc {
  [self invalidate];
}
```

Creating Synchronization Points with Dispatch Barriers

GCD offers a rich system of serial and concurrent queues. With some thought, you can use these to create many things other than simple thread management. For instance, GCD queues can be used to solve many common locking problems at a fraction of the overhead.

A *dispatch barrier* creates a synchronization point within a concurrent queue. While it's running, no other block on the queue is allowed to run, even if it's concurrent and other cores are available. If that sounds like an exclusive (write) lock, it is. Non-barrier blocks can be thought of as shared (read) locks. As long as all access to the resource is performed through the queue, barriers provide very cheap synchronization.

For comparison, you could manage multithreaded access with `@synchronize`, which takes an exclusive lock on its parameter, as shown in the following code.

```
- (id)objectAtIndex:(NSUInteger)index {
  @synchronized(self) {
    return [self.array objectAtIndex:index];
  }
}

- (void)insertObject:(id)obj atIndex:(NSUInteger)index {
  @synchronized(self) {
    [self.array insertObject:obj atIndex:index];
  }
}
```

`@synchronized` is easy to use, but very expensive even when there is little contention. There are many other approaches. Most are either complicated and fast or simple and slow. GCD barriers offer a nice tradeoff.

```
- (id)objectAtIndex:(NSUInteger)index {
  __block id obj;
  dispatch_sync(self.concurrentQueue, ^{
    obj = [self.array objectAtIndex:index];
  });
  return obj;
}

- (void)insertObject:(id)obj atIndex:(NSUInteger)index {
  dispatch_barrier_async(self.concurrentQueue, ^{
    [self.array insertObject:obj atIndex:index];
  });
}
```

All that is required is a `concurrentQueue` property, created by calling `dispatch_queue_create` with the `DISPATCH_QUEUE_CONCURRENT` option. In the reader (`objectAtIndex:`), you use `dispatch_sync` to wait for it to complete. Creating and dispatching blocks in GCD has very little overhead, so this approach is much faster than using a mutex. The queue can process as many reads in tandem as it has cores available. In the writer, you use `dispatch_barrier_async` to ensure exclusive access to the queue while writing. By making the call asynchronous, the writer returns quickly, but any future reads on the same thread are guaranteed to return the value the writer set. GCD queues are FIFO, so all requests on the queue before the write are completed first, then the write runs alone, and only then are requests that were placed on the queue after the write processed. This prevents writer starvation and ensures that immediately reading after a write always yields the correct result.

Queue Targets and Priority

Queues are hierarchical in GCD. Only the global system queues are actually scheduled to run. You can access these queues with `dispatch_get_global_queue` and one of the following priority constants:

- DISPATCH_QUEUE_PRIORITY_HIGH
- DISPATCH_QUEUE_PRIORITY_DEFAULT

- DISPATCH_QUEUE_PRIORITY_LOW

- DISPATCH_QUEUE_PRIORITY_BACKGROUND

All of these queues are concurrent. GCD schedules as many blocks as there are threads available from the HIGH queue. When the HIGH queue is empty, it moves on to the DEFAULT queue, and so on. The system creates and destroys threads as needed, based on the number of cores available and the system load.

When you create your own queue, it is attached to one of these global queues (its *target*). By default, it is attached to the DEFAULT queue. When a block reaches the front of your queue, the block is effectively moved to the end of its target queue. When it reaches the front of the global queue, it's executed. You can change the target queue with dispatch_set_target_queue.

Once a block is added to a queue, it runs in the order it was added. There is no way to cancel it, and there is no way to change its order relative to other blocks on the queue. But what if you want a high-priority block to "skip to the head of the line"? As shown in the following code, you create two queues, a high-priority queue and a low-priority queue, and make the high-priority queue the target of the low-priority queue.

```
dispatch_queue_t
low = dispatch_queue_create("low", DISPATCH_QUEUE_SERIAL);

dispatch_queue_t
high = dispatch_queue_create("high", DISPATCH_QUEUE_SERIAL);

dispatch_set_target_queue(low, high);
```

Dispatching to the low-priority queue is normal:

```
dispatch_async(low, ^{ /* Low priority block */ });
```

To dispatch to the high-priority queue, suspend the low queue and resume it after the high-priority block finishes:

```
dispatch_suspend(low);
dispatch_async(high, ^{
  /* High priority block */
  dispatch_resume(low);
});
```

Suspending a queue prevents scheduling any blocks that were initially put on that queue, as well as any queues that target the suspended queue. It won't stop currently executing blocks, but even if the low-priority block is next in line for the CPU, it won't be scheduled until dispatch_resume is called.

You need to balance dispatch_suspend and dispatch_resume exactly like retain and release. Suspending the queue multiple times requires an equal number of resumes.

Dispatch Groups

A dispatch group is similar to dependencies in NSOperation. First, you create a group:

```
dispatch_group_t group = dispatch_group_create();
```

Notice that groups have no configuration of their own. They're not tied to any queue. They're just a group of blocks. You usually add blocks by calling `dispatch_group_async`, which behaves like `dispatch_async`.

```
dispatch_group_async(group, queue, block);
```

You then register a block to be scheduled when the group completes using `dispatch_group_notify`:

```
dispatch_group_notify(group, queue, block);
```

When all blocks in the group complete, `block` will be scheduled on `queue`. You can have multiple notifications for the same group, and they can be scheduled on different queues if you like. If you call `dispatch_group_notify` and there are no blocks on the queue, the notification will fire immediately. You can avoid this situation by using `dispatch_suspend` on the queue while you configure the group and then `dispatch_resume` to start it.

Groups aren't actually tracking blocks as much as they're tracking *tasks*. You can increment and decrement the number of tasks directly using `dispatch_group_enter` and `dispatch_group_leave`. So, in effect, `dispatch_group_async` is really no more than this:

```
dispatch_async(queue, ^{
  dispatch_group_enter(group);
  dispatch_sync(queue, block);
  dispatch_group_leave(group);
});
```

Calling `dispatch_group_wait` blocks the current thread until the entire group completes, which is similar to a thread *join* in Java.

Queue-Specific Data

Much like associative references discussed in Chapter 3, queue-specific data allows you to attach a piece of data directly to a queue. This approach can sometimes be a useful and extremely fast way to pass information in and out of a queue. This is combined with *dispatch data*, discussed in the next section, to allow extremely high-performance data passing that reduces memory copying and allocation/deallocation churn.

Like associative references, queue-specific data uses a unique address as its key, rather than a string or other identifier. This unique address is usually achieved by passing the address of a `static char`. Unlike associative references, queue-specific data does not know how to retain and release. You have to pass it a destructor function that it calls when the value is replaced. For memory you've allocated with `malloc`, the destructor is `free`. It's cumbersome to use queue-specific data with Objective-C objects under ARC, but Core Foundation objects are a bit easier to use, as demonstrated here. (See the earlier section "GCD and ARC" for more information on using ARC-managed objects.) In this example, `value` is released automatically when the queue is destroyed or if another value is set for `kMyKey`.

```
static char kMyKey;
CFStringRef *value = CFStringCreate...;
dispatch_queue_set_specific(queue,
                            &kMyKey,
                            (void*)value,
                            (dispatch_function_t)CFRelease);
...

dispatch_sync(queue, ^{
  CFStringRef *string = dispatch_get_specific(&kMyKey);
  ...
});
```

One nice thing about queue-specific data is that it respects queue hierarchies. So if the current queue doesn't have the given key assigned, `dispatch_get_specific` automatically checks the target queue, then that queue's target queue, and on up the chain.

Dispatch Data

Dispatch data is the foundation of one of the most powerful low-level performance advances added in iOS 5, and you'll likely never need to use it directly. Dispatch data are blocks of noncontiguous, immutable memory buffers that can be joined very quickly and split between blocks with minimal copying. Buffers can also be incrementally released as they're processed, improving memory usage.

This is an incredibly robust system and is the basis for a feature called *dispatch I/O*, which promises significant I/O performance improvements on multicore iOS devices, and particularly on the Mac. However, in most cases, you'll get most of the benefit for free by using the higher-level abstractions without taking on the complexity of using dispatch I/O directly. My recommendation is to leave this technology alone while it finishes maturing and Apple works out the best patterns for using it. You may want to start looking at it now if your application needs to process very large amounts of data very quickly and you've found that memory allocation or disk access are your major bottlenecks. These types of problems are common for the OS, but less common at the application layer. See the "Further Reading" section for links to more information.

Summary

The future of iOS development is multitasking. Apps will need to do more operations in parallel to leverage multicore hardware and provide the best experience for users. Traditional threading techniques are still useful, but operation queues and Grand Central Dispatch are more effective and promise greater performance with less contention and less locking. Learning to manage your internal multitasking and behaving appropriately when multitasking with other applications are fundamental parts of today's iOS development.

Further Reading

Apple Documentation

The following documents are available in the iOS Developer Library at `developer.apple.com` or through the Xcode Documentation and API Reference.

iOS App Programming Guide, "Executing Code in the Background"

iOS App Programming Guide, "State Preservation and Restoration"

Concurrency Programming Guide

File System Programming Guide. "Techniques for Reading and Writing Files." The section "Processing a File Using GCD" includes example code explaining dispatch I/O channels.

Threading Programming Guide

WWDC Sessions

The following session videos are available at `developer.apple.com`.

WWDC 2011, "Session 320: Adopting Multitasking in Your App."

WWDC 2011, "Session 210: Mastering Grand Central Dispatch."

WWDC 2012, "Session 208: Saving and Restoring Application State on iOS."

WWDC 2012, "Session 712: Asynchronous Design Patterns with Blocks, GCD, and XPC."

Other Resources

Ash, Mike. *NSBlog*. Mike Ash is one of the most prolific writers on low-level threading issues. Although some of this information is now dated, many of his blog posts are still required reading.
`http://mikeash.com/pyblog/`

- Friday Q&A 2010-01-01: "NSRunLoop Internals"

- Friday Q&A 2009-07-10: "Type Specifiers in C, Part 3"

- Friday Q&A 2010-07-02: "Background Timers"

- Friday Q&A 2010-12-03: "Accessors, Memory Management, and Thread Safety"

CocoaDev, "LockingAPIs." CocoaDev collects much of the accumulated wisdom of the Cocoa developer community. The "Locking APIs" page includes links and discussion of the available technologies and tradeoffs.
`www.cocoadev.com/index.pl?LockingAPIs`

Chapter 14
REST for the Weary

At some point, most iOS applications have to communicate with a remote web server in one way or another. Some apps can run and be useful without a network connection, and web server communication might be short-lived (or even optional) for the application. Apps that fall into this category are those that sync data with a remote server when a connection is present, such as to-do lists.

Another set of apps needs nearly continuous network connectivity to provide any meaningful value to the user. These apps typically act as a mobile client for a web service. Twitter clients, foursquare, Facebook, and most apps you write fall into this category. This chapter presents some techniques for writing apps the right way for consuming a web service. We discuss caching data offline or synchronizing data with a remote server in Chapter 24.

As of this writing, a quick search for Twitter in Apple's App Store turns up nearly 650 iPad apps and more than 3,000 iPhone apps. Today, if you want to create the next Twitter client, you don't have to know anything about web services or the Twitter API. There are more than a dozen implementations of the Twitter API in Objective-C. The same is true for most public services like Facebook's Graph API or Dropbox. So, rather than explaining how to build your next Twitter client, this chapter provides insights and best practices for designing your next iPhone app that consumes a generic, simple, and hypothetical web service.

Because the ideas and techniques presented here are generic, you can easily apply them to any projects you undertake. If you've been an iOS developer for at least one year, you may have already implemented a project like this, where your customer sends you a documentation of his server APIs. You've been introduced to the server developer and probably have some control over negotiating the output format and error-handling stuff. In most cases, both the client and the server code were developed in tandem.

In addition to discussing the REST implementation on iOS, this chapter provides some guidelines for the server that will help you achieve the following:

- Improve the code quality.
- Reduce development time.
- Improve code readability and maintainability.
- Increase the perceived performance of the app.

The Worldwide Web Consortium has identified two major classes of web services (W3C Web Services Architecture 2004): RESTful services that manipulate XML representation of web resources using a uniform set of stateless operations, and arbitrary services that might expose any operation. SOAP and WSDL are in the second category. Web services used in 2011 are mostly RESTful, including but not limited to Twitter APIs, foursquare, and Dropbox. This chapter focuses on consuming a RESTful service in your application.

The REST Philosophy

The three most important features of a RESTful server are its *statelessness, uniform resource identification,* and *cacheability.*

A RESTful server is always stateless. This means every API is treated as a new request and no client context is remembered on the server. Clients do maintain the state of the server, which includes but isn't limited to caching responses and login access tokens.

Resource identification on a RESTful server is done through URLs. For instance, instead of accepting a resource ID as a parameter, a REST server accepts it as a part of the URL. For example, `http://example.com/resource?id=1234` becomes `http://example.com/resources/1234`.

This method of resource identification and the fact that a RESTful server doesn't maintain the state of the client together allow clients to cache responses based on the URL, just as a browser caches web pages.

Response from a RESTful server is usually sent in a uniform, agreed-upon format, usually to decouple the client/server interface. The client iOS app communicates with a RESTful server through this agreed-upon data exchange format. As of today, the most commonly used formats are XML and JSON. The next section discusses the differences among the formats and the ways you can parse them in your app.

Choosing Your Data Exchange Format

Web services traditionally support two major kinds of data exchange format: JSON (JavaScript Object Notation) and XML (eXtensible Markup Language). Microsoft pioneered XML as the default data exchange format for its SOAP services, whereas JSON became an open standard described in RFC 4627. Although there are debates about which is superior, as an iOS developer, you need to be able to handle both kinds of data format on your app.

Several parsers are available for both XML and JSON for Objective-C. The following sections discuss some of the most commonly used toolkits.

Parsing XML on iOS

You can do XML parsing using two kinds of parsers, a DOM parser or a SAX parser. A SAX parser is a sequential parser and returns parsed data on a callback as it steps through the XML document. Most SAX parsers work by taking in a URL as a parameter and giving you data as it becomes available. For example, the `NSXMLParser` foundation class has a method called `initWithContentsOfURL:`.

```
(id)initWithContentsOfURL:(NSURL *)url;
```

You essentially initialize a parser object with the URL, and the `NSXMLParser` does the rest. Parsed data becomes available through callback via delegate methods defined in `NSXMLParserDelegate`. The most commonly handled methods are

```
parserDidStartDocument:

parserDidEndDocument:

parser:didStartElement:namespaceURI:qualifiedName:attributes:

parser:didEndElement:namespaceURI:qualifiedName:

parser:foundCharacters:
```

Because the parser uses delegation to return data, you need a NSObject subclass conforming to `NSXMLParserDelegate` for every object you're handling. This tends to make your code base a bit more verbose compared to a DOM parser.

> While you can use the one class (even your controller) to conform to the `NSXMLParserDelegate`, your code will be highly unmanageable when the XML format changes. Create a separate subclass of `NSObject` and make that class conform to `NSXMLParserDelegate` for clarity.

A DOM parser, on the other hand, loads the complete XML before it starts parsing. The advantage of using a DOM parser is its capability to access data at random using XPath queries, and there's no delegation as in the SAX model.

Mac OS X SDK has an Objective-C based DOM parser, `NSXMLDocument`. iOS doesn't have an Objective-C based DOM parser built in. You can use libxml2 or third party Objective-C wrappers like KissXML, TouchXML or GDataXML built around libxml2. Some third party libraries cannot write XML. A web service that responds in XML will mostly expect the post body in XML. In that case, you will require a library that can write XML (say from a NSObject or NSDictionary). KissXML and GDataXML are two good libraries. For a complete comparsion, check the link to "How to Choose The Best XML Parser for Your iPhone Project" in the "Further Reading" section.

If you are using KissXML, the most commonly used classes are `DDXMLDocument` and `DDXMLNode` and the commonly used methods are `initWithXMLString:options:error:` in `DDXMLDocument` and `elementWithName:stringValue:` in `DDXMLNode`.

Using a DOM parser makes your code cleaner and easier to read. Although this comes at the expense of execution time for handling web service requests, the effect is minor because DOM parsers become slower only for documents larger than a megabyte or so. A web service response generally is less than that. Any performance gain you get is negligible compared to the time of the network operation. These performance gains make a lot of sense when you're parsing XML from your resource bundle. For handling web service requests, I would always recommend a DOM parser.

To learn more about XML performance, download and test the XML Performance app published by Apple and Ray Wenderlich (See the "Further Reading" section at the end of this chapter).

Parsing JSON on iOS

The second data exchange format is JSON, which is much more commonly used than XML. Although Apple has a JSON processing framework, it was a private API in iOS 4, and the Mac Snow Leopard and wasn't available for general use. With iOS 5, Apple introduced NSJSONSerialization (Apple's JSON parsing and serializing framework) that you can use for parsing.

There are also plenty of third-party JSON processing frameworks to choose from. The most commonly used frameworks by far are SBJson, TouchJSON, YAJL, and JSONKit. (See the "Further Reading" section for the links to download these frameworks.) Almost all frameworks have category extensions on NSString, NSArray, and NSDictionary to convert to and from JSON. The code samples in this chapter use Apple's own NSJSONSerialization.

Apple's classes are fast compared to other frameworks. As of this writing, most other third-party offerings, including the most popular JSONKit, are not ARC-ready and therefore I recommend using NSJSONSerialization. However NSJSONSerialization lacks the capability to serialize custom objects, something JSONKit can do. JSONKit has a couple of convenient methods that accomplish this: serial izeUnsupportedClassesUsingDelegate:selector:error: and serializeUnsupportedClas sesUsingBlock:error:.

If your JSON parsing needs unsupported class handling, you will not be able to use NSJSONSerialization.

> When you're choosing a library for your app, you might have to do some performance evaluation. (You can compare the frameworks using the open source test project json-benchmarks on GitHub (see the "Further Reading" section for the link to this tool). Because all five (SBJson, TouchJSON, YAJL, JSONKit, and NSJSONSerialization) are actively developed, every library is equally good, and there's no one best library as of this writing. Keep a close eye on them and be ready to swap frameworks if one seems superior to another. Usually, swapping a JSON library won't require monumental refactoring, because in most cases, it involves changing the class category extension methods. As of this writing, the two most popular ones are JSONKit and Apple's NSJSONSerialization with JSONKit slightly faster than NSJSONSerialization.

XML Versus JSON

Source code fragments in this chapter are based on using JSON. You'll discover how to design your classes to make it easy to add XML support without affecting the rest of the code base. In every case, JSON processing on iOS is an order of magnitude easier than XML. So if your server supports both XML and JSON formats, choosing JSON is a wise decision. If your server code is not yet developed, start by supporting JSON initially.

Designing the Data Exchange Format

It's essential to keep in mind that we're talking about data exchange between client and server. The most common mistake iOS developers make is to think of JSON as some arbitrary data sent by the server in response to an API call. Although that's true to some extent, a quick look at what happens on the server will give you a better picture of what JSON actually is.

Internally, most servers are coded using some object-oriented programming language. Whether it's Java, Scala, Ruby, or C# (even PHP and Python support objects to some extent), any data you need on your iOS app will likely be an object on the server as well. Whether the object is an ORM-mapped entity (ORM stands for *object relational mapping*) or a business object is of little importance. Just call them model objects, and these objects are serialized to JSON only at the transport level. Most object-oriented languages provide interfaces to serialize objects, and developers usually harness this capability to convert their objects to JSON. This means the JSON you see on the response is just a different representation of the objects (or object list) on the server.

Keep this concept in mind while writing your code, and you will probably create model objects for every equivalent server model object. When you do that, you need not worry about changes affecting your code later. Refactoring will be far easier.

Rather than thinking in terms of JSON strings, it makes more sense to think in terms of objects and resources. Design and develop your code such that you always reconstruct model objects for every object on the server. When the reconstructed objects on your iOS app match 100 percent with the objects on the server, the goal of data exchange is attained, and it becomes easier to write an error-free app.

In short, think of JSON as a data exchange format instead of a language with a bunch of syntax. Consider documenting the data exchange contracts on an object/resource basis rather than as primitive data types. foursquare's developer documentation is a very good example. We, in fact, recommend using the foursquare's documentation as the starting point. These objects in turn become the model objects for your app. You see this in detail a little later in this chapter (Key Value Coding JSONs), and you look at how to convert JSON dictionaries into models by using Objective-C's key-value coding/observing (KVC/KVO) mechanism.

Model Versioning

In the past, at least from the late 1990s or early 2000s until the first iPhone was launched in 2007, most client/server development happened in tandem with a web-based interface. Native clients were not commonly used. The client app running on the web browser was always deployed together with the server. As such, it wasn't really necessary to handle versioning in your models. However, on iOS, deploying the client requires that the app be physically installed on your user's device. This could take days or months, so you should also handle situations when the server is accessed with an older client. How many older versions of the client you want to support depends on your business goals. As an iOS developer, you should probably build support for catering to those business needs. Using class clusters on your iOS app is one way to do that. You find out more about this design later in this chapter.

A Hypothetical Web Service

We start by describing a hypothetical app concept, the iHotelApp, and develop the iOS code for it. Later, in Chapter 24, you will revisit this app and add a caching layer.

Assume that you're in charge of developing an iOS app for a restaurant. The restaurant uses iPads to take orders. Orders can be placed directly with waiters who enter the orders into their iPads. Customers can also directly place orders using the kiosks (a dedicated iPad running your app) at their tables. Here's a brief description of the top-level functionalities of the app:

- Customer orders are sent to the remote servers based on the customers' table numbers, whereas waiters pick a table number along with every order they send through their own login accounts. So, it's clear that there are two kinds of login/authentication mechanisms. One is the traditional username/password-based type, and the other is based on customer table numbers. In all cases, the server will exchange an access token for a given authentication information. The important point is that you need to develop one code base that caters to both types of login. After logging in, every web service requires you to send an access token with every subsequent call you make.

   This requirement translates to the `/loginwaiter` and `/logintable` web service endpoints.

   Both these endpoints return an access token. In the iOS client implementation, I'll show you how to "remember" this access token and send it along with every request.

- Customers should be able to see the menu, along with the details of every menu item including the photos/videos of the food and ratings left by other customers.

   This requirement translates to a web service `/menuitems` endpoint and a `/menuitem/<itemid>` that returns a JSON object that will be modeled as a `MenuItem` object.

   In the iOS implementation, you will discover how to map the JSON keys to your model object with as little code as possible by making use of Objective-C's most powerful technique, key-value coding (KVC).

- Customers should be able to submit reviews of an item.

   This requirement translates to a web service endpoint `/menuitem/<itemid>/review`.

   In these cases, some iOS apps show a floating heads-up display (commonly known as HUD) that prevents users from doing any operation until the review is posted. This is clearly bad from a user experience perspective. You'll see how to post reviews in the background without showing a modal HUD.

Although there are other requirements for this app, these three cover the most commonly used patterns when talking to a web service.

Important Reminders

Keep these essential points in mind as you build your app:

- **Never make a synchronous network call.** Even if they're on a background thread, synchronous calls don't report progress. Another reason is that to cancel a synchronous request running on a background thread, you have to kill the thread, which is again not a good idea. Additionally, you will not be able to control the number of network calls in your app, which is very critical to the performance of your app. You find more tips to improve performance on iOS in this chapter.

- **Avoid using NSThread or GCD-based threading directly for network operations (unless your project is small and has just a couple of API calls).** Running your own threads or using GCD has some caveats, as just explained.

- **Use** `NSOperationQueue`-**based threading instead.** `NSOperationQueue` helps with controlling the queue length and the number of concurrent network operations. GCD-based threads also cannot be cancelled after the block has been dispatched.

Now, it's time to start designing the iOS app's web service architecture.

RESTfulEngine Architecture (iHotelApp Sample Code)

iOS apps traditionally use model-view-controller (MVC) as the primary design pattern. When you're developing a REST client in your app, you need to isolate the REST calls to their own class. The stateless nature of REST and its cacheable nature can be best applied when it's written in its own class. Moreover, it also provides a layer of isolation (which is also good for unit testing) and helps keep your controller code cleaner.

To get started, you choose a network management framework.

NSURLConnection Versus Third-Party Frameworks

Apple provides classes in `CFNetwork.framework`, and a Foundation-based `NSURLConnection`, for making asynchronous requests. However, for developing RESTful services, you need to customize those classes by subclassing them. Rather than reinvent what's already available for the development of web services, I recommend using `MKNetworkKit` (see the "Further Reading" section at the end of this chapter). `MKNetworkKit` encapsulates many often-used features such as basic or digest authentication, form posts, and uploading or downloading files. Another important feature it provides is an `NSOperationQueue` encapsulation, which you can use to queue network requests.

> My advice is generally to refrain from using third-party code when you're developing for iOS. However, some components and frameworks are worth using. As far as possible, avoid third-party code that is heavily interdependent. `MKNetworkKit` is a block-based wrapper around `NSURLConnection` that doesn't bloat your code base while providing powerful features, most importantly, caching. We show you the benefits of caching your responses in Chapter 23. We recommended `ASIHttpRequest` in the previous edition of this book in Chapter 9. Because the original author of `ASIHttpRequest` has stopped development and support, it makes sense to move your code base to another framework. Other similar frameworks that you can consider are AFNetworking, LRResty, and RestKit. However, none of these frameworks is fully compatible with ARC. LRResty has an ARC branch, but it was still experimental at the time of this writing.

The code sample provided in the download files for this chapter uses `MKNetworkKit`. You can find this code in the Chapter 14 directory (iHotelApp) on the book's website.

Note that the code download for this chapter is quite large. The chapter provides important code snippets, and you should look at the corresponding files. Open the project in Xcode to better understand the code and the architecture.

The necessary server component for the RESTfulEngine is already hosted on the book's website at `iosptl.com`.

Creating the RESTfulEngine

The `RESTfulEngine` is the heart of the iHotelApp. This class encapsulates every call to the web service standalone class, which handles your network calls. Data should be passed from `RESTfulEngine` to view controllers only as `Model` objects instead of JSON or `NSDictionary` objects. (I discuss the process of creating model classes in the next section.) Now, what should happen when there's a back-end-related error? Communicating errors from `RESTfulEngine` to the view controller will be covered in the subsequent section. Here are the first two important steps:

1. Add `MKNetworkKit` code to your project. You can add it as a submodule and drag the relevant files or add a cocoapod dependency to your project.

2. Create a `RESTfulEngine` object of type `MKNetworkEngine` in your project's `AppDelegate`. For a demo implementation, refer to this chapter's source code on the book's website.

   This `RESTfulEngine` encapsulates most of the network-related mundane tasks, such as managing the concurrent queue, showing the activity indicator, and so on.

   The `RESTfulEngine` object automatically changes the maximum number of concurrent operations to 6 when on WiFi and to 2 when on carrier data. This improves the performance of your REST client considerably. I explain this later in the section "Tips to Improve Performance on iOS."

Refrain from writing or storing state variables in your application delegate. Doing so is almost as bad as using global state variables. However, pointers to commonly used modules like `managed ObjectContext` or `persistentStoreCoordinator` or our own `networkEngine` are fine.

Adding Authentication to the RESTfulEngine

Now that the class is ready, you will add methods to handle web service calls: first and foremost, authentication. `MKNetworkKit` provides wrapper methods for a variety of authentication schemes, including, but not limited to, HTTP Basic Authentication, HTTP Digest Authentication scheme, NIL Authentication, and so on. I won't go through the details of the authentication mechanisms in this book, but for the sake of simplicity, assume that you exchange an access token with the server by sending the username and password to the `/loginwaiter` request or to the `/logintable` request. You need to define macros for these URL endpoints. Add the following code to the `RESTfulEngine` class header file:

The Constants in RESTfulEngine.h

```
#define LOGIN_URL @"loginwaiter"
#define MENU_ITEMS_URL @"menuitem"
```

Next, create a property in `RESTfulEngine` to hold the access token and then create a new method, `loginWithName:password:onSucceeded:onError:`, as in the following code:

The init Method (and Property Declaration) in RESTfulEngine.h

```
@property (nonatomic, strong) NSString *accessToken;

-(id) loginWithName:(NSString*) loginName
          password:(NSString*) password
       onSucceeded:(VoidBlock) succeededBlock
           onError:(ErrorBlock) errorBlock;
```

The init Method (and Property Declaration) in RESTfulEngine.m

```
-(RESTfulOperation*) loginWithName:(NSString*) loginName
          password:(NSString*) password
       onSucceeded:(VoidBlock) succeededBlock
           onError:(ErrorBlock) errorBlock
{
  RESTfulOperation *op = (RESTfulOperation*) [self
operationWithPath:LOGIN_URL];

  [op setUsername:loginName password:password basicAuth:YES];
  [op onCompletion:^(MKNetworkOperation *completedOperation) {

    NSDictionary *responseDict = [completedOperation responseJSON];
    self.accessToken = [responseDict objectForKey:@"accessToken"];
    succeededBlock();
  } onError:^(NSError *error) {

    self.accessToken = nil;
    errorBlock(error);
  }];

  [self enqueueOperation:op];
  return op;
}
```

That completes your web service call. Now you need to notify the caller, (which is usually the view controller) about the outcome of the web service call. You use blocks for this.

Adding Blocks to the RESTfulEngine

Every web service call in this `RESTfulEngine` mandates the caller to implement two block methods, one for notifying a successful call and another for error notification.

Block Definitions

```
typedef void (^VoidBlock)(void);
typedef void (^ModelBlock)(JSONModel* aModelBaseObject);
typedef void (^ArrayBlock)(NSMutableArray* listOfModelBaseObjects);
typedef void (^ErrorBlock)(NSError* engineError);
```

You use the first block type to notify success without passing additional information. The second block type is to notify success and send a model object.

> If you're not familiar with blocks yet, you may want to skip to Chapter 23 and then come back here. Chapter 23 discusses in depth how blocks work internally. Blocks have advantages and drawbacks that we talk about in Chapter 23.

The `RESTfulEngine` class implementation is now complete for the first method, `loginWithName:pass word:onSucceeded:onFailure`. You can call this method from the view controller (which is usually the login page that shows the username and password fields):

Login Button Event Handling in iHotelAppViewController.m

```
-(IBAction) loginButtonTapped:(id) sender
{
   [AppDelegate.engine loginWithName:@"mugunth"
                            password:@"abracadabra"
                          onSucceeded:^{

                          [[[UIAlertView alloc]
                             initWithTitle:NSLocalizedString(@"Success",
@"")
             message:NSLocalizedString(@"Login successful", @"")
           delegate:self
cancelButtonTitle:NSLocalizedString(@"Dismiss", @"")
otherButtonTitles: nil] show];
} onError:^(NSError *engineError){
   [UIAlertView showWithError:engineError];
   }];
}
```

Thus, with just a few lines of code, you're able to implement the login functionality of the web service.

Remember the access token? If your access token is simply a string, you can store it in Keychain or in `NSUserDefaults`. Storing it in Keychain is more secure than `NSUserDefaults`. You learn more about security in Chapter 15. The easiest and probably the cleanest way to store your keychain is to write a custom synthesizer for `accessToken` like the one below:

Access Token Custom Accessor in RESTfulEngine.m

```
-(NSString*) accessToken
{
    if(!_accessToken)
    {
        _accessToken = [[NSUserDefaults standardUserDefaults]
                        stringForKey:kAccessTokenDefaultsKey];
    }

    return _accessToken return_accessToken;
}
-(void) setAccessToken:(NSString *) aAccessToken
{
    _accessToken = _accessToken = aAccessToken;

    [[NSUserDefaults standardUserDefaults] setObject:self.accessToken
      forKey:kAccessTokenDefaultsKey];
    [[NSUserDefaults standardUserDefaults] synchronize];
}
```

If your web server sends user profile information at login, to cache the data, you may need a bit more sophisticated mechanism than NSUserDefaults. Keyed Archiving or Core Data can be used for that.

Whew! That completes your first endpoint, but you're not done yet! Next, you create a second endpoint, /menuitems, which is used to download a list of menu items from the server.

Authenticating Your API Calls with Access Tokens

In most web services, every call after login is probably protected and can be accessed only by passing the access token. Instead of sending the access token in every method, a cleaner way is to write a factory method in your RestfulEngine that creates a request object. This request object can then be filled with parameters specific to the call.

Overriding Methods to Add Custom Authentication Headers in RESTfulEngine.m

MKNetworkKit (and most other third-party networking frameworks) already provides a factory method, operationWithURLString:params:httpMethod: that internally calls the prepareHeaders: and allows you to pass additional headers to a request. You override the method prepareHeaders: and add your custom HTTP Header fields including the Authorization header to add your access token to the request.

Every network operation created on an engine calls this method. Be sure to call the superclass implementation after you're done with it. With techniques like this, you will never again have a buggy API call because you accidentally forgot to set the access token.

```
-(void) prepareHeaders:(MKNetworkOperation *)operation {
```

```
if(self.accessToken)
    [operation setAuthorizationHeaderValue:self.accessToken
     forAuthType:@"Token"];
[super prepareHeaders:operation];
}
```

Note that this `prepareHeaders:` method can also have additional headers set depending on your web service requirements. If your web service requires you to turn on gzip encoding for all calls or to send the application version number and the device-related information on the HTTP header, this method is the best place to add code for adding these additional header parameters.

Now, it's time to add a method to the `RESTfulEngine` class for fetching menu items from the server:

Method to Fetch the List of Menu Items in RESTfulEngine.m

```
-(RESTfulOperation*) fetchMenuItemsOnSucceeded:(ArrayBlock) succeededBlock
                                       onError:(ErrorBlock) errorBlock
{
  RESTfulOperation *op = (RESTfulOperation*) [self
  operationWithPath:MENU_ITEMS_URL];
[op onCompletion:^(MKNetworkOperation *completedOperation) {
    // convert the response to model objects and invoke succeeded block
} onError:errorBlock];
  [self enqueueOperation:op];
  return op;
}
```

Custom parameters passed to your API should be added to this method. Your view controller code remains clean of unnecessary strings/dictionaries and URLs.

Canceling Requests

View controllers that need to display the information from your web service call methods like `fetchMenuItems:` on the `RESTfulEngine`. To ensure that the view controllers plays nicely with system resources, it's the view controller's responsibility to cancel any network operation it creates when the user navigates out of the view. For example, tapping the Back button means that even if the request returns, the response is not used. Canceling the request at this point means that other requests queued in the `RESTfulEngine` get a chance to run, and your subsequent views' requests are executed faster.

To enable this behavior, every method that is written on your `RESTfulEngine` class should return the operation object back to the view controller. Canceling a running operation speeds up the execution waiting time for the request submitted by the next view. A good example of this scenario on the foursquare app is the user tapping on a profile view and then tapping on the Mayorships button. In this case, the profile view submits a request to fetch the user's profile but the user has already navigated to the Mayorship view without viewing the profile. It's now the responsibility of the profile view to cancel its request. Canceling the profile fetch request naturally speeds up the Mayorship fetch request by freeing up the bandwidth. This is applicable not only to foursquare but also to most web service apps you develop.

Request Responses

When you call the `fetchMenuItems:` method, the server's response is a list of menu items. In the previous web service call example, the response was an access token, a simple string, so you didn't need to design a model. In this case, you create a model class. Assume that the JSON returned by the server is in the following format:

```
{
"menuitems" : [{
  "id": "JAP122",
  "image": "http://d1.myhotel.com/food_image1.jpg",
  "name": "Teriyaki Bento",
  "spicyLevel": 2,
  "rating" : 4,
  "description" : "Teriyaki Bento is one of the best lorem ipsum dolor
sit",
  "waitingTime" : "930",
  "reviewCount" : 4
}]
```

One easy way to create a model from a JSON is to write verbose code to fill in your model class with the JSON. The other, much more elegant way is to piggyback on Objective-C's arguably most important feature: key-value coding. The JSONKit classes (or any other JSON parsing framework, including Apple's `NSJSONSerialization`) discussed earlier, converts a JSON-formatted string into a `NSMutableDictionary` (or a `NSMutableArray`). In this case, you get a dictionary with a entry, `menuitems`. The call shown in the following code can extract the menu items' dictionary from the response.

```
NSMutableDictionary *responseDict = [[request responseString]
                                      mutableObjectFromJSONString];
NSMutableArray *menuItems = [responseDict objectForKey:@"menuitems"];
```

Now that you have an array of menu items, you can iterate through them, extract the JSON dictionary of every `menuitem,` and use KVO to convert them into model objects. This process is covered in the next section. The server sends another dictionary entry called `status`. I come back to that entry in the "Error Handling" section later in this chapter.

Key Value Coding JSONs

Before you start writing your first model class, you need to know a bit about the model class inheritance architecture. Any web service-based app includes more than one model. In fact, a count of ten models for a single app is not uncommon. Instead of writing the KVC code in ten different classes, you write a base class that does the bulk of KVC and delegates very little work to the subclasses. Call this base class `JSONModel`. Any model class in the app that models a JSON and needs JSON observing will inherit from this `JSONModel`.

Because you'll be making copies and/or mutable copies of your model classes, implement `NSCopying` and `NSMutableCopying` in this base class. Derived classes must override this base class implementation and provide their own deep copy methods.

To start, add a method called `initWithDictionary:` to the base class. Your `JSONModel.h` should look similar to the following:

JSONModel.h

```
@interface JSONModel : NSObject <NSCopying, NSMutableCopying>
-(id) initWithDictionary:(NSMutableDictionary*) jsonDictionary;
@end
```

Then implement the `initWithDictionary:` method:

JSONModel.m

```
-(id) initWithDictionary:(NSMutableDictionary*) jsonObject
{
    if((self = [super init]))
    {
        [self init];
        [self setValuesForKeysWithDictionary:jsonObject];
    }
    return self;
}
```

The important part of this procedure is the method `setValuesForKeysWithDictionary:`. This method is a part of Objective-C KVC that matches each property in the class that has the same name as a key in the dictionary, and sets the property's value to the value of that entry. Most importantly, if `self` is a derived class, it automatically matches the derived class properties and sets their values. There are some exception cases to be handled, which are covered shortly.

Voila! With just one line of code, you've "mapped" the JSON into your model class. But will everything work automatically when you have a derived class? Isn't there a catch here? Before going into the details, you need to understand how the method `setValuesForKeysWithDictionary:` works. Your `MenuItem` dictionary looks like this:

```
"id": "JAP122",
"image": "http://d1.myhotel.com/food_image1.jpg",
"name": "Teriyaki Bento",
"spicyLevel": 2,
"rating" : 4,
"description" : "Teriyaki Bento is one of the best lorem ipsum dolor sit",
"waitingTime" : "930",
"reviewCount" : 4
```

When you pass this dictionary to the `setValuesForKeysWithDictionary:` method, it sends the following messages along with their corresponding values: `setId`, `setImage`, `setName`, `setSpicyLevel`, `setRating`, `setDescription`, `setWaitingTime`, and `setReviewCount`. So a class modeling this JSON will implement these methods. The easiest way to implement this is to use Objective-C's built-in `@property`, so your `MenuItem.h` model class looks like the following:

Menuitem.h

```
@interface MenuItem : JSONModel
@property (nonatomic, strong) NSString *itemId;
@property (nonatomic, strong) NSString *image;
@property (nonatomic, strong) NSString *name;
@property (nonatomic, strong) NSString *spicyLevel;
@property (nonatomic, strong) NSString *rating;
@property (nonatomic, strong) NSString *itemDescription;
@property (nonatomic, strong) NSString *waitingTime;
@property (nonatomic, strong) NSString *reviewCount;

@end
```

Note that the property names for id and `description` have been changed to `itemId` and `itemDescription`. That's because id is a reserved keyword and `description` is a method in `NSObject` that prints out the address of the object. To avoid conflicts you must rename them. However, you need to handle these exception cases because the default implementation of the `setValuesForKeysWithDictionary:` method crashes with a familiar error message stating, "This class is not key value coding-compliant for the `key:id`." To handle this case, KVC provides a method called `setValue:forUndefinedKey:`.

In fact, the default implementation of this method raises the `NSUndefinedKeyException`. Override this method in your derived class and set the values accordingly.

Your `MenuItem.m` now looks like this:

Menuitem.m

```
- (void)setValue:(id)value forUndefinedKey:(NSString *)key
{
    if([key isEqualToString:@"id"])
        self.itemId = value;
    if([key isEqualToString:@"description"])
        self.itemDescription = value;
    else
        [super setValue:value forKey:key];
}
```

To avoid crashes in the future because of spurious keys in JSON, and be a bit more defensive in your programming style, you can override this `setValue:forUndefinedKey:` method in the base class, `JSONModel.m`, like this:

```
- (void)setValue:(id)value forUndefinedKey:(NSString *)key {
    NSLog(@"Undefined Key: %@", key);
}
```

Now, in your `RESTfulEngine`'s `fetchMenuItems:onSuccceeded:onError:` method, add the following code to convert the JSON responses to `MenuItem` model objects.

RESTfulEngine.m

```
NSMutableDictionary *responseDictionary = [completedOperation
responseJSON];
    NSMutableArray *menuItemsJson = [responseDictionary
                                    objectForKey:@"menuitems"];
    NSMutableArray *menuItems = [NSMutableArray array];
    [menuItemsJson enumerateObjectsUsingBlock:^(id obj, NSUInteger idx,
BOOL
                                               *stop) {
      [menuItems addObject:[[MenuItem alloc] initWithDictionary:obj]];
    }];
    succeededBlock(menuItems);
```

As you see, you call the `MenuItem init` method with a JSON dictionary to initialize itself from the dictionary keys. In short, by overriding a method *only* for special cases, you have successfully mapped a JSON dictionary to your custom model, and this model is clean of any JSON key strings! That's the power of KVC. The code is also inherently defensive, in the sense that whenever there's a change in JSON keys that the server sends (probably arising from a bug on the server-side), you see `NSLog` statements displaying the wrong undefined key on the console, and you can probably notify the server developers or make changes to your client to support the new keys.

It's also a good idea to add methods for performing deep copy to the derived class. Just override methods in `NSCopying` and `NSMutableCopying` and you're done. Tools like Accessorizer available from the Mac App Store can help you with that. (See the "Further Reading" section for a link to the app.)

List Versus Detail JSON Objects

A JSON object is a payload that gets transferred from the server to the client. To improve performance and reduce payload size, it's common for server developers to use two kinds of payload for the same object. One is a large payload format that contains all information about the object; the second is a small payload that contains information needed just to display the information on a list. For the example in this chapter, a minimum amount of information about the menu item is displayed on the listing page, and most of the other content, including images, photos, and reviews, is displayed on the detail page.

This technique goes a long way toward improving an iOS app's perceived performance. On the implementation side, the iOS app doesn't have to be changed for mapping two kinds of JSON. You get either a complete JSON or a JSON that fills your object partially. The code written to map the detailed JSON should work without any modification in this scenario. For example, the server can send the small payload JSON for `/menuitems` and a detailed payload for `/menuitems/<menuitemid>`. The detailed payload will contain exactly the same data plus the first page of reviews and links to the photos of the dishes, and so on.

Nested JSON Objects

In the example, every menu item will have an array of reviews left by the user. If you depend on the default implementation of KVC and declare an `NSMutableArray` property on your model, the KVC binding will set the array's value to an array of `NSMutableDictionary`. But what you actually want is to map it as a array of models, which means you have to map the dictionaries as well in a recursive fashion. You can handle this case by the overriding the `setValue:forKey:` method.

Assume that the following represents the format of JSON sent by the `/menuitems/<itemid>` method:

```
{
"menuitems" : [{
  "id": "JAP122",
  "image": "http://d1.myhotel.com/food_image1.jpg",
  "name": "Teriyaki Bento",
  "spicyLevel": 2,
  "rating" : 4,
  "description" : "Teriyaki Bento is one of the best lorem ipsum dolor
sit",
  "waitingTime" : "930",
  "reviewCount" : 4,
  "reviews": [{
    "id": "rev1",
    "reviewText": "This is an awesome place to eat",
      "reviewerName": "Awesome Man",
    "reviewedDate": "10229274633",
    "rating": "5"
  }]
}],
"status" : "OK"
}
```

This code is very similar to what you've already seen, but it has one additional payload: an array of reviews. In a real-life scenario, there might be multiple such additions, such as a list of photos, a list of "likes," and so on. But for the sake of simplicity, just assume that the detailed listing of a menu item has only one additional piece of information, which is the array of reviews. Now, before overriding the `setValue:forKey:` method, create a model object for a review entry. This class's header file will look similar to the following one. The implementation contains nothing but synthesizers and overridden `NSCopying` and `NSMutableCopying` (deep copy) methods.

Review.m

```
@property (nonatomic, strong) NSString *rating;
@property (nonatomic, strong) NSString *reviewDate;
@property (nonatomic, strong) NSString *reviewerName;
@property (nonatomic, strong) NSString *reviewId;
@property (nonatomic, strong) NSString *reviewText;
```

Again, you can generate these accessors using tools like Accessorizer and Objectify, both available on the Mac App Store site. Because the JSON data for Review doesn't have special keys that might be in conflict with Objective-C's reserved list, you don't even have to write explicit code for converting JSON to a review model. The initialization code is in the base class, and the KVC-compliant code is generated by the properties. That's the power of KVC.

Next, override the `setValue:forKey:` method in the `MenuItem` model to convert review dictionaries to `Review` models:

Custom Handling of KVC's setValue:forKey: Method in MenuItem.m

```
-(void) setValue:(id)value forKey:(NSString *)key
{
  if([key isEqualToString:@"reviews"])
  {
    for(NSMutableDictionary *reviewArrayDict in value)
    {
      Review *thisReview = [[[Review alloc] initWithDictionary:reviewArrayDict]
                                       autorelease];
      [self.reviews addObject:thisReview];
    }
  }
  else
    [super setValue:value forKey:key];
}
```

The idea behind this code is to handle the `reviews` key of the JSON in a specialized way and to let the default superclass implementation handle the other keys.

Less Is More

You may have heard how great KVC and KVO are via veteran Objective-C developers' blogs. Now that you understand them, you can put these concepts to use in your next app. You'll realize how powerful they are and how easily they enable you to write less code more efficiently.

Next, you find how to handle server-side errors gracefully on the iOS client.

Error Handling

Recall that you saw a key called `status` in the JSON payload. Every web service has some way to communicate error messages to the client. In some cases, they're sent through a special key, such as `status`. In other cases, the web server sends an `error` key with more information about the actual error, and no such `error` key is sent when the API call is successful. This section shows you how to model both these scenario on iOS so that you write as little code as possible, yet write it in a way that's easy to read and understand.

The first thing to understand is that not all API errors can be mapped to a custom HTTP error code. In fact, a server might throw errors even when everything is perfectly fine, but the user input is wrong. A website registration web service might throw an error if the user tries to register with an e-mail address that's already taken. This is just one example, and in most cases, you need specialized error handling for handling your own internal business logic errors. In this example, for instance, a missing menu item results in a 404 error. Most web services send a custom error message along with the 404 notice so that clients can understand what caused that 404.

A client implementation should not only report the HTTP error as an error message to the user, but also understand the internal business logic error for elegant error reporting and do proper error reporting.

Otherwise, the only error you can ever show is, "Sorry, something bad happened, please try again later." and no one, especially your customer, is interested in seeing that kind of vague message. This section shows you how to handle such cases elegantly.

In the following steps, you subclass `MKNetworkOperation` to handle custom API errors.

1. Create a subclass of `MKNetworkOperation`. Name it RESTfulOperation. This subclass will have a property to store the business logic errors thrown from the server.

2. Create an `NSError*` property called `restError` in the subclass.

3. Override two methods to handle error conditions. The first method to override is the `operationFailedWithError:`

Code in RESTfulOperation.m That Illustrates Error Handling

```
-(void) operationFailedWithError:(NSError *)theError
{
  NSMutableDictionary *errorDict = [[self responseJSON]
objectForKey:@"error"];

  if(errorDict == nil)
  {
    self.restError = [[RESTError alloc] initWithDomain:kRequestErrorDomain
          code:[theError code]
        userInfo:[theError userInfo]];
  }
  else
  {
    self.restError = [[RESTError alloc] initWithDomain:kBusinessErrorDoma
in
                                        code:[[errorDict
                                        objectForKey:@"code"]
intValue]
userInfo:errorDict];
  }

  [super operationFailedWithError:theError];
}
```

Using this class, you check for the presence of the `"error"` JSON key and process it appropriately. The `failWithError` method will be called when there is a HTTP error. You handle non-HTTP, business logic errors in the same manner. As you saw earlier, not every business logic error can be mapped to an equivalent HTTP error code. Moreover, in some cases, a benign error may be sent along with your response, and the server may delegate the responsibility of treating that as an error or normal condition to the client. For handling both these cases, you have to override another method, `operationSucceeded:`, as shown in the following code:

Code in RESTRequest.m That Illustrates Request Handling for Successful Conditions and Report Business Logic Error if Any

```
- (void)operationSucceeded
{
  // even when request completes without a HTTP Status code, it might be a
     benign error

  NSMutableDictionary *errorDict = [[self responseJSON]
objectForKey:@"error"];

  if(errorDict)
  {
    self.restError = [[RESTError alloc] initWithDomain:kBusinessErrorDoma
in
                                         code:[[errorDict
                         objectForKey:@"code"] intValue]
                                   userInfo:errorDict];
    [super operationFailedWithError:self.restError];
  }
  else
  {
    [super operationSucceeded];
  }
}
```

Both these methods remember the business logic errors in the `restError` property of your subclassed request object. This enables the client to know both the HTTP error (by accessing the `RESTfulOperation`'s superclass's error object) and the business layer error, from the local property `restError`.

Because this handling is done on a subclass, the class `RESTfulEngine` doesn't have to do any additional error handling. All it gets is a nicely wrapped `NSError` object for both kinds of errors, HTTP or business logic. The view controller implementation will now be as simple as checking whether the error is `nil`; if it's not `nil`, show the message inside the `[[request restError] userInfo]`.

With that, I move on to a discussion of localization.

Localization

This section is about localizing web service-related error messages and not localizing your app. Adding internationalization and localization support to your app is explained in detail in Chapter 14.

Some implementations require you to localize error messages in multiple languages. For errors generated within the app, this is simple and can be handled using the Foundation classes and macros. For server-related errors, the previous implementation just showed the server errors on the UI. The best way to show localized errors is for the server to return errors in agreed-upon codes. The iOS client can then look into a localized string table and show the correct error for a given code.

RESTError.m

```
+ (void) initialize
{
   NSString *fileName = [NSString stringWithFormat:@"Errors_%@", [[NSLocale
                       currentLocale] localeIdentifier]];
   NSString *filePath = [[NSBundle mainBundle] pathForResource:fileName
                                             ofType:@"plist"];

   if(filePath != nil)
   {
      errorCodes = [[NSMutableDictionary alloc] initWithContentsOfFile:fileP
ath];
   }
   else
   {
      // fall back to English for unsupported languages
      NSString *filePath = [[NSBundle mainBundle] pathForResource:@"Errors_
en_US"
                               ofType:@"plist"];
      errorCodes = [[NSMutableDictionary alloc] initWithContentsOfFile:fileP
ath];
   }
}
```

This RESTError class can again be initialized with the error dictionary you get from the server using the KVC technique covered earlier in this chapter. Override NSError's localizedDescription and localizedRecoverySuggestion methods to provide proper user-readable error methods. If your web service provides error codes along with error messages, handling and showing error messages this way is better than showing the server error from the userInfo dictionary. Now replace the NSError you created in RESTRequest with this RESTError. This will ensure that for custom error codes sent from the server, the localizedDescription and localizedRecoverySuggestion, come from the Errors_en_US.plist file.

Handling Additional Formats Using Category Classes

Assume that you've written and delivered your app, and for some reason, your client wants to move the server implementation to a Windows-based system, and the server now sends you XML data instead of JSON. With this current architecture in place, it's easy to add an additional format parsing to your model. The recommended way to do so is to write a category extension on your model that has a method to convert XML to dictionaries. In short, write a method in your category extension to convert an XML tree into a NSMutableDictionary and pass this dictionary to the initWithDictionary: method, which you previously wrote. Category classes like this provide a very powerful way to extend and add features to your existing implementation without creating unwanted side effects.

Tips to Improve Performance on iOS

The best tip for improving performance for a web service-based app is to avoid sending data that's not immediately necessary. Unlike a web-based app, an iPhone app has very limited bandwidth, and in most cases, it will be connected to a 3G network. Trying to implement techniques like prefetching contents for what could be the user's next page will only slow down your app.

Avoid multiple small AJAX-like API calls. In the "Creating the RESTfulEngine" section, earlier in this chapter, you learned that `MKNetworkEngine` sets the `networkQueue` to run six concurrent operations because most servers don't allow more than six parallel HTTP connections from a single IP address. Running more than six operations will only result in the seventh and subsequent operations timing out.

On a 3G network, at least at the time of this writing, most network operators throttle the bandwidth and limit the number of outbound connections from a mobile device to two, usually one on an EDGE connection. `MKNetworkKit` automatically changes the number of concurrent operations based on the network that the device is connected to. In case you aren't using MKNetworkKit, you should check for reachability notifications using the Reachability classes provided by Apple and change the queue size dynamically as and when the connectivity changes. Again, this count of two on 3G and one on EDGE is not absolute, and you need to test your customer base's network and use the results accordingly.

If you have control over the server development, the following tips can help to get the best out of the iOS app you develop:

- A server that caters to a web-based client should almost always have multiple small web service calls that are usually performed using AJAX. On iOS, it's best to avoid these APIs and possibly use or develop a custom API that gives more customized data per call.

- Unlike a browser, most carrier networks throttle the number of parallel data connections. Again, it's safe to assume that you run not more than one network operation on an EDGE connection, not more than two parallel network operations on a 3G network, and not more than six on a Wi-Fi connection.

Summary

In this chapter you learned how to architect an iOS application that uses a web service. The chapter also presented the different data exchange formats and ways to parse them in Objective-C, and you learned a very powerful method of processing responses from a RESTful service using Objective-C's powerful method, KVC. You then learned about using queues for handing concurrent requests and how to maximize performance by altering the maximum concurrent operations on the queue-based available network.

Further Reading

Apple Documentation

The following documents are available in the iOS Developer Library at `developer.apple.com` or through the Xcode Documentation and API Reference.

Reachability

Apple XMLPerformance Sample Code

NSXMLDocument Class Reference

Other Resources

Kumar, Mugunth. "MKNetworkKit Documentation"
`http://mknetworkkit.com`

Callahan, Kevin. "Mac App Store" (2011)
`http://itunes.apple.com/gb/app/accessorizer/id402866670?mt=12`

Cocoanetics. "JSON versus Plist, the Ultimate Showdown" (2011)
`www.cocoanetics.com/2011/03/json-versus-plist-the-ultimate-showdown/`

How To Choose The Best XML Parser for Your iPhone Project
`http://www.raywenderlich.com/553/how-to-chose-the-best-xml-parser-for-your-iphone-project`

Crockford, Douglas. "RFC 4627. 07 01" (2006)
`http://tools.ietf.org/html/rfc4627`

W3C. "Web Services Architecture. 2 11" (2004)
`www.w3.org/TR/ws-arch/#relwwwrest`

Brautaset, Stig. "JSON Framework 1 1" (2011)
`http://stig.github.com/json-framework/`

Wight, Jonathan. "TouchCode/TouchJSON. 1 1" (2011)
`https://github.com/TouchCode/TouchJSON`

Gabriel. "YAJL-ObjC" (2011)
`https://github.com/gabriel/yajl-objc`

Johnezang. "JSONKit" (2011)
`https://github.com/johnezang/JSONKit`

"mbrugger json-benchmarks" on GitHub
`https://github.com/mbrugger/json-benchmarks/`

Chapter 15

Batten the Hatches with Security Services

iOS is likely the first platform most developers encounter that employs a true least-privilege security model. Most modern operating systems employ some kind of privilege separation, allowing different processes to run with different permissions, but it is almost always used in a very rough way. Most applications on UNIX, OS X, and Windows either run as the current user, or run as an administrative user that can do nearly anything. Attempts to segment privileges further, whether with Security Enhanced Linux (SELinux) or Windows User Account Control (UAC), have generally led developers to revolt. The most common questions about SELinux are not how to best develop for it, but how to turn it off.

> With the Mac App Store, and particularly OS X 10.8, Apple has expanded some of iOS's least privilege approach to the desktop. Time will tell if it's successful.

Coming from these backgrounds, developers tend to be shocked when encountering the iOS security model. Rather than ensure maximal flexibility, Apple's approach has been to give developers the fewest privileges it can and see what software developers are *incapable* of making with those privileges. Then Apple provides the fewest additional privileges that allow the kinds of software it wants for the platform. This approach can be very restrictive on developers, but it's also kept iOS quite stable and free of malware. Apple is unlikely to change its stance on this approach, so understanding and dealing with the security model is critical to iOS development.

This chapter shows the way around the iOS security model, dives into the numerous security services that iOS offers, and provides the fundamentals you need to really understand Apple's security documentation. Along the way, you'll gain a deeper understanding of how certificates and encryption work in practice, so that you can leverage these features to really improve the security of your products.

The code for this chapter is available in the online sample code. A clear-cut project called `FileExplorer` is also available that enables you to investigate the public parts of the file system.

Understanding the iOS Sandbox

The heart of the iOS security model is the *sandbox*. When an application is installed, it's given its own home directory in the file system, readable only by that application. This makes it difficult to share information between applications, but also makes it difficult for malicious or poorly written software to read or modify your data.

Applications are not separated from each other using standard UNIX file permissions. All applications run as the same user ID (501, `mobile`). Calling `stat` on another application's home directory fails, however, because of operating system restrictions. Similar restrictions prevent your application from reading `/var/log` while allowing access to `/System/Library/Frameworks`.

Within your sandbox, there are four important top-level directories: your `.app` bundle, `Documents`, `Library`, and `tmp`. Although you can create new directories within your sandbox, it's not well defined how iTunes will deal with them. I recommend keeping everything in one of these top-level directories. You can always create subdirectories under `Library` if you need more organization.

Your `.app` bundle is the package built by Xcode and copied to the device. Everything within it is digitally signed, so you can't modify it. In particular, this includes your `Resources` directory. If you want to modify files that you install as part of your bundle, you'll need to copy them elsewhere first, usually somewhere in `Library`.

The `Documents` directory is where you store user-visible data, particularly files like word-processing documents or drawings that the user assigns a filename. These files can be made available to the desktop through file sharing if `UIFileSharingEnabled` is turned on in `Info.plist`.

The `Library` directory stores files that shouldn't be directly visible to users. Most files should go into `Library/Application Support`. These files are backed up, so if you want to avoid that, you can attach the attribute `NSURLIsExcludedFromBackupKey` to the files using the method `NSURL setResourceValue:forKey:error:`.

> The addition of `setResourceValues:error:` to `NSURL` was a very strange move by Apple in iOS 5. `NSURL` represents a URL, not the resource at that URL. This method makes more sense as part of `NSFileManager`. I've opened a radar on this, and hopefully it will become more consistent.

The `Library/Caches` directory is special because it isn't backed up, but it is preserved between application upgrades. This is where you want to put most things you don't want copied to the desktop.

The `tmp` directory is special because it's neither backed up nor preserved between application upgrades, which makes it ideal for temporary files, as the name implies. The system may also delete files in a program's `tmp` directory when that program isn't running.

When considering the security of the user's data, backups are an important consideration. Users may choose whether to encrypt the iTunes backup with a password. If there is data that shouldn't be stored unencrypted on the desktop machine, you store it in the keychain (see the "Using Keychains" section, later in this chapter). iTunes backs up the keychain only if backup encryption is enabled.

If you have information you don't want the user to have access to, you can store it in the keychain or in `Library/Caches` because these are not backed up. This is weak protection, however, because the user can always jailbreak the phone to read any file or the keychain. There is no particular way to prevent the owner of a device from reading data on that device. iOS security is about protecting the user from attackers, not about protecting the application from the user.

Securing Network Communications

The greatest risk to most systems is their network communication. Attackers don't need access to the device, only to the device's network. The most dangerous areas are generally coffee shops, airports, and other public Wi-Fi networks. It's your responsibility to make sure that the user's information is safe, even on hostile networks.

The first and easiest solution is to use Hypertext Transfer Protocol Secure (HTTPS) for your network communication. Most iOS network APIs automatically handle HTTPS, and the protocol eliminates many of the easiest attacks. In the simplest deployment, you put a self-signed certificate on the web server, turn on HTTPS, and configure `NSURLConnection` to accept untrusted certificates, as discussed shortly. This is still vulnerable to several kinds of attacks, but it's easy to deploy and addresses the most basic attacks.

How Certificates Work

Hopefully, you have encountered public-private key infrastructure (PKI) systems before. This section gives a quick overview of the technology, and then discusses how it affects the security of your application.

Asymmetric cryptography is based on the mathematical fact that you can find two very large numbers (call them A and B) that are related in such a way that anything encrypted with one can be decrypted with the other, and vice versa. Key A cannot decrypt things that key A encrypted, nor can key B decrypt things that key B encrypted. Each can decrypt only the other's encrypted data (called *ciphertext*). There is no real difference between key A and key B, but for the purposes of public key cryptography, one is called the *public key*, which generally everyone is allowed to know, and the other is called the *private key*, which is secret. You can use a public key to encrypt data such that only a computer with the private key can decrypt it. This is an important property that is used repeatedly in public key systems. If you want to prove that some entity (person or machine) has the private key, you make up a random number, encrypt it with the entity's public key, and send it. That entity decrypts the message with the entity's private key and sends it back to you. Because only the private key could have decrypted the message, the entity you're communicating with must have the private key.

This property also allows you to *digitally sign* data. Given some data, you first hash it with some well-known hashing algorithm and then encrypt it with your private key. The resulting ciphertext is the signature. To validate the signature, you hash the data again with the same algorithm, decrypt the signature using the public key, and compare the hashes. If they match, you know the signature was created by some entity that had access to the private key.

Just because an entity has access to the private key doesn't prove it's who it says it is. You need to ask two questions. First, how well is the private key protected? Anyone with access to the private key can forge a signature. Second, how do you know that the public key you have is related to the entity you care about? If I approach you on the street and hand you a business card that says I'm the President of the United States, it hardly proves anything. I'm the one who handed you the business card. Similarly, if a server presents you with a public key that claims to be for `www.apple.com`, why should you believe it? This is where a *certificate chain* comes in, and it's relevant to both questions.

A certificate is made up of a public key, some metadata about the certificate (more on that later), and a collection of signatures from other certificates. In most cases, there is a short chain of certificates, each signing the one below it. In very rare cases, there may be multiple signatures on one certificate. An example of a certificate chain is shown in Figure 15-1.

Figure 15-1 The certificate chain for daw.apple.com

In this example, the server `daw.apple.com` presents a certificate that includes its own public key, signed by an intermediate certificate from VeriSign, which is signed by a root certificate from VeriSign. Mathematically, you can determine that the controllers of each of these certificates did sign the next certificate in the chain, but why would you trust any of them? You trust them because Apple trusts the VeriSign root certificate, which has signed the intermediate certificate, which has signed the Apple certificate. Apple ships the VeriSign root certificate in the trusted root store of every iOS device, along with more than a hundred other trusted root certificates. The *trusted root store* is a list of certificates that is treated as explicitly trustworthy. Explicitly trusted certificates are called *anchors*. You can set your own anchors if don't want to trust Apple's list.

This brings you to the much-misused term *self-signed certificate*. For cryptographic reasons, every certificate includes a signature from itself. A certificate that has only this signature is called self-signed. Often, when people talk about a self-signed certificate, they mean a certificate that you shouldn't trust. But the VeriSign root certificate is a self-signed certificate, and it's one of the most trusted certificates in the world. Every root certificate, by definition, is a self-signed certificate. What's the difference? It isn't how many signatures a certificate has in its chain that matters, but how well all of the private keys in the chain are protected, and whether the identity of the owner has been authenticated.

If you generate your own self-signed certificate and protect the private key very well, then that's more secure than a certificate that VeriSign issues you. In both cases, you're dependent on protecting your private key, but in the latter case, you also have to worry about VeriSign protecting *its* private key. VeriSign spends a lot of money and effort doing that, but protecting two keys is always more risky than protecting just one of them.

This isn't to say that commercial certificates from VeriSign, DigiTrust, and other providers are bad. But you don't get a commercial certificate to improve the security of your system. You get one for convenience because the commercial certs are already in the root key store. But remember, you control the root key store in your own application. This leads to a surprising fact: *There is no security reason to purchase a commercial certificate to secure your application's network protocol to your own server.*

Commercial certificates are valuable only for websites visited by browsers or other software you don't control. Generating your own certificate and shipping the public key in your application is marginally *more* secure than using a commercial certificate. If you already have a commercial certificate for your server, using it for your application's network protocol is somewhat more convenient; it's just not more secure. This is not to say that it's okay to trust random certificates (that is, turning off certificate validation). It's to say that it's slightly better to trust only *your* certificates than to trust commercial certificates.

Certificates can be corrupt or not, valid or not, and trusted or not. These are separate attributes that need to be understood individually. The first question is whether a certificate is corrupt. A certificate is *corrupt* if it doesn't conform to the X.509 data format or if its signatures are incorrectly computed. A corrupt certificate should never be used for anything, and the iOS certificate function generally rejects them automatically.

> **X.509 refers to the data format specification and semantics originally defined by ITU-T** (`www.itu.int/ITU-T`). **The current version (v3) is defined by IETF RFC 5280** (`www.ietf.org/rfc/rfc5280.txt`).

Checking Certificate Validity

Given that a certificate is not corrupt, is it valid? Certificates contain a great deal of metadata about the public key they contain. The public key is just a very large number. It doesn't represent anything by itself. It's the metadata that gives that number meaning. A certificate is *valid* if the metadata it presents is consistent and appropriate for the requested use.

The most important piece of metadata is the subject. For servers, this is generally the fully qualified domain name (FQDN), such as `www.example.org`. The first test of validity is a name match. If you walk into a bank and identify yourself as "John Smith," you might be asked for your driver's license. If you hand over a license that says "Susan Jones," that would not help in identifying you no matter how authentic the driver's license. Similarly, if you're visiting a site named `www.example.org` and the site presents a certificate with a common name `www.badguy.com`, you should generally reject it. Unfortunately, it's not always that simple.

What if you visit `example.org` and it presents a certificate that says `www.example.org`? Should you accept that certificate? Most people would assume that `example.org` and `www.example.org` refer to the same server (which may or may not be true), but certificates use a simple string match. If the strings don't match, the certificate is invalid. Some servers present wild card certificates with subjects like `*.example.org`, and iOS will accept those, but there are still some cases when it will reject a certificate because of a name mismatch you believe it should accept. Unfortunately, iOS doesn't make this easy to manage, but it can be done.

The primary tool for determining whether to accept a certificate is the `NSURLConnection` delegate method `connection:willSendRequestForAuthenticationChallenge:`. In this method, you're supposed to determine whether you're willing to authenticate to this server and if so to provide the credentials. The following code authenticates to any server that presents a non-corrupt certificate, whether or not the certificate is valid or trusted:

```
- (void)connection:(NSURLConnection *)connection
  willSendRequestForAuthenticationChallenge:
  (NSURLAuthenticationChallenge *)challenge
{
  SecTrustRef trust = challenge.protectionSpace.serverTrust;
  NSURLCredential *cred;
  cred = [NSURLCredential credentialForTrust:trust];
  [challenge.sender useCredential:cred
        forAuthenticationChallenge:challenge];
}
```

This code extracts the trust object, discussed later, and creates a credential object for it. HTTPS connections always require a credential object, even if you're not passing credentials to the server.

In the next example, you're trying to connect to the IP address 72.14.204.113, which is `encrypted.google.com`. The certificate you receive is `*.google.com`, which is a mismatch. The string `72.14.204.113` doesn't include the string `.google.com`. You decide to accept any trusted certificate that includes `google.com` in its subject. To compile this example, you need to link `Security.framework` into your project.

ConnectionViewController.m (Connection)

```
- (void)connection:(NSURLConnection *)connection
  willSendRequestForAuthenticationChallenge:
  (NSURLAuthenticationChallenge *)challenge
{
  NSURLProtectionSpace *protSpace = challenge.protectionSpace;
  SecTrustRef trust = protSpace.serverTrust;
  SecTrustResultType result = kSecTrustResultFatalTrustFailure;

  OSStatus status = SecTrustEvaluate(trust, &result);
  if (status == errSecSuccess &&
      result == kSecTrustResultRecoverableTrustFailure) {
    SecCertificateRef cert = SecTrustGetCertificateAtIndex(trust,
                                                           0);
    CFStringRef subject = SecCertificateCopySubjectSummary(cert);

    NSLog(@"Trying to access %@. Got %@.", protSpace.host,
          subject);
    CFRange range = CFStringFind(subject, CFSTR(".google.com"),
                                 kCFCompareAnchored|
                                 kCFCompareBackwards);
    if (range.location != kCFNotFound) {
      status = RNSecTrustEvaluateAsX509(trust, &result);
    }
    CFRelease(subject);
```

```
   }

   if (status == errSecSuccess) {
      switch (result) {
        case kSecTrustResultInvalid:
        case kSecTrustResultDeny:
        case kSecTrustResultFatalTrustFailure:
        case kSecTrustResultOtherError:
// We've tried everything:
        case kSecTrustResultRecoverableTrustFailure:
           NSLog(@"Failing due to result: %lu", result);
           [challenge.sender cancelAuthenticationChallenge:challenge];
           break;

        case kSecTrustResultProceed:
        case kSecTrustResultConfirm:
        case kSecTrustResultUnspecified: {
           NSLog(@"Success with result: %lu", result);
           NSURLCredential *cred;
           cred = [NSURLCredential credentialForTrust:trust];
           [challenge.sender useCredential:cred
                 forAuthenticationChallenge:challenge];
        }
        break;

        default:
           NSAssert(NO, @"Unexpected result from trust evaluation:%ld",
                 result);
           break;
      }
   }
   else {
      // Something was broken
      NSLog(@"Complete failure with code: %lu", status);
      [challenge.sender cancelAuthenticationChallenge:challenge];
   }
}
```

In this routine, you're passed a challenge object and extract the trust object. You evaluate the trust object (`SecTrustEvaluate`) and receive a recoverable failure. Typically, a recoverable failure is something like a name mismatch. You fetch the certificate's subject and determine whether it's "close enough" (in this case, checking if it includes `.google.com`). If you're okay with the name you were passed, you reevaluate the certificate as a simple X.509 certificate rather than as part of an SSL handshake (that is, you evaluate it while ignoring the hostname). This is done with a custom function `RNSecTrustEvaluateAsX509`, shown here.

```
static OSStatus RNSecTrustEvaluateAsX509(SecTrustRef trust,
                                         SecTrustResultType *result
                                         )
{
   OSStatus status = errSecSuccess;

   SecPolicyRef policy = SecPolicyCreateBasicX509();
```

(continued)

```
SecTrustRef newTrust;
CFIndex numberOfCerts = SecTrustGetCertificateCount(trust);
CFMutableArrayRef certs;
certs = CFArrayCreateMutable(NULL,
                             numberOfCerts,
                             &kCFTypeArrayCallBacks);
for (NSUInteger index = 0; index < numberOfCerts; ++index) {
  SecCertificateRef cert;
  cert = SecTrustGetCertificateAtIndex(trust, index);
  CFArrayAppendValue(certs, cert);
}

status = SecTrustCreateWithCertificates(certs,
                                        policy,
                                        &newTrust);
if (status == errSecSuccess) {
  status = SecTrustEvaluate(newTrust, result);
}

CFRelease(policy);
CFRelease(newTrust);
CFRelease(certs);

return status;
}
```

This function creates a new trust object by copying all the certificates from the original trust object created by the URL loading system . This trust object uses the simpler X.509 policy, which only checks the validity and trust of the certificate itself, without considering the hostname as the original SSL policy does.

A certificate may also be invalid because it has expired. Unfortunately, while you can reevaluate the certificate using any date you want using `SecTrustSetVerifyDate`, there is no easy, public way to determine the validity dates for the certificate. The following private methods allow you to work out the valid range:

```
CFAbsoluteTime SecCertificateNotValidBefore(SecCertificateRef);
CFAbsoluteTime SecCertificateNotValidAfter(SecCertificateRef);
```

> As with all private methods, these may change at any time, and may be rejected by Apple. The only other practical way to parse the certificate is to export it with `SecCertificateCopyData` and parse it again using OpenSSL. Building and using OpenSSL on iOS is beyond the scope of this book. Search the Web for "OpenSSL iOS" for several explanations on how to build this library.

After evaluating the trust object, the final result will be a `SecTrustResultType`. Several results represent "good" or "possibly good" certificates:

▪ `kSecTrustResultProceed`—The certificate is valid, and the user has explicitly accepted it.

▪ `kSecTrustResultConfirm`—The certificate is valid, and you should ask the user whether to accept it.

- `kSecTrustResultUnspecified`—The certificate is valid, and the user has not explicitly accepted or rejected it. Generally, you accept it in this case.

- `kSecTrustResultRecoverableTrustFailure`—The certificate is invalid, but in a way that may be acceptable, such as a name mismatch, expiration, or lack of trust (such as a self-signed certificate).

The following results indicate that the certificate should not be accepted:

- `kSecTrustResultDeny`—The certificate is valid, and the user has explicitly rejected it.

- `kSecTrustResultInvalid`—The validation was unable to complete, likely because of a bug in your code.

- `kSecTrustResultFatalTrustFailure`—The certificate itself was defective or corrupted.

- `kSecTrustResultOtherError`—The validation was unable to complete, likely because of a bug in Apple's code. You should never see this error.

Determining Certificate Trust

So far, you've found out how to determine whether a certificate is valid, but that doesn't mean it's trusted. Returning to the example of identifying yourself at the bank, if you present your Metallica fan club membership card, it probably won't be accepted as identification. The bank has no reason to believe that your fan club has done a good job making sure you're who you say you are. That's the same situation an application faces when presented with a certificate signed by an unknown authority.

To be trusted, a certificate must ultimately be signed by one of the certificates in the trust object's list of *anchor certificates*. Anchor certificates are those certificates that are explicitly trusted by the system. iOS ships with more than a hundred of them from companies and government agencies. Some are global names like VeriSign and DigiTrust; others are more localized like QuoVadis and Vaestorekisterikeskus. Each of these organizations went through a complex audit process and paid significant amounts of money to be in the root store, but that doesn't mean your application needs to trust them.

If you generate your own certificate, you can embed the public key in your application and configure your trust object to accept only that certificate or certificates signed by it. This gives you greater control over your security and can save you some money.

For this example, you create a self-signed root certificate.

1. Open Keychain Access.

2. Select Keychain Access menu⇨Certificate Assistant⇨Create a Certificate.

3. Enter any name you like, set the Identity Type to Self Signed Root, set the Certificate Type to SSL Client, and create the certificate. You receive a warning that this is a self-signed certificate. That is the intent of this process, so you click Continue. Your newly created certificate displays a warning that "This root certificate is not trusted." That's also as expected because it isn't in the root keychain.

4. Back in the Keychain Access window, select the login keychain and select the category Certificates.

5. Find your certificate and drag it to the desktop to export it. This file includes only the public key. Keychain does not export the private key by default. Drag the public key file into your Xcode project.

You can test that a certificate you've received is signed by your certificate as follows:

```
SecTrustRef trust = ...; // A trust to validate
NSError *error;
NSString *path = [[NSBundle mainBundle] pathForResource:@"MyCert"
                                                 ofType:@"cer"];
NSData *certData = [NSData dataWithContentsOfFile:path
                                          options:0
                                            error:&error];

SecCertificateRef certificate;
certificate = SecCertificateCreateWithData(NULL,
                              (__bridge CFDataRef)certData);
CFArrayRef certs = CFArrayCreate(NULL,
                          (const void**)&certificate,
                          1,
                          &kCFTypeArrayCallBacks);

SecTrustSetAnchorCertificates(trust, certs);

CFRelease(certs);
CFRelease(certificate);
```

You load the certificate from your resource bundle into an `NSData`, convert it into a `SecCertificate`, and set it as the anchor for the trust object. The trust object will now accept only the certificates passed to `SecTrustSetAnchorCertificates` and will ignore the system's anchors. If you want to accept both, you can use `SecTrustSetAnchorCertificatesOnly` to reconfigure the trust object.

Using these techniques, you can correctly respond to any certificate in your `connection:willSendReques tForAuthenticationChallenge:` method and control which certificates you accept or reject.

Employing File Protection

iOS provides hardware-level encryption of files. Files marked for protection are encrypted using a per-device key, which is encrypted using the user's password or PIN. Ten seconds after the device is locked, the unencrypted per-device key is removed from memory. When the user unlocks the device, the password or personal identification number (PIN) is used to decrypt the per-device key again, which is then used to decrypt the files.

The weakest link in this scheme is the user's password. On an iPhone, users almost exclusively use a 4-digit PIN, which offers only 10,000 combinations (far fewer are used in practice). In May 2011, ElcomSoft Co. Ltd demonstrated that it could brute-force a 4-digit PIN in about 20–40 minutes. This doesn't protect against forensics or device theft, but does protect against attackers who have access to the device only for a few minutes. On iPad, typing a real password is much more convenient, so the security is similar to file encryption on a laptop.

For a developer, the specifics of the iOS encryption scheme aren't critical. The scheme is effective enough for users to expect it on any application that holds sensitive information.

You can configure the protection of individual files that you create with NSFileManager or NSData. The options, shown in the following list, have slightly different names. NSFileManager applies string attributes to the file, whereas NSData uses numeric options during creation, but the meanings are the same. The FileManager constants begin with NSFileProtection..., and the NSData constants begin with NSDataWritingFileProtection....

- ...None—The file is not protected and can be read or written at any time. This is the default value.

- ...Complete—Any file with this setting is protected ten seconds after the device is locked. This is the highest level of protection, and the setting you should generally use. Files with this setting may not be available when your program is running in the background. When the device is unlocked, these files are unprotected.

- ...CompleteUnlessOpen—Files with this setting are protected ten seconds after the device is locked unless they're currently open. This allows your program to continue accessing the file while running in the background. When the file is closed, it will be protected if the device is locked.

- ...CompleteUntilFirstUserAuthentication—Files with this setting are protected only between the time the device boots and the first time the user unlocks the device. The files are unprotected from that point until the device is rebooted. This allows your application to open existing files while running in the background. You can create new files using ...CompleteUnlessOpen. This is better than the None setting, but should be avoided if at all possible because it provides very limited protection.

To create a new file with file protection turned on, convert it to an NSData and then use writeToFile:options:error:. This is preferable to creating the file and then using NSFileManager to set its protection attribute.

```
[data writeToFile:dataPath
        options:NSDataWritingFileProtectionComplete
          error:&writeError];
```

To create a protected file in the background, you can apply the option ...CompleteUnlessOpen, which allows you to read as long as it's open when the device locks. You should generally avoid this option unless you're actually in the background. The easiest way to create a protected file when you may or may not be in the background is as follows:

```
[data writeToFile:path
        options:NSDataWritingFileProtectionComplete
          error:&error] ||
[data writeToFile:path
        options:NSDataWritingFileProtectionCompleteUnlessOpen
          error:&error];
```

If you use this technique, upgrade your file protection at startup with a routine like this:

```
- (void)upgradeFilesInDirectory:(NSString *)dir
                          error:(NSError **)error {
  NSFileManager *fm = [NSFileManager defaultManager];
  NSDirectoryEnumerator *dirEnum = [fm enumeratorAtPath:dir];
  for (NSString *path in dirEnum) {
    NSDictionary *attrs = [dirEnum fileAttributes];
```

(continued)

```
    if (![[attrs objectForKey: NSFileProtectionKey]
        isEqual:NSFileProtectionComplete]) {
      attrs = [NSDictionary dictionaryWithObject:
            NSFileProtectionComplete forKey:NSFileProtectionKey];
      [fm setAttributes:attrs ofItemAtPath:path error:error];
    }
  }
}
```

If your application needs to know whether protected data is available, you can use one of the following:

- Implement the methods `applicationProtectedDataWillBecomeUnavailable:` and `applicationProtectedDataDidBecomeAvailable:` in your application delegate.

- Observe the notifications `UIApplicationProtectedDataWillBecomeUnavailable` and `UIApplicationProtectedDataDidBecomeAvailable` (these constants lack the traditional `Notification` suffix).

- Check `[[UIApplication sharedApplication] protectedDataAvailable]`.

For foreground-only applications, file protection is very easy. Because it's so simple and it's hardware-optimized, unless you have a good reason not to, you'll generally want to protect your files. If your application runs in the background, you need to give more careful thought to how to apply file protection, but still be sure to protect all sensitive information as well as possible.

Using Keychains

File protection is intended to protect *data*. Keychain is intended to protect *secrets*. In this context, a secret is a small piece of data used to access other data. The most common secrets are passwords and private keys.

The keychain is protected by the operating system and is encrypted when the device is locked. In practice, it works very similarly to file protection. Unfortunately, the Keychain API is anything but friendly. Many people have written wrappers around the Keychain API. However, my recommendation is Apple's `KeyChainItemWrapper` from the `GenericKeychain` sample code. This is what I'll discuss in this section after a brief introduction to the low-level data structures.

`KeyChainItemWrapper` has some problems, most significantly that it's not ARC-compliant. You can fix that by turning off ARC for the `KeyChainItemWrapper.m` or by adding the few `__bridge` casts that are required. If you search the Web for "GenericKeyChain ARC," you'll find several posts by people who have already done this. It also isn't as flexible as we'd like. A number of potential replacements are available. None of them are good enough for us to recommend yet, but this will likely change. Watch `iosptl.com` for updates.

An item in the keychain is called a `SecItem`, but it's stored in a `CFDictionary`. There is no `SecItemRef` type. There are five classes of `SecItem`: generic password, Internet password, certificate, key, and identity. In most cases, you want to use a generic password. Many problems come from developers trying to use an

Internet password, which is more complicated and provides little benefit unless you're implementing a web browser. `KeyChainItemWrapper` uses only generic password items, which is one reason I like it. Storing private keys and identities is rare in iOS applications and won't be discussed in this book. Certificates that contain only public keys should generally be stored in files rather than in the keychain.

You eventually need to search the keychain for the item you want. There are many pieces of the key to search for, but the best way is to assign your own identifier and search for that. Generic password items include the attribute `kSecAttrGeneric`, which you can use to store your identifier. This is how `KeyChainItemWrapper` operates.

Keychain items have several searchable *attributes* and a single encrypted *value*. For a generic password item, some of the more important attributes are the account (`kSecAttrAccount`), service (`kSecAttrService`), and identifier (`kSecAttrGeneric`). The value is generally the password.

So with that background, it's time to see how to use `KeychainItemWrapper`. First, as shown in the following code, you create one with `initWithIdentifier:accessGroup:`. I discuss access groups in the section "Sharing Data with Access Groups," but for now leave it `nil`.

```
KeychainItemWrapper *
wrapper = [[KeychainItemWrapper alloc]
                          initWithIdentifier:@"MyKeychainItem"
                                accessGroup:nil];
```

You can now read from and write to `wrapper` like you would an `NSDictionary`. It automatically synchronizes with the keychain. The `__bridge` casts are to allow you to pass Core Foundation constants to a Cocoa method under ARC.

```
id kUsernameKey = (__bridge id)kSecAttrAccount;
id kPasswordKey = (__bridge id)kSecValueData;

NSString *username = [wrapper objectForKey:kUsernameKey];
[wrapper setObject:password forKey:kPasswordKey];
```

> Note that I am using a Core Foundation type (`kSecAttrAccount`) where an `id` is expected (`objectForKey:`), with no type casting and no `__bridge`. This is a new feature.

`KeychainItemWrapper` caches reads, but doesn't write. Writing to the keychain can be expensive, so you don't want to do it too often. The keychain is not a place to store sensitive data that changes often. That should be written in an encrypted file, as described in the earlier section, "Employing File Protection."

Sharing Data with Access Groups

The iOS sandbox creates a significant headache for application suites. If you have multiple applications that work together, there is no easy way to share information between them. Of course, you can save the data on a server, but the user still needs to enter credentials for each of your applications.

iOS offers a solution to this with access groups. Multiple applications can share keychain data as long as they share an access group. To create an access group, open the target in Xcode. At the bottom of the Summary pane, enable Entitlements. Then add a new keychain access group, as shown in Figure 15-2.

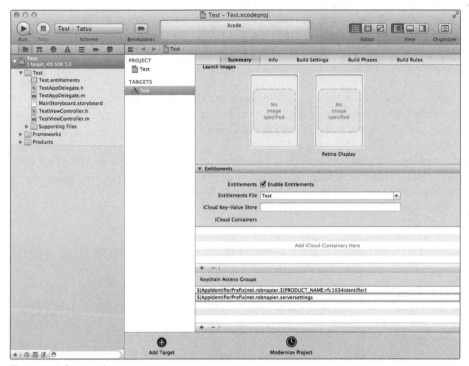

Figure 15-2 Creating the serversettings access group

Utilizing `KeychainItemWrapper`, you can use this access group by passing the identifier to `initWith Identifier:accessGroup:`. For more information on this feature, see the documentation for `SecItemAdd` in the *Keychain Services Reference* at `developer.apple.com`.

Storing small pieces of sensitive information in the keychain is quite easy with `KeychainItemWrapper`. Unless you have a very good reason not to, I recommend using it instead of directly accessing the Keychain API, which is much more complicated.

Using Encryption
Most of the time, iOS handles all your encryption needs for you. It automatically encrypts and decrypts HTTPS for network traffic and manages encrypted files using file protections. If you have certificates, `SecKeyEncrypt` and `SecKeyDecrypt` handle asymmetric (public/private key) encryption for you.

But what about simple, symmetric encryption using a password? iOS has good support for this, but limited documentation. The available documentation is in `/usr/include/CommonCrypto`. Most of it assumes that you have some background in cryptography and doesn't warn you of common mistakes. This unfortunately has led to a lot of bad AES code examples on the Internet. In an effort to improve things, I developed `RNCryptor` as an easy-to-use encryption library based on CommonCrypto. In this section, I discuss many of the underlying details of how `RNCryptor` works and why. This will let you modify `RNCryptor` with confidence, or enable you to build your own solution. If you just want a prebuilt solution, skip this section and read the `RNCryptor` docs. You can find the source for `RNCryptor` at `github.com/rnapier/RNCryptor`.

Overview of AES

The *Advanced Encryption Standard*, or AES, is a symmetric encryption algorithm. Given a key, it converts *plaintext* into *ciphertext*. The same key is used to convert ciphertext back into plaintext. Originally named Rijndael, in 2001 the algorithm was selected by the U.S. government as its standard for encryption.

It's a very good algorithm. Unless you need another algorithm for compatibility with an existing system, always use AES for symmetric encryption. The best cryptographers in the world have carefully scrutinized it, and it's hardware-optimized on iOS devices, making it extremely fast.

AES offers three key lengths: 128, 192, and 256 bits. There are slight differences in the algorithm for each length. Unless you have very specialized needs, I recommend AES-128. It offers an excellent tradeoff of security and performance, including time performance and battery life performance.

At its heart, AES is just a mathematical function that takes a fixed-length key and a 16-byte block and returns a different 16-byte block. This has several implications:

- AES uses a fixed-length key, not a variable-length password. You must convert passwords to keys in order to use them in AES. We explain this further in the next section, "Converting Passwords to Keys with PBKDF2."

- AES can only transform a 16-byte block into another 16-byte block. In order to work on plaintext that isn't exactly 16-bytes long, some mechanism outside of AES must be applied. See the sections "AES Mode and Padding" and "The Initialization Vector (IV)" for how AES works with data that is not exactly 16 bytes long.

- Every possible block of 16 bytes is legal as plaintext or as ciphertext. This means that AES provides no authentication against accidental or intentional corruption of the ciphertext. See the section "Authentication with HMAC" for how to add error detection to AES.

- Every possible key can be applied to every possible block. This means that AES provides no mechanism for detecting that the key was incorrect. See the section "Bad Passwords" for more information on this problem.

- Although AES supports three key lengths, it has only one block size. This sometimes creates confusion over the constant `kCCBlockSizeAES128`. The "128" in this constant refers to the block size, not the key size. So when using AES-192 or AES-256, you still pass `kCCBlockSizeAES128` as the block size.

Converting Passwords to Keys with PBKDF2

A *key* is not the same thing as a *password*. A key is a very large number used to encrypt and decrypt data. All possible keys for an encryption system are called its *key space*. A password is something a human can type. Long passwords that include spaces are sometimes called *passphrases*, but for simplicity, I just use the word "password" no matter the construction. If you try to use a password as an AES key, you significantly shrink the number of available keys. If the user selects a random 16-character password using the 94 characters on a standard keyboard, that only creates about a 104-bit key space, approximately one ten-millionth the size of the full AES key space. Real users select passwords from a much smaller set of characters. Worse yet, if the user has a password longer than 16 bytes (16 single-byte characters or 8 double-byte characters), you will throw away part of it when using AES-128.

You need a way to convert a password into a useable key that makes it as hard as possible on the attacker to search every possible password. The answer is a *password-based key derivation function*. Specifically, you will use PBKDF2, which is defined by RSA Laboratories' Public-Key Cryptography Standards (PKCS) #5. You don't need to know the internals of PBKDF2 or PKCS #5, but it's important to know the names because they show up in the documentation. What is important is that PBKDF2 converts a password into a key.

To use PBKDF2, you need to generate a *salt*, which is just a large random number. The standard recommends at least 64 bits. The salt is combined with the password to prevent identical passwords from generating identical keys. You then iterate through the PBKDF2 function a specific number of times, and the resulting data is your key. This process is called *stretching*. To decrypt the data, you need to preserve the salt and the number of iterations. Typically, the salt is saved with the encrypted data, and the number of iterations is a constant in your source code, but you can also save the number of iterations with the encrypted data.

The important fact here is that the salt, the number of iterations, and the final ciphertext are all public information. Only the key and the original password are secrets.

Generating the salt is easy. It's just a large random number. You can create it with a method like `randomDataOfLength:`, shown in the following code:

RNCryptor.m (RNCryptor)

```
const NSUInteger kPBKDFSaltSize = 8;

+ (NSData *)randomDataOfLength:(size_t)length {
  NSMutableData *data = [NSMutableData dataWithLength:length];

  int result = SecRandomCopyBytes(kSecRandomDefault,
                                  length,
                                  data.mutableBytes);
  NSAssert(result == 0, @"Unable to generate random bytes: %d",
           errno);

  return data;
}
...
NSData *salt = [self randomDataOfLength:kPBKDFSaltSize];
```

Originally, the standard called for 1,000 iterations of PBKDF2, but this has gone up as CPUs have improved. I recommend between 10,000 and 100,000 iterations on an iPhone 4 and 50,000 to 500,000 iterations on a modern MacBook Pro. The reason for the large number is to slow down brute-force attacks. An attacker generally tries passwords rather than raw AES keys because the number of practical passwords is much smaller. By requiring 10,000 iterations of the PBKDF2 function, the attacker must waste about 80ms per attempt on an iPhone 4. That adds up to 13 minutes of search time for a 4-digit PIN, and months or years to search for even a very simple password. The extra 80ms for a single key generation is generally negligible. Going up to 100,000 iterations adds nearly a second to key generation on an iPhone, but provides much better protection if the password guessing is done on a desktop, even if the password is very weak.

PBKDF2 requires a *pseudorandom function* (PRF), which is just a function that can generate a very long series of statistically random numbers. iOS supports various sizes of SHA-1 and SHA-2 for this. SHA256, which is a specific size of SHA-2, is often a good choice.

> RNCryptor uses SHA-1 as its PRF for historical compatibility reasons. This will likey change to SHA256 in the future.

Luckily, it's easier to use PBKDF2 than it is to explain it. The following method accepts a password string and salt data and returns an AES key.

RNCryptor.m (RNCryptor)

```
+ (NSData *)keyForPassword:(NSString *)password salt:(NSData *)salt
                settings:(RNCryptorKeyDerivationSettings)keySettings
{
  NSMutableData *derivedKey = [NSMutableData
                            dataWithLength:keySettings.keySize];

  int result = CCKeyDerivationPBKDF(keySettings.PBKDFAlgorithm,
                               password.UTF8String,
                               password.length,
                               salt.bytes,
                               salt.length,
                               keySettings.PRF,
                               keySettings.rounds,
                               derivedKey.mutableBytes,
                               derivedKey.length);

  // Do not log password here
  NSAssert(result == kCCSuccess,
    @"Unable to create AES key for password: %d", result);

  return derivedKey;
}
```

The password salt and number of iterations must be stored along with the cipher text.

AES Mode and Padding

AES is a block cipher, which means that it operates on a fixed-sized block of data. AES works on exactly 128 bits (16 bytes) of input at a time. Most things you want to encrypt, however, are not exactly 16 bytes long. Many things you want to encrypt aren't even a multiple of 16 bytes. In order to address this issue, you need to choose an appropriate mode.

A *mode* is an algorithm for chaining blocks together so that arbitrary data can be encrypted. Modes are applicable to any block cipher, including AES.

> **If you don't need the details on how AES modes work, just trust me that the right answer on iOS devices is CBC with PKCS#7 padding and skip to the section, "The Initialization Vector (IV)."**

The simplest mode is *electronic codebook* (ECB). *Never use this mode in an iOS app.* ECB is appropriate only for cases when you need to encrypt a huge number of random blocks. This is such a rare situation that you shouldn't even consider it for iOS apps. ECB simply takes each block and applies the block cipher (AES) and outputs the result. It is extremely insecure. The problem is that if two plaintext blocks are the same, the ciphertext will be the same. This leaks a lot of information about the plaintext and is subject to numerous attacks. Do not use ECB mode.

The most common block cipher mode is *cipher-block chaining* (CBC). The ciphertext of each block is XORed with the next plaintext block prior to encryption. This is a very simple mode, but very effective in solving the problems with ECB. It has two major problems: Encryption cannot be done in parallel, and the plaintext must be an exact multiple of the block size (16 bytes). Parallel encryption is not a major concern on iOS platforms. iPhone and iPad perform AES encryption on dedicated hardware and don't have sufficient hardware to parallelize the operation. The block size problem is dealt with by padding.

Padding is the practice of expanding plaintext to the correct length prior to encryption. It needs to be done in a way that the padding can be unambiguously removed after decryption. There are several ways to achieve this, but the only way supported by iOS is called PKCS #7 padding. This is requested with the option `kCCOptionPKCS7Padding`. In PKCS #7 padding, the encryption system appends *n* copies of the byte *n*. So, if your last block were 15 bytes long, it would append one 0x01 byte. If your last block were 14 bytes long, it would append two 0x02 bytes. After decryption, the system looks at the last byte and deletes that number of padding bytes. It also performs an error check to make sure that the last *n* bytes are actually the value *n*. This has surprising implications that we discuss in the section "Bad Passwords."

There is a special case if the data is exactly the length of the block. In order to be unambiguous, the PKCS #7 has to append a full block of 0x10.

PKCS #7 padding means that the ciphertext can be up to a full block longer than the plaintext. Most of the time this is fine, but in some situations, this is a problem. You may need the ciphertext to fit in the same space as the plaintext. There is a CBC-compatible method called ciphertext stealing that can do this, but it's not supported on iOS with CBC.

CBC is the most common cipher mode, making it the easiest to exchange with other systems. Unless you have a strong reason to use something else, use CBC. This is the mode recommended by Apple's security team, based on my discussions with them.

If padding is impossible in your situation, I recommend cipher feedback mode (CFB). It's less widely supported than CBC, but it is a fine mode and doesn't require padding. Output feedback (OFB) is also fine for the same reasons. I have no particular advice on choosing one over the other. They are both good.

Counter (CTR) mode is useful if you need to avoid the overhead of an initialization vector, but is easy to use insecurely. I discuss CTR, when to use it and how, in the section "The Initialization Vector (IV)."

XTS is a specialized mode for random-access data, particularly encrypted file systems. I don't have enough relevant information to recommend it. FileVault on Mac uses it, so Apple clearly has an interest in it.

All of the modes offered by iOS are unauthenticated. This means that modifying the ciphertext will not necessarily create errors. In most cases, it will just change the resulting plaintext. It's possible for attackers who know some of the contents of an encrypted document to modify it in predictable ways without the password. There are encryption modes (called *authenticated modes*) that protect against this, but none are supported by iOS. See the section "Authentication with HMAC," for more information.

The many encryption modes can be confusing, but in almost all cases, the most common solution is best: CBC with PKCS #7 padding. If padding is problematic for you, then I recommend CFB.

The Initialization Vector (IV)

As discussed in the earlier section "AES Mode and Padding," in chaining modes such as CBC, each block influences the encryption of the next block. This ensures that two identical blocks of plaintext will not generate identical blocks of ciphertext.

The first block is a special case because there's no previous block. Chaining modes allow you to define an extra block called the *initialization vector* (IV) to begin the chain. This is often labeled optional, but you need to always provide one. Otherwise, an all-zero block is used, and that leaves your data vulnerable to certain attacks.

As with the salt, the IV is just a random series of bytes that you save with the ciphertext and use during decryption.

```
iv = [self randomDataOfLength:kAlgorithmIVSize];
```

You then store this IV along with the ciphertext and use it during decryption. In some cases, adding this IV creates unacceptable overhead, particularly in high-performance network protocols and disk encryption. In that case, you still cannot use a fixed IV (such as NULL). You must use a mode that doesn't require a random IV. Counter mode (CTR) uses a nonce rather than an IV. A *nonce* is essentially a predictable IV. While chaining modes like CBC require that the IV be random, nonce-based modes require that the nonce be unique. This requirement is very important. In CTR mode, a given nonce/key pair must never, ever be reused. This is usually implemented by making the nonce a monotonically increasing counter starting at 0. If the nonce counter ever needs to be reset, then the key must be changed.

The advantage of a nonce is that it doesn't have to be stored with the ciphertext the way an IV must. If a key is chosen randomly at the beginning of a communication session, then both sides of the communication can use the current message number as the nonce. As long as the number of messages cannot exceed the largest possible nonce, this is a very efficient way to communicate.

The nonce uniqueness requirement is often more difficult to implement than it seems. Randomly selecting a nonce, even in a large nonce-space, may not be sufficient if the random number generator is not sufficiently random. A nonce-based mode with a non-unique nonce is completely insecure. IV-based modes, on the other hand, are weaker if the IV is reused, but they aren't completely broken.

In most iOS applications, the correct solution is CBC with PKCS #7 padding and a random IV sent along with the data.

Authentication with HMAC

None of the modes discussed so far protect against accidental or malicious modification of the data. AES does not provide any authentication. If an attacker knows that a certain string occurs at a certain location in your data, the attacker can change that data to anything else of the same size. This could change "$100" to "$500" or "bob@example.com" to "sue@example.net".

Many block cipher modes provide authentication. Unfortunately, iOS does not support any of them. To authenticate data, you need to do it yourself. The best way to do this is to add an *authentication code* to the data. The best choice on iOS is a hash-based message authentication code (HMAC).

An HMAC requires its own key, just like encryption. If you're using a password, you can use PBKDF2 to generate the HMAC key using a different salt. If you're using a random encryption key, you'll need to also generate a random HMAC key. If you use a salt, you'll include the HMAC salt with the ciphertext, just like the encryption salt. See the section "Converting Passwords to Keys with PBKDF2" for more information.

Because a HMAC can hash anything, it can be computed on the plaintext or on the ciphertext. It is best to encrypt first and then HMAC the ciphertext. This allows you to verify the data prior to decrypting it. It also protects against certain kinds of attacks.

In practice, the code required to HMAC is very similar to the encryption code. You either pass all the ciphertext to a single HMAC call, or you call the HMAC routines multiple times with blocks of ciphertext.

Bad Passwords

The lack of authentication has a commonly misunderstood side effect. AES cannot directly detect when you pass the wrong key (or password). Systems that use AES often appear to detect this, but that's actually just luck.

Most AES systems use CBC encryption with PKCS #7 padding. As I explained in the earlier section "AES Mode and Padding," the PKCS #7 padding scheme adds a predictable pattern to the end of the plaintext. When you decrypt, the padding provides a kind of error detection. If the plaintext does not end in a legal padding value, the decryptor returns an error. But if the last decrypted byte happens to be 0x01, this appears to be legal padding, and you won't be able to detect a bad password. For systems without padding (such as CFB or CTR), even this check is not possible.

If you use a password and PBKDF2 to generate your encryption key and HMAC key, the HMAC will give you password validation. Because the HMAC key is based on the password, and the password is wrong, the HMAC won't validate. But if you use a random encryption key and random HMAC key, you still won't be able to detect a bad password.

To some extent, this is a security feature. If the attacker cannot easily detect the difference between a correct and incorrect password, it's harder to break the encryption. In almost all cases where you want to report a bad password, it's because you're using PBKDF2 and HMAC will solve the problem. Otherwise, you may have to add error detection inside your plaintext to be certain.

Note that even with HMAC, you cannot easily distinguish a corrupted ciphertext from a bad password.

Performing One-Shot Encryption

That's most of the theory you need. iOS provides all the math functions, so you can ignore the implementation details.

The first example is one-shot encryption and decryption routines. These take an NSData and return an NSData. They use the convenience function CCCrypt from CommonCryptor.

The encryption routine accepts plaintext data and a password and returns ciphertext data, an IV, a salt, a HMAC salt, and a HMAC.

RNCryptManager.m (CryptPic)

```objc
#import <CommonCrypto/CommonCryptor.h>
const CCAlgorithm kAlgorithm = kCCAlgorithmAES128;
const NSUInteger kAlgorithmKeySize = kCCKeySizeAES128;
const NSUInteger kAlgorithmBlockSize = kCCBlockSizeAES128;
const NSUInteger kAlgorithmIVSize = kCCBlockSizeAES128;
const NSUInteger kPBKDFSaltSize = 8;
const NSUInteger kPBKDFRounds = 10000;   // ~80ms on an iPhone 4

+ (NSData *)encryptedDataForData:(NSData *)data
                        password:(NSString *)password
                              iv:(NSData **)iv
                            salt:(NSData **)salt
                        HMACSalt:(NSData **)HMACSalt
                            HMAC:(NSData **)HMAC
                           error:(NSError **)error {
  NSAssert(iv, @"IV must not be NULL");
  NSAssert(salt, @"salt must not be NULL");

  *iv = [self randomDataOfLength:kAlgorithmIVSize];
  *salt = [self randomDataOfLength:kPBKDFSaltSize];

  NSData *key = [self AESKeyForPassword:password salt:*salt];

  size_t outLength;
  NSMutableData *
  cipherData = [NSMutableData dataWithLength:data.length +
              kAlgorithmBlockSize];

  CCCryptorStatus
  result = CCCrypt(kCCEncrypt, // operation
                  kAlgorithm, // Algorithm
                  kCCOptionPKCS7Padding, // options
                  key.bytes, // key
                  key.length, // keylength
                  (*iv).bytes,// iv
                  data.bytes, // dataIn
```

(continued)

```
                      data.length, // dataInLength,
                      cipherData.mutableBytes, // dataOut
                      cipherData.length, // dataOutAvailable
                      &outLength); // dataOutMoved

    if (result == kCCSuccess) {
      cipherData.length = outLength;
    }
    else {
     // ... Handle Error ...
      return nil;
    }

    if (HMAC) {
      NSAssert(HMACSalt, @"HMAC salt must not be NULL if HMAC is passed.");
      *HMACSalt = [self randomDataOfLength:kPBKDFSaltSize];
      NSData *HMACKey = [self AESKeyForPassword:password salt:*HMACSalt];
      NSMutableData *
        HMACOut = [NSMutableData dataWithLength:CC_SHA256_DIGEST_LENGTH];
      CCHmac(kCCHmacAlgSHA256,
             [HMACKey bytes], [HMACKey length],
             [cipherData bytes], [cipherData length],
             [HMACOut mutableBytes]);
      *HMAC = HMACOut;
    }
    return cipherData;
}
```

The decryption routine accepts ciphertext data, a password, IV, salt, HMAC salt, and HMAC.

RNCryptManager.m (CryptPic)

```
+ (NSData *)decryptedDataForData:(NSData *)data
                        password:(NSString *)password
                              iv:(NSData *)iv
                            salt:(NSData *)salt
                        HMACSalt:(NSData *)HMACSalt
                            HMAC:(NSData *)HMAC
                           error:(NSError **)error {

    if (HMAC) {
      NSData *HMACKey = [self AESKeyForPassword:password salt:HMACSalt];
      NSMutableData *
        HMACOut = [NSMutableData dataWithLength:CC_SHA256_DIGEST_LENGTH];
      CCHmac(kCCHmacAlgSHA256,
             [HMACKey bytes],
             [HMACKey length],
             [data bytes],
             [data length],
             [HMACOut mutableBytes]);
      if (! [HMAC isEqualToData:HMACOut]) {
        // ... Handle error ...
```

```
      return nil;
   }
}
NSData *key = [self AESKeyForPassword:password salt:salt];

size_t outLength;
NSMutableData *
decryptedData = [NSMutableData dataWithLength:data.length];
CCCryptorStatus
result = CCCrypt(kCCDecrypt, // operation
                 kAlgorithm, // Algorithm
                 kCCOptionPKCS7Padding, // options
                 key.bytes, // key
                 key.length, // keylength
                 iv.bytes,// iv
                 data.bytes, // dataIn
                 data.length, // dataInLength,
                 decryptedData.mutableBytes, // dataOut
                 decryptedData.length, // dataOutAvailable
                 &outLength); // dataOutMoved

if (result == kCCSuccess) {
   [decryptedData setLength:outLength];
}
else {
   if (result != kCCSuccess) {
      // ... Handle Error ...
      return nil;
   }
}

return decryptedData;
}
```

Improving CommonCrypto Performance

The CCCrypt function is fairly straightforward. It has a lot of parameters, and you need to generate a key, but once you have your data in place, it's just one function call. As presented in the section "Performing One-Shot Encryption," however, CCCrypt requires enough memory to hold two copies of your plaintext. It also requires that all the plaintext be available when it gets started.

You can save half the memory by reusing the buffer in CCCrypt. The dataIn and dataOut parameters can point to the same buffer as long as it's as large as the ciphertext. For AES, that's the size of the plaintext plus one 16-byte block.

This still requires that all the plaintext be available in memory at the same time. That can be expensive for large files, especially on a mobile device. It also prevents you from decrypting as the data is read from the network. This is particularly useful in cases when you want to store data on an untrusted server. HTTPS protects it on the network but that doesn't help if you don't trust the server. It's better to use file protection locally on the device and use AES to protect the file remotely.

CCCrypt is just a convenience function around the normal CommonCrypto routines: CCCryptorCreate, CCCryptorUpdate, and CCCryptorFinal. In this example, you use these to handle encryption and decryption with NSStream objects. The full source code is available in the CryptPic sample code for this chapter.

This routine handles either encrypting or decrypting, based on the operation parameter (kCCEncrypt or kCCDecrypt). First, it reads or writes the IV and salt at the beginning of the stream. The _CM...Data methods are helpers for dealing with NSStream. They're available in the sample code.

RNCryptManager.m (CryptPic)

```
switch (operation) {
  case kCCEncrypt:
    // Generate a random IV for this file.
    iv = [self randomDataOfLength:kAlgorithmIVSize];
    salt = [self randomDataOfLength:kPBKDFSaltSize];

    if (! [outStream _CMwriteData:iv error:error] ||
        ! [outStream _CMwriteData:salt error:error]) {
      return NO;
    }
    break;
  case kCCDecrypt:
    // Read the IV and salt from the encrypted file
    if (! [inStream _CMgetData:&iv
                      maxLength:kAlgorithmIVSize
                          error:error] ||
        ! [inStream _CMgetData:&salt
                      maxLength:kPBKDFSaltSize
                          error:error]) {
      return NO;
    }
    break;
  default:
    NSAssert(NO, @"Unknown operation: %d", operation);
    break;
}
```

Next, it generates the key from the password and creates the CCCryptor object. This is the object that performs the encryption or decryption.

```
NSData *key = [self AESKeyForPassword:password salt:salt];

// Create the cryptor
CCCryptorRef cryptor = NULL;
CCCryptorStatus result;
result = CCCryptorCreate(operation,                 // operation
                         kAlgorithm,                // algorithim
                         kCCOptionPKCS7Padding,      // options
                         key.bytes,                  // key
                         key.length,                 // keylength
```

```
                        iv.bytes,              // IV
                        &cryptor);             // OUT cryptorRef
```

Next, it allocates some buffers to use. According to the documentation, you can use a single buffer to manage the plaintext and ciphertext, but there is a bug in CCCryptorUpdate that prevents this (radar://9930555). If you use padding and call CCCryptorUpdate multiple times, you can't do "in place" encryption. That isn't a major problem in this case because the buffer size is small.

CCCryptorGetOutputLength returns the size of the buffer required to process the requested number of bytes, including any extra data that may be needed for the final block. You could also use kMaxReadSize + kAlgorithmBlockSize, which is always greater than or equal to the result of CCCryptorGetOutputLength. There's no problem with allocating a little too much memory here. Using NSMutableData rather than malloc lets ARC take care of the memory management for you, even if there's an error.

```
        dstBufferSize = CCCryptorGetOutputLength(cryptor, // cryptor
                                       kMaxReadSize, // input length
                                               true); // final

        NSMutableData *
        dstData = [NSMutableData dataWithLength:dstBufferSize];

        NSMutableData *
        srcData = [NSMutableData dataWithLength:kMaxReadSize];

        uint8_t *srcBytes = srcData.mutableBytes;
        uint8_t *dstBytes = dstData.mutableBytes;
```

Now the routine reads a block of data, encrypts or decrypts it, and writes it to the output stream. processR esult:bytes:length:toStream:error: just checks the result and handles the file writing in a way that simplifies error handling. The important call is CCCryptorUpdate. This reads data from srcBytes and writes them to dstBytes. It updates dstLength with the number of bytes written.

```
        ssize_t srcLength;
        size_t dstLength = 0;

        while ((srcLength = [inStream read:srcBytes
                            maxLength:kMaxReadSize]) > 0 ) {
          result = CCCryptorUpdate(cryptor,        // cryptor
                          srcBytes,        // dataIn
                          srcLength,       // dataInLength
                          dstBytes,        // dataOut
                          dstBufferSize,   // dataOutAvailable
                          &dstLength);     // dataOutMoved

        if (![self processResult:result
                        bytes:dstBytes
                      length:dstLength
                       toStream:outStream
                      error:error]) {
```

(continued)

```
            CCCryptorRelease(cryptor);
            return NO;
        }
    }
```

When you've read the entire file (`srcLength == 0`), there may still be some unprocessed data in the `CCCryptor`. `CCCryptorUpdate` processes only data in block-sized units (16 bytes for AES). If padding was enabled, you need to call `CCCryptorFinal` to deal with whatever's left over. If you did not enable padding, you can skip this step, but it's generally not worth writing special code to avoid it.

```
    result = CCCryptorFinal(cryptor,          // cryptor
                            dstBytes,          // dataOut
                            dstBufferSize,     // dataOutAvailable
                            &dstLength);       // dataOutMoved
    if (![self processResult:result
                        bytes:dstBytes
                       length:dstLength
                     toStream:outStream
                        error:error]) {
        CCCryptorRelease(cryptor);
        return NO;
    }

    CCCryptorRelease(cryptor);
    return YES;
```

Note the calls to `CCCryptorRelease`. Unlike other `...Release` functions, this immediately frees the memory. There's no retain counting on `CCCryptor`. `CCCryptorRelease` also overwrites the memory with zeros, which is good security practice for sensitive data structures.

> The fact that `CCCryptorRelease` **overwrites the memory with zeros is not documented in** `CCCommonCryptor.h`, **but it can be verified in the source code.** `CommonCrypto` **is open source, and the source is available from** `http://opensource.apple.com`. **Look in the OS X tree, not the iOS tree.**

You can see this code in action in the CryptPic sample project. For a more advanced version of this approach, including handling HMAC, see "RNCryptor" in the "Further Reading" section.

Combining Encryption and Compression

It's sometimes a good idea to compress data before encrypting it. There's a theoretical security benefit to doing so, but generally it's just to make the data smaller. The important thing to remember is that you must compress before you encrypt. You can't compress encrypted data. If you could, that would suggest patterns in the ciphertext, which would indicate a poor encryption algorithm. In most cases, encrypting and then compressing leads to a larger output than the original plaintext.

Summary

iOS provides a rich collection of security frameworks to make it as easy as possible to secure your users' data. This chapter showed you how to secure network communications, files, and passwords. You also found out how to properly validate certificates so that you can ensure that your application communicates only with trusted sources. Securing your application requires a few extra lines of code, but taking care of the basics is generally not difficult using the code provided in this chapter.

Further Reading

Apple Documentation

The following documents are available in the iOS Developer Library at `developer.apple.com` or through the Xcode Documentation and API Reference.

Certificate, Key, and Trust Services Programming Guide

iOS Application Programming Guide: "The Application Runtime Environment"

Secure Coding Guide (/usr/lib/CommonCrypto)

WWDC Sessions

The following session videos are available at `developer.apple.com`.

WWDC 2011, "Session 208: Securing iOS Applications"

WWDC 2012, "Session 704: The Security Framework"

WWDC 2012, "Session 714: Protecting the User's Data"

Other Resources

Aleph One. *Phrack, Volume 7, Issue Forty-Nine*, "Smashing The Stack For Fun And Profit" (1996). Fifteen years later, this is still one of the best introductions to buffer overflows available, with examples. `www.phrack.org/issues.html?issue=49&id=14#article`

Boneh, Dan. Stanford "Cryptography" Course. There is no better free resource for learning how to do cryptography correctly than this course. This is not a quick overview. You'll discover the math behind cryptosystems and how they're put together correctly and incorrectly. You'll learn how to reason about the security of systems and mathematically prove security theorems about them. It's more math than computer programming. I highly recommend it, but it is a significant commitment. It's broken into two six-week sections. `https://www.coursera.org/course/crypto`

Granoff, Mark. *Lockbox*. While not yet ideal, this is the most promising Keychain wrapper in my opinion. `https://github.com/granoff/Lockbox`

Napier, Rob. "RNCryptor." This is my framework for AES encryption, based on CommonCryptor. Its purpose is to make it easy to use AES correctly, and it implements all the features discussed in this chapter. `https://github.com/rnapier/RNCryptor`

Schneier, Bruce, *Applied Cryptography* (John Wiley & Sons 1996). Anyone interested in the guts of cryptography should read this book. The main problem is that after reading it, you may think you can create your own cryptography implementations. You shouldn't. Read this book as a fascinating, if dated, introduction to cryptography. Then put it down and use a well-established implementation.

Chapter 16

Running on Multiple iPlatforms and iDevices

The iOS SDK was announced to the public in February 2008. At that time, there were only two devices using it: iPhone and iPod touch. Apple has since been innovating vigorously, and in 2010 it introduced another bigger brother to the family, the iPad. In 2010, another new device running iOS was introduced: the Apple TV. Who knows what the future might hold—Apple might even announce an SDK for Apple TV development and may even enable running games from Apple TV controlled by your iPhone on iPod touch.

Every year, a new version of the SDK comes out along with at least two or three new device updates, and these new devices often come with additional sensors. The GPS sensor debuted with iPhone 3G, the magnetometer—a sensor used to show the direction of magnetic north (more commonly known as a compass)—debuted in iPhone 3GS, and the gyroscope (for life-like gameplay) in iPhone 4. The iPad was introduced later with a whole new UI, a far bigger screen than the iPhone, but without a camera. iPad added a couple of cameras (including a front-facing camera) in the second iteration, iPad 2. The iPad 2 was superceded by the new iPad, which has a better camera and features like face detection and video stabilization. Recently, Apple announced the iPhone 5, which has a larger screen than its predecessors. iPhone 5 poses another challenge in designing your user interfaces. I show you how to support iPhone 5 in your app later in this chapter.

Similarly, every version of the SDK comes with powerful new features: In App Purchases, Push Notification Service, Core Data, and MapKit support in iOS 3; multitasking, blocks, and Grand Central Dispatch in iOS 4; iCloud, Twitter integration, and Storyboards in iOS 5; and, PassKit in iOS 6, to name a few. When you use one of these features, you might be interested in providing backward compatibility to users running an older version of the operating system. Keep in mind, however, that if you're using a feature available in a newer version of the SDK, you must either forget about old users (not a good idea) or write code that adapts to both users (either by supporting an equivalent feature for older users or by prompting them that additional features are available if they run a newer version).

As a developer, you need to know how to write code that easily adapts to any device (known or unknown) and platform. For that purpose, it's easier to depend on Cocoa framework APIs to detect capabilities than writing code assuming that a certain sensor will be present on a given hardware. In short, developers need to avoid making assumptions about hardware capabilities based on device model strings.

This chapter looks at some strategies that can help you write code that adapts easily to multiple platforms and devices using the various APIs provided by Cocoa framework. You also learn how to adapt your code to support the new, taller, iPhone 5 In the course of this chapter, you write a category extension on the `UIDevice` class and add methods that check for features that aren't readily exposed by the framework.

Developing for Multiple Platforms

The iOS debuted with a public SDK in version 2.0, and version 6.0 is the fifth iteration that is available for developers. One important advantage of iOS over competing platforms is that users don't have to wait for carriers to "approve" their OS updates, and because the updates are free of charge, most users (more than 75 percent) get the latest available OS within a month. It's usually fine for iOS developers to support just the two latest iterations of the SDK. That is, in late 2011 and early 2012, it was enough to support iOS 4 and iOS 5; now, in late 2012 to early 2013, it should be enough to support iOS 5 and iOS 6, all of which makes life easier for developers.

Configurable Target Settings: Base SDK Versus Deployment Target

To customize the features your app can use and the devices and OS versions your app can run, Xcode provides two configurable settings for the target you build. The first is your base SDK setting and the second is the iOS Deployment Target.

Configuring the Base SDK Setting

The first configurable setting is called Base SDK. You can configure this setting by editing your target. To do so, follow these steps:

1. Open your project and select the project file on the project navigator.

2. On the editor pane, select the target and select the Build Settings tab. The Base SDK setting is usually the third option here, but the easiest way to look for a setting in this pane is to search for it in the search bar.

You can change the value to "Latest iOS SDK" or any version of SDK installed on your development machine. The Base SDK setting instructs the compiler to use that version of SDK to compile and build your app, and this means it directly controls which APIs are available for your app. By default, new projects created with Xcode always use the latest-available SDK, and Apple handles API deprecation. Unless you have very specific reasons not to, stick to this default value.

Configuring the Deployment Target Setting

The second setting is the Deployment Target, which governs the minimum required OS version necessary for using your app. If you set this to a particular version, say 5.0, the AppStore app automatically prevents users running previous operating systems from downloading or installing your app. To cater to a wider audience, I recommend providing backward compatibility for at least one previous version of the OS. For example, if iOS 6 is the latest version, you should also support at least iOS 5. You can set the Deployment Target on the same Build Settings tab as the Base SDK setting.

When you're using a feature available in iOS 6 SDK, but still want to support older versions, set your Base SDK setting to the latest SDK (or iOS 6) and your Deployment Target to at least iOS 5. However, when your app is running on iOS 5 devices, some frameworks and features may not be available. It's your responsibility as a developer to adapt your app to work properly without crashing.

Considerations for Multiple SDK Support: Frameworks, Classes, and Methods

You need to handle three cases when you support multiple SDKs: frameworks, classes, and methods. In the following sections, you find out about the ways to make this possible.

Framework Availability

Sometimes a new SDK may add a whole new framework, which means that a complete framework is not available on older operating systems. An example from iOS 5 is the `Twitter.framework`. This framework is available only to users running iOS 5 and above. You have two choices here. Either set the deployment target to iOS 5 and build your app only for customers running iOS 5 and above or check whether the given framework is present on the user's operating system and hide necessary UI elements that invoke a call to this framework. Clearly, the second choice is the optimal one.

When you use a symbol that's defined in a framework that is not available on older versions, your app will not load. To avoid this and to selectively load a framework, you must weak-link it. To weak-link a framework, open the target settings page from the project settings editor. Then open the Build Phases tab and expand the fourth section (Link Binary With Libraries). You will see a list of frameworks that are currently linked to your target. If you haven't yet changed a setting here, all the frameworks are set to Required by default. Click the Required combo box and change it to Optional. This will weak-link the said framework.

When you weak-link a framework, missing symbols automatically become null pointers, and you can use this null check to enable or disable UI elements.

An example on iOS 6 is the `PassKit.Framework`. When you use the built-in PassKit framework for storing user's coupons, weak-link it and do a runtime check to see if it is available. If not, you have to implement your own methods to mimic that functionality.

> When you link a framework that is present only on a newer version of the SDK, but you still specify the iOS Deployment target to a SDK older than that, your application will fail to launch and crash almost immediately. This will cause your app to be rejected. When you receive a crash report from the Apple review team stating that the app crashes immediately on launch (mostly without any useful crash dumps), this is what you have to look for. The fix for this crash is to "weak link" the framework. To learn more about debugging, read Chapter 19 in this book.

Class Availability

Sometimes a new SDK might add new classes to an existing framework. This means that even if the framework gets linked, not all symbols will be available to you on older operating systems. An example from iOS 4 is the `UILocalNotification` class defined in `UIKit.Framework`. This framework is linked with every iOS app, so when you're using this class, you need to check for its presence by instantiating an object using the `NSClassFromString` method. If it returns `nil`, that class is not present on the target device. An example from iOS 5 is the `UIStepper` control. If you're using this class, check for its existence.

Another method to check for class availability is to use the `class` method instead of `NSClassFromString`, as shown in the following code.

Checking for Availability of the UIStepper Control

```
if ([UIStepper class])  {
    // Create an instance and add it to the subview
} else {
    // create instance of a equivalent control and add it to subview
}
```

> To use the `class` method, use the LLVM Clang compiler, and the deployment target should be 3.1 or later.

Method Availability

In some cases, new methods are added to an existing class in the new SDK. A classic example from iOS 4 is multitasking support. The class `UIDevice` has a method called `isMultitaskingSupported`. The following code checks for this class.

Code for Checking Whether a Method Is Available in a Class

```
if ([[UIDevice currentDevice] respondsToSelector:@selector(isMultitaskingS
upported)])  {
if([UIDevice currentDevice].isMultitaskingSupported)  {
    // Code to support multitasking goes here
  }
}
```

To check whether a method is available in a given class, use the `respondsToSelector:` method. If it returns `YES`, you can use the method you checked for.

If the method you're checking is a global C function, equate it to `NULL` instead, as shown in the following code.

Checking Availability of a C Function

```
if (CFunction != NULL) {
  CFunction(a);
}
```

> You have to equate the function name explicitly to `NULL`. Implicitly assuming pointers as `nil` or `NULL` will not work. Do note that it is not `CFunction()`. It's just `CFunction` without the parenthesis. Checking the condition will not invoke the method.

Checking the Availability of Frameworks, Classes, and Methods

Although it's quite easy to remember framework availability, it can be challenging to remember the availability of every single class and method. Equally difficult is reading through the complete iOS documentation to find out which method is available and which method is not. I recommend two different ways to check the availability of a framework, class, or method.

Developer Documentation

The straightforward way to check the availability of symbols or frameworks is to search in the Availability section of the developer documentation. Figure 16-1 is a screenshot from the developer documentation showing how to look for multitasking availability.

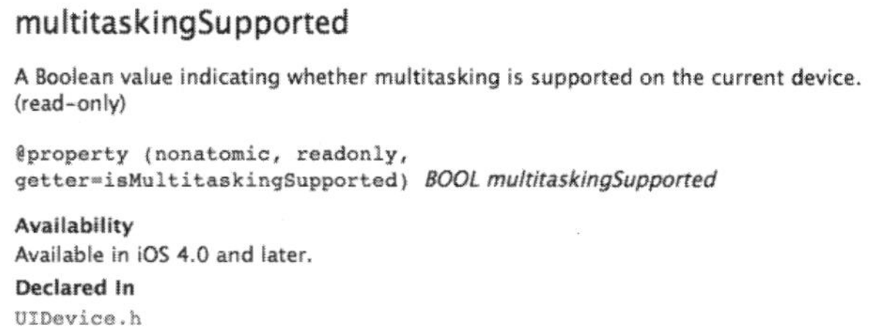

multitaskingSupported

A Boolean value indicating whether multitasking is supported on the current device. (read-only)

```
@property (nonatomic, readonly,
getter=isMultitaskingSupported) BOOL multitaskingSupported
```

Availability
Available in iOS 4.0 and later.
Declared In
`UIDevice.h`

Figure 16-1 Multitasking availability in developer documentation

Macros in iOS Header Files

The other method for checking the availability of a method or class is to read through the header files. I find this easier than fiddling through the documentation. Just Cmd-click the symbol from your source code, and Xcode opens the header file where it's defined. Most newly added methods have either one of the macro decorations shown in Figure 16-2.

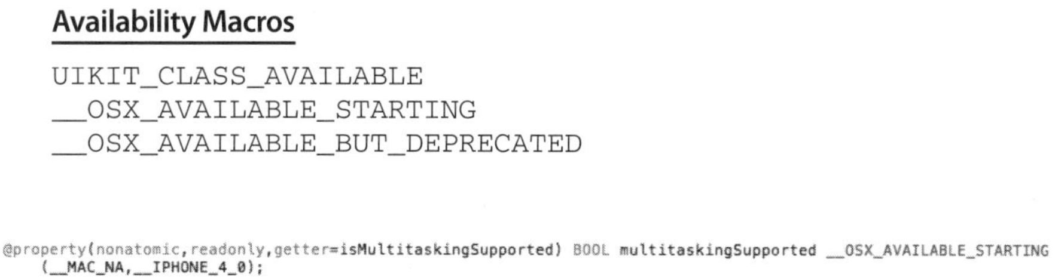

Availability Macros

```
UIKIT_CLASS_AVAILABLE
__OSX_AVAILABLE_STARTING
__OSX_AVAILABLE_BUT_DEPRECATED
```

```
@property(nonatomic,readonly,getter=isMultitaskingSupported) BOOL multitaskingSupported __OSX_AVAILABLE_STARTING
    (__MAC_NA,__IPHONE_4_0);
```

Figure 16-2 Multitasking availability in header file

It's usually easier and faster to check availability of a class or method for a given SDK version from the header file. But not all methods will have this macro decoration. If it doesn't, you have to look at the developer documentation.

> **If a method doesn't have a macro decoration, it probably means that the method was added ages ago to the SDK, and you normally don't have to worry if you're targeting the two most recent SDKs.**

Now that you know how to support multiple SDK versions, it's time to focus on the meat of the chapter: supporting multiple devices. In the next section, you discover the subtle differences between the devices and learn the correct way to check for availability of a particular feature. In parallel, you also write a category extension class on `UIDevice` that adds methods and properties for checking features not exposed by the framework.

Detecting Device Capabilities

The first and most common mistake that developers made in the past, when there were only two devices (iPod touch and iPhone), was to detect the model name and check whether it was an "iPhone," thereby assuming capabilities. This worked well for a year or so. But soon, when new devices with new hardware sensors became available, the method became highly prone to error. For example, the initial version of the iPod touch didn't have a microphone; however, after the iPhone OS 2.2 software update, users could add one by connecting an external microphone/headset. If your code assumes device capabilities based on model name, it will still work, but it's not correct and not the right thing to do.

Detecting Devices and Assuming Capabilities

Consider the following code fragment, which assumes the capabilities of the iPhone.

Detecting a Microphone the Wrong Way

```
if(![[UIDevice currentDevice].model isEqualToString:@"iPhone"])  {
        UIAlertView *alertView = [[UIAlertView alloc]
initWithTitle:@"Error"
message:@"Microphone not present"
delegate:self
        cancelButtonTitle:@"Dismiss"
otherButtonTitles: nil];
        [alertView show];
    }
```

The problem with the preceding code is that the developer has made a broad assumption that only iPhones will ever have microphones. This code worked well initially. But with the iOS software 2.2 update, when Apple added external microphone capability to iPod touch, the preceding code prevents users from using the app. Another problem is that this code shows an error for any new device introduced later—for example, iPad.

You should use some other method for detecting hardware or sensor availability rather than assume devices' capabilities. Fortunately or unfortunately, these methods are scattered around on various frameworks.

Now, it's time to start looking at various methods for checking device capabilities the right way and grouping them under a `UIDevice` category class.

Detecting Hardware and Sensors

The first thing to understand is that instead of assuming capabilities, you need to check for the presence of the exact hardware or sensor you need. For example, instead of assuming that only iPhones have a microphone, use APIs to check for the presence of a microphone. The first advantage of the following code is that it automatically works for new devices to be introduced in the future and for externally connected microphones.

What's the second advantage? The code is a one-liner.

Correct Way to Check for Microphone Availability

```
- (BOOL) microphoneAvailable   {
 AVAudioSession *session = [AVAudioSession sharedInstance];
 return session.inputIsAvailable;
}
```

In the case of a microphone, you also need to consider detecting input device change notifications. That is, enable your Record button on the UI when the user plugs in a microphone, in addition to `viewDidAppear`. Sounds cool, right? Here's how to do that.

Detecting Whether a Microphone Is Being Plugged In

```
void audioInputPropertyListener(void* inClientData,
AudioSessionPropertyID inID, UInt32 inDataSize, const void *inData)   {

    UInt32 isAvailable = *(UInt32*)inData;
    BOOL micAvailable = (isAvailable > 0);
    // update your UI here
}

- (void)viewDidLoad   {
    [super viewDidLoad];
AudioSessionAddPropertyListener(
kAudioSessionProperty_AudioInputAvailable,
audioInputPropertyListener, nil);
}
```

All you need to do here is to add a property listener for `kAudioSessionProperty_AudioInputAvailable` and on the callback check for the value.

With just few extra lines of code, you're able to write the correct version of device detection code. Next, you extend this for other hardware and sensors.

`AudioSessionPropertyListeners` **behave much like observing** `NSNotification` **events. When you add a property listener to a class, it's your responsibility to remove it at the right time. In the preceding example, because you added the property listener in** `viewDidLoad`, **you need to remove it in the** `didReceiveMemoryWarning` **method.**

Detecting Camera Types

The iPhone shipped with a single camera originally and added a front-facing camera later in iPhone 4. The iPod touch had no camera until the fourth generation. While the iPhone 4 has a front-facing camera, the iPad 1 (its bigger brother) doesn't have one, whereas the newer iPad 2 has both a front-facing and a back-facing camera. All this means that you should not write code with device-based assumptions. It's actually far easier to use the API.

The `UIImagePickerController` class has class methods to detect source type availability.

Checking for Camera Presence

```
- (BOOL) cameraAvailable  {
  return [UIImagePickerController isSourceTypeAvailable:
UIImagePickerControllerSourceTypeCamera];
}
```

Checking for a Front-Facing Camera

```
- (BOOL) frontCameraAvailable
{
#ifdef __IPHONE_4_0
  return [UIImagePickerController isCameraDeviceAvailable:
UIImagePickerControllerCameraDeviceFront];
#else
  return NO;
#endif
}
```

For detecting a front-facing camera, you need to be running on iOS 4 and above. The enumeration `UIImagePickerControllerCameraDeviceFront` is available only on iOS 4 and above because any device that has a front-facing camera (iPhone 4 and iPad 2) always runs iOS 4 and above. So you use a macro and return NO if the device runs iOS 3 or below.

Similarly, you can check whether the camera attached has video-recording capabilities. Cameras on iPhone 3GS and above can record videos. You can check that using the following code.

Checking for a Video-Recording Capable Camera

```
- (BOOL) videoCameraAvailable  {
  UIImagePickerController *picker =
[[UIImagePickerController alloc] init];
// First call our previous method to check for camera presence.
if(![self cameraAvailable])  return NO;
NSArray *sourceTypes = [UIImagePickerController
availableMediaTypesForSourceType:
UIImagePickerControllerSourceTypeCamera];

  if (![sourceTypes containsObject:(NSString *)kUTTypeMovie]){

    return NO;
```

```
    }

    return YES;
}
```

This enumerates the available media types for a given camera and determines whether it contains `kUTTypeMovie`.

Detecting Whether a Photo Library Is Empty

If you're using a camera, you will almost always use the user's photo library. Before calling `UIImagePicker` to show the user's photo album, ensure that there are photos in it. You can check this the same way you check for camera presence. Just pass `UIImagePickerControllerSourceTypePhotoLibrary` or `UIImagePickerControllerSourceTypeSavedPhotosAlbum` for the source type.

Detecting the Presence of a Camera Flash

So far, the only device to have a camera flash is the iPhone 4. In coming years, more and more devices will have a camera flash. It's easy to check for camera flash presence using `UIImagePickerController`'s class method.

Checking for a Camera Flash

```
-  (BOOL) cameraFlashAvailable   {
#ifdef __IPHONE_4_0
    return [UIImagePickerController isFlashAvailableForCameraDevice:
UIImagePickerControllerCameraDeviceRear];
#else
    return NO;
#endif
}
```

Detecting a Gyroscope

The gyroscope is an interesting addition to the iPhone 4. Devices introduced after iPhone 4, including the new iPad and iPhone 5, also have a gyroscope. It allows developers to measure relative changes to the physical position of the device. By comparison, an accelerometer can measure only force. Twisting movements cannot be measured by an accelerometer. Using a gyroscope, it's possible for game developers to implement 6-axis control like that found in Sony's PlayStation 3 controller or Nintendo's Wii controller. You can detect the presence of a gyroscope using an API provided in the `CoreMotion.framework`.

Code to Detect the Presence of a Gyroscope

```
-  (BOOL) gyroscopeAvailable   {
#ifdef __IPHONE_4_0
    CMMotionManager *motionManager = [[CMMotionManager alloc] init];
    BOOL gyroAvailable = motionManager.gyroAvailable;
    return gyroAvailable;
#else
    return NO;
#endif
}
```

> If a gyroscope is a core feature of your app but your target device doesn't have a gyroscope, you have to design your app with alternative input methods, or you can specify them in the `UIRequiredDeviceCapablities` **key in your app's** `info.plist`, **preventing devices without a gyroscope from installing the app. You learn more about this key later in the chapter.**

Detecting a Compass or Magnetometer

Compass availability can be checked using the `CoreLocation.framework` class `CLLocationManager`. Call the method `headingAvailable` in `CLLocationManager`, and if it returns true, you can use a compass in your app. A compass is more useful in a location-based application and augmented reality-based applications.

Detecting a Retina Display

As an iOS developer, you already know that catering to a retina display is as easy as adding an @2x image file for every resource you use in the app. But in cases where you download the image from a remote server, you need to download images at twice the resolution on devices with retina display.

A good example of this is a photo browser app like, say, a Flickr viewer or Instagram. When your user launches the app in iPhone 4 or the new iPad or iPhone 5, you should be downloading images of double the resolution you do for non-retina display devices. Some developers choose to ignore this and download higher resolution images for all devices, but that is a waste of bandwidth and might even be slower to download over EDGE. Instead, download higher-resolution files after determining that the device has a retina display. Checking for this is easy.

Retina Display Capable

```
- (BOOL) retinaDisplayCapable  {
 int scale = 1.0;
 UIScreen *screen = [UIScreen mainScreen];
 if([screen respondsToSelector:@selector(scale)])
  scale = screen.scale;
 if(scale == 2.0f) return YES;
 else return NO;
}
```

With this code, you look for the `mainScreen` of the device and check whether the device is capable of showing high-resolution retina display-capable graphics. This way, if Apple introduces an external retina display (maybe the newer Apple Cinema Displays) and allows the current generation iPads to project to it in retina mode, your app will still work without changes.

Detecting Alert Vibration Capability

As of this writing, only iPhones are capable of vibrating to alert the user. Unfortunately, there is no public API for checking whether a device is vibration-capable. However, the `AudioToolbox.framework` has two methods to selectively vibrate only iPhones:

```
AudioServicesPlayAlertSound(kSystemSoundID_Vibrate);

AudioServicesPlaySystemSound(kSystemSoundID_Vibrate);
```

The first method vibrates the iPhone and plays a beep sound on iPod touch/iPad. The second method just vibrates the iPhone. On devices not capable of vibrating, it doesn't do anything. If you're developing a game that vibrates the device to signify danger or a Labyrinth game where you want to vibrate whenever the player hits the wall, use the second method. The first method is for alerting the user, which includes vibration plus beeps, whereas the second is just for vibrations.

Detecting Remote Control Capability

iOS apps can handle remote control events generated by buttons pressed on the external headset. To handle these events, use the following method to start receiving notifications:

```
[[UIApplication sharedApplication] beginReceivingRemoteControlEvents];
```

Implement the following method in your `firstResponder`:

```
remoteControlReceivedWithEvent:
```

Be sure to turn this off when you no longer need the events by calling

```
[[UIApplication sharedApplication] endReceivingRemoteControlEvents];
```

Detecting Phone Call Capability

You can check whether a device can make phone calls by checking if it can open URLs of type `tel:`. The `UIApplication` class's `canOpenURL:` method is handy for checking whether a device has an app that can handle URLs of a specific type. `tel:` URLs are handled by the phone app on iPhone. The same method can also be used to check whether a specific app that can handle a given URL is installed on a device.

Phone Call Capabilities

```
- (BOOL) canMakePhoneCalls  {
  return [[UIApplication sharedApplication]
canOpenURL:[NSURL URLWithString:@"tel://"]];
}
```

A word about usability: Developers should completely hide features specific to phones on iPod touch devices. For example, if you're developing a Yellow Pages app that lists phone numbers from the Internet, show the button to place a call only on devices that are capable of making phone calls. Do not simply disable it (because nothing can be done by the user to enable it) or show an error alert. There have been cases where showing a "Not an iPhone" error on an iPod touch leads to rejection of the app by the app review team.

In App Email and SMS

Although In App email and In App SMS are technically not sensors or hardware, not every device can send e-mails or SMSs. This includes iPhones as well—even those that run iOS 4 and above. Although `MFMessageViewController` and `MFMailComposeViewController` are available from iOS 4, and even if your app's minimum deployment target is set to iOS 4, you still need to know and understand the common pitfalls when using these classes.

A common case is an iOS device that has no configured e-mail accounts and therefore cannot send e-mail, even when it's technically capable of sending one. The same applies to SMS/MMS. An iPhone that doesn't have a SIM card cannot send text messages. You need to be aware of this and always check for capabilities before attempting to use this feature.

Checking for this capability is easy. Both `MFMessageComposeViewController` (for In app SMS) and `MFMailComposeViewController` (for In App email) have class methods `canSendText` and `canSendMail`, respectively, that can be used.

Checking Multitasking Awareness

Checking whether a device can multitask is straightforward. As you saw earlier in this chapter, you have to check whether the method `isMultitaskingSupported` is available, as shown in the following code. If it returns `YES`, you can write multitasking-related code. Otherwise, remember your app's state and continue when the app is launched again.

Is Multitasking Available?

```
if ([[UIDevice currentDevice] respondsToSelector:
@selector(isMultitaskingSupported)])   {
  if([UIDevice currentDevice].isMultitaskingSupported)   {
    // Code to support multitasking goes here
  }
}
```

But there is something more. On devices that don't support multitasking, your application delegate will not receive the following callbacks:

```
- applicationDidEnterBackground:
- applicationWillEnterForeground:
```

This means that any part of your startup code and initialization sequence you write in `applicationWillEnterForeground:` should be written in `applicationDidFinishLaunchingWithOptions:` as well as for nonmultitasking capable devices.

Similarly, the teardown code (including your Core Data-managed context save methods) that you write in `applicationDidEnterBackground:` needs to be written in `applicationWillTerminate:` as well.

Obtaining the UIDevice+Additions Category

The code fragments you've seen so far in this chapter are available as a UIDevice category addition. You can download them from the book's website.

It has just two files: `UIDevice+Additions.h` and `UIDevice+Additions.m`. You have to link necessary frameworks to avoid those pesky linker errors because this class links to various Apple library frameworks. But don't worry; they are dynamically loaded, so they don't bloat your app.

Supporting the iPhone 5

iPhone 5 was announced in September 2012, and it poses a new challenge to developers: a bigger screen. iOS developers have never been required to support multiple device resolutions in the past. Fret not, Apple has made things easy for us. The first step is to add a launch image (`Default-568h@2x.png`). As shown in Figure 16-3, when you build your project with Xcode 4.5, you will see a warning: "Missing Retina 4 launch image." Click Add to add a default launch image to your project.

Figure 16-3 Xcode 4.5 prompting for addition of a launch image for iPhone 5

The app will then launch in full screen without letter boxing. However, most of your nib files will still not scale properly. The next step is to check the auto resizing mask of every nib file and ensure that the view inside the nib file automatically sizes based on the super view's height. Figure 16-4 illustrates this.

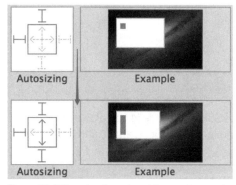

Figure 16.4 Changing the autoresizing mask property using Interface Builder

The properties to use are `UIViewAutoresizingFlexibleTopMargin`, `UIViewAutoresizingFlexibleBottomMargin`, and `UIViewAutoresizingFlexibleHeight`.

You use the `UIViewAutoresizingFlexibleHeight` for the top-most view so that it autosizes with the main window. You use the `UIViewAutoresizingFlexibleTopMargin` and/or `UIViewAutoresizingFlexibleBottomMargin` for subviews. Use `UIViewAutoresizingFlexibleTopMargin` property when you want the subview to be "pinned" to the bottom (the top margin is flexible). Use the `UIViewAutoresizingFlexibleBottomMargin` when you want the subview to be "pinned" to the top (the bottom margin is flexible).

> Remember, a flexible bottom margin pins the UI element to the top and a flexible top margin pins the UI element to the bottom. Note too that when you use Cocoa Auto Layout, you don't have to deal with auto resize masks.

If you have hard-coded the height of a view to 460 or 480, you might have to change this using bounds. For example, you use

```
self.window = [[UIWindow alloc] initWithFrame:[[UIScreen mainScreen]
bounds]];
```

instead of

```
self.window = [[UIWindow alloc] initWithFrame:CGRectMake(0, 0, 320,
480)];
```

Finally, any `CALayer` that you added to the view will have to be manually resized. The following code shows how to do this. This code assumes you have a "patternLayer" for your all your view controllers.

Resizing a CALayer for iPhone 5 compatiblity

```
-(void)viewWillLayoutSubviews {

self.patternLayer.frame = self.view.bounds;
[super viewWillLayoutSubviews];
}
```

iPhone 5 requires a new instruction set—the armv7s. Only the latest version of Xcode (4.5 as of this writing) supports generating an armv7s instruction set. Note that Xcode 4.5 no longer supports armv6 and deprecates iPhone 3G and older devices.

UIRequiredDeviceCapabilities

So far, you've found out how to conditionally check a device for specific capabilities and use them if they are present. In some cases, your app depends solely on the presence of particular hardware, and without that hardware your app will be unusable. Examples include a camera app like Instagram or Camera+. The core functionality of the app doesn't work without a camera. In this case, you need something more than just checking for device capabilities and hiding specific parts of your app. You normally won't need devices without a camera to use or download your app.

Apple provides a way to ensure this using the `UIRequiredDeviceCapabilities` key in the Info plist file. The following values are supported for this key:

`telephony`	`wifi`	`sms, still-camera`
`auto-focus-camera`	`front-facing-camera`	`camera-flash`
`video-camera`	`accelerometer`	`gyroscope`
`location-services`	`gps`	`magnetometer`
`gamekit`	`opengles-1`	`opengles-2`
`armv6`	`armv7`	`peer-peer`
`accelerometer`	`Bluetooth-le`	`microphone.`

You can explicitly require particular device capabilities or prohibit installation of your app on devices without a specific capability. For example, you can prevent your apps from running on devices with `video-camera` by setting the `video-camera` key to NO. Alternatively, you can mandate the presence of `video-camera` by setting the `video-camera` key to YES.

Apple doesn't allow you to submit an update to an existing app and prevent it from running on a specific device that was supported before the update. For example, if your app supported both iPhone and iPod touch in version 1.0, you cannot submit an update that prevents it from running on either device. Put another way, you cannot introduce a mandate for the presence of particular hardware later in your app's product life cycle. The submission process on iTunes Connect will fail and show you an error. The opposite is allowed, however. That is, if you have been excluding a device previously, you can allow installations on it in a subsequent version. In other words, if version 1 of your app supported only iPhones, you can submit a version 2 to support all devices.

There is one exception to this rule. If you have been supporting armv6 in your current application, you can submit a new app that removes support for armv6 and hence the original iPhone and iPhone 3G. In fact, Xcode 4.5 doesn't allow you to build for armv6 devices (iPhone 3G, iPod touch 2nd generation, the original iPhone) and versions below Xcode 4.5 don't allow you to build for armv7s devices (iPhone 5). So if you support iPhone 5, you have to forego support for iPhone 3G.

Adding values to the `UIRequiredDeviceCapablities` key will prohibit your app from being installed on devices without the capabilities you requested. If you specify that telephony is needed, users cannot even download the app on their iPod touch or iPad. You must be certain that this is your expected behavior before using this key.

Summary

This chapter discussed various techniques and tricks to help run your app on multiple platforms. It also looked at the various hardware and sensors available for iOS developers and how to detect their presence the right way. You incrementally wrote a category extension on `UIDevice` that could be used for detecting most device capabilities. You also learned about supporting the new iPhone 5's screen size. Finally, you learned about the `UIRequiredDeviceCapablities` key and how to completely exclude devices without a required capability. My recommendation is to depend on the methods explained in this chapter and use the `UIRequiredDeviceCapablities` key sparingly.

Further Reading

Apple Documentation

The following documents are available in the iOS Developer Library at `developer.apple.com` or through the Xcode Documentation and API Reference.

Understanding the UIRequiredDeviceCapablities Key

iOS Build Time Configuration Details

Other Resources

MK blog. (Mugunth Kumar) "iPhone Tutorial: Better way to check capabilities of iOS devices"
`http://blog.mugunthkumar.com/coding/iphone-tutorial-better-way-to-check-capabilities-of-ios-devices/`

Github. "MugunthKumar/DeviceHelper"
`https://github.com/MugunthKumar/DeviceHelper`

## Chapter 17

Internationalization and Localization

Localization is a key concern for any application with a global market. Users want to interact in their own languages, with their familiar formatting. Supporting this in your application is called *internationalization* (sometimes abbreviated "i18n" for the 18 characters between the "i" and the "n") and *localization* ("L10n"). The differences between i18n and L10n aren't really important or consistently agreed upon. Apple says, "Internationalization is the process of designing and building an application to facilitate localization. Localization, in turn, is the cultural and linguistic adaptation of an internationalized application to two or more culturally-distinct markets." (See "Internationalization Programming Topics" at `developer.apple.com`.) This chapter uses the terms interchangeably.

After reading this chapter, you will have a solid understanding of what localization is and how to approach it. Even if you're not ready to localize your application yet, this chapter provides easy steps to dramatically simplify localization later. You find out how to localize strings, numbers, dates, and nib files, and how to regularly audit your project to make sure it stays localizable.

What Is Localization?

Localization is more than just translating strings. Localization means making your application culturally appropriate for your target audience. That includes translating language in strings, images, and audio, and formatting numbers and dates. iOS 6 adds auto layout, which makes localization much simpler. This is discussed in Chapter 7.

Here are some general things you can do to improve your iOS localizations:

- **Keep nib files simple.** This isn't difficult on iOS because there aren't as many complicated things you can do in a nib as you can on a Mac. But just remember that every IBOutlet and IBAction connection you make has to be made in every localized nib file.

- **Separate nib files that require localization from ones that don't.** Many iOS nib files have no strings or localized images. You don't need to localize these. If you just need a localized title, then make it an IBOutlet and set the localized value at runtime rather than localizing the nib file. String localization is much easier to maintain than nib file localization.

- **Remember right-to-left languages.** This is one of the hardest things to fix later, especially if you have custom text views.

- **Don't assume that a comma is the thousands separator or a dot is the decimal point.** These are different in different languages, so build your regular expressions using NSLocale.

■ **Glyphs (drawn symbols) and characters do not always map one-to-one.** If you're doing custom text layout, this can be particularly surprising. Apple's frameworks generally handle this automatically, but don't try to circumvent systems like Core Text when they force you to calculate the number of glyphs in a string rather than using `length`. This issue is particularly common in Thai, but exists in many languages (even occasionally in English, as we discuss in Chapter 26).

In my experience, it is best to do all of your development up to the point of release and then translate rather than try to translate as you go. The cost of localization is best absorbed at fixed points during development, generally at the end. It's expensive to retranslate things every time you tweak the UI.

Although translation is best done near the time of release, you should line up your localization provider fairly early in the development cycle and prepare for localization throughout the process. A good localization provider does more than just translate a bunch of strings. Ideally, your localization provider will provide testing services to make sure your application "makes sense" in the target culture. Getting the provider involved early in the process can save expensive rework later if your interface is hard to localize.

An example of a "hard-to-localize" application is one that includes large blocks of text. Translating large blocks of text can play havoc with layout, even when using auto-layout. Remembering that you will often pay by the word for translation may help you focus on reducing the number of words you use. Redesign your UI so it doesn't need so much text to let the user know what to do. Rely on Apple's UI elements and icons as much as possible. Apple's done the hard and expensive work for you to make them internationally appropriate. Don't waste that. For example, when using a `UIToolBarItem`, use a system item whenever appropriate rather than drawing your own icons. If the icon's meaning matches your intent, always try to use the system icon, even if you believe you could create a better one. In our opinion, the "action" icon (an arrow coming out of a box) is incomprehensible, but users are used to it. Apple has trained them in what it means, so you should use it. Never use a system icon for something other than its intended meaning, however. For example, do not use `UIBarButtonSystemItemReply` to mean "go left" or "go back."

Another frequent localization problem is icons that assume a cultural background, such as a decorated tree to indicate "winter." Check marks can also cause problems, because they are not used in all cultures (French for instance), and in some cultures a check mark means "incorrect" (Finnish, for instance). Involving a good localization provider before producing your final artwork can save you a lot of money re-creating your assets.

Localizing Strings

The most common tool for localizing strings is `NSLocalizedString`. This function looks up the given key in `Localizeable.strings` and returns the value found there, or the key itself if no value is found. `Localizeable.strings` is a localized file, so there is a different version for each language, and `NSLocalizedString` automatically selects the correct one based on the current locale. A command-line tool called `genstrings` automatically searches your files for calls to `NSLocalizedString` and writes your initial `Localizeable.strings` file for you.

The easiest approach is to use the string as its own key (the second parameter is a comment to the localizer):

```
NSString *string =
    NSLocalizedString(@"Welcome to the show.",
                      @"Welcome message");
```

To run `genstrings`, you open a terminal, change to your source code directory, and run it as shown here (assuming an English localization):

```
genstrings -o en.lproj *.m
```

This creates a file called `en.lproj/Localizeable.string` that contains the following:

```
/* Welcome message */
"Welcome to the show." = "Welcome to the show.";
```

Even if you don't run `genstrings`, this works in the developer's language because it automatically returns the key as the localized string.

In most cases, I recommend using the string as its own key and automatically generating the `Localizeable.strings` file when you're ready to hand the project off to localizers. This approach simplifies development and helps keep the `Localizeable.strings` file from accumulating keys that are no longer used.

Auditing for Nonlocalized Strings

During development, be sure to periodically audit your program to make sure that you're using `NSLocalizedString` as you should. I recommend a script like this:

find_nonlocalized

```
#!/usr/bin/perl -w
# Usage:
#     find_nonlocalized [<directory> ...]
#
# Scans .m and .mm files for potentially nonlocalized
#    strings that should be.
# Lines marked with DNL (Do Not Localize) are ignored.
# String constant assignments of this form are ignored if
#    they have no spaces in the value:
#    NSString * const <...> = @"...";
# Strings on the same line as NSLocalizedString are
#    ignored.
# Certain common methods that take nonlocalized strings are
#    ignored
# URLs are ignored
```

(continued)

```perl
#
# Exits with 1 if there were strings found

use File::Basename;
use File::Find;
use strict;

# Include the basenames of any files to ignore
my @EXCLUDE_FILENAMES = qw();

# Regular expressions to ignore
my @EXCLUDE_REGEXES = (
    qr/\bDNL\b/,
    qr/NSLocalizedString/,
    qr/NSString\s*\*\s*const\s[^@]*@"[^ ]*";/,
    qr/NSLog\(/,
    qr/@"http/, qr/@"mailto/, qr/@"ldap/,
    qr/predicateWithFormat:@"/,
    qr/Key(?:[pP]ath)?:@"/,
    qr/setDateFormat:@"/,
    qr/NSAssert/,
    qr/imageNamed:@"/,
    qr/NibNamed?:@"/,
    qr/pathForResource:@"/,
    qr/fileURLWithPath:@"/,
    qr/fontWithName:@"/,
    qr/stringByAppendingPathComponent:@"/,
);

my $FoundNonLocalized = 0;

sub find_nonlocalized {
    return unless $File::Find::name =~ /\.mm?$/;
    return if grep($_, @EXCLUDE_FILENAMES);

    open(FILE, $_);

    LINE:
    while (<FILE>) {
        if (/@"[^"]*[a-z]{3,}/) {
            foreach my $regex (@EXCLUDE_REGEXES) {
                next LINE if $_ =~ $regex;
            }
            print „$File::Find::name:$.:$_";
            $FoundNonLocalized = 1;
        }
    }
    close(FILE);
}

my @dirs = scalar @ARGV ? @ARGV : („.");
find(\&find_nonlocalized, @dirs);
exit $FoundNonLocalized ? 1 : 0;
```

Periodically run this script over your source to make sure that there are no nonlocalized strings. If you use Jenkins at (`jenkins-ci.org`) or another continuous-integration tool, you can make this script part of the build process, or you can add it as a script step in your Xcode build. Whenever it returns a new string, you can decide whether to fix it, update the regular expressions to ignore it, or mark the line with `DNL` (Do Not Localize).

Formatting Numbers and Dates

Numbers and dates are displayed differently in different locales. This is generally straightforward using `NSDateFormatter` and `NSNumberFormatter`, which you are likely already familiar with.

> For an introduction to `NSDateFormatter` and `NSNumberFormatter`, see the "Data Formatting Guide" in Apple's documentation at `developer.apple.com`.

There are a few things to keep in mind, however. First, formatters are needed for input as well as output. Most developers remember to use a formatter for date input, but may forget to use one for numeric input. The decimal point is not universally used to separate whole from fractional digits on input. Some countries use a comma or an apostrophe. It's best to validate number input using an `NSDateFormatter` rather than custom logic.

Digit groupings have a bewildering variety. Some countries split thousands groups with space, comma, or apostrophe. China sometimes groups ten thousands (four digits). Don't guess. Use a formatter. Remember that this can impact the length of your string. If you leave room for only seven characters for one hundred thousand ("100,000") you may overflow in India, which uses eight ("1,00,000" or one *lakh*).

Percentages are another place where you need to be careful because different cultures place the percent sign at the beginning or end of the number, and some use a slightly different symbol. Using `NSNumberFormatterPercentStyle` will behave correctly.

Be especially careful with currency. Don't store currency as a float because that can lead to rounding errors as you convert between binary and decimal. Always store currency as an `NSDecimalNumber`, which does its math in decimal. Keep track of the currency you're working in. If your user switches locale from the U.S. to France, don't switch his $1 purchase to @@eu1. Generally, you need to persist in using the currency in which a given value is expressed. The `RNMoney` class is an example of how to do this. First, the following code demonstrates how to use the class to store Rubles and Euros.

main.m (Money)

```
NSLocale *russiaLocale = [[NSLocale alloc]
                        initWithLocaleIdentifier:@"ru_RU"];

RNMoney *money = [[RNMoney alloc]
                    initWithIntegerAmount:100];
NSLog(@"Local display of local currency: %@", money);
NSLog(@"Russian display of local currency: %@",
```

(continued)

```
                [money localizedStringForLocale:russiaLocale]);


    RNMoney *euro =[[RNMoney alloc] initWithIntegerAmount:200
                                    currencyCode:@"EUR"];
    NSLog(@"Local display of Euro: %@", euro);
    NSLog(@"Russian display of Euro: %@",
            [euro localizedStringForLocale:russiaLocale]);
```

RNMoney is an immutable object that stores an amount and a currency code. If you do not provide a currency code, it defaults to the current locale's currency. It is a very simple data class designed to be easy to initialize, serialize, and format. Here is the code.

RNMoney.h (Money)

```objc
#import <Foundation/Foundation.h>

@interface RNMoney : NSObject <NSCoding>
@property (nonatomic, readonly, strong)
                                    NSDecimalNumber *amount;
@property (nonatomic, readonly, strong)
                                    NSString *currencyCode;

- (RNMoney *)initWithAmount:(NSDecimalNumber *)anAmount
        currencyCode:(NSString *)aCode;
- (RNMoney *)initWithAmount:(NSDecimalNumber *)anAmount;

- (RNMoney *)initWithIntegerAmount:(NSInteger)anAmount
                    currencyCode:(NSString *)aCode;
- (RNMoney *)initWithIntegerAmount:(NSInteger)anAmount;

- (NSString *)localizedStringForLocale:(NSLocale *)aLocale;
- (NSString *)localizedString;

@end
```

RNMoney.m (Money)

```objc
#import "RNMoney.h"

@implementation RNMoney

static NSString * const kRNMoneyAmountKey = @"amount";
static NSString * const kRNMoneyCurrencyCodeKey =
                                    @"currencyCode";

- (RNMoney *)initWithAmount:(NSDecimalNumber *)anAmount
                currencyCode:(NSString *)aCode {
  if ((self = [super init])) {
    _amount = anAmount;
    if (aCode == nil) {
      NSNumberFormatter *formatter = [[NSNumberFormatter alloc] init];
```

```
      _currencyCode = [formatter currencyCode];
    }
    else {
      _currencyCode = aCode;
    }
  }
  return self;
}

- (RNMoney *)initWithAmount:(NSDecimalNumber *)anAmount {
  return [self initWithAmount:anAmount
                 currencyCode:nil];
}

- (RNMoney *)initWithIntegerAmount:(NSInteger)anAmount
                      currencyCode:(NSString *)aCode {
    return [self initWithAmount:
             [NSDecimalNumber decimalNumberWithDecimal:
              [[NSNumber numberWithInteger:anAmount]
               decimalValue]]
                   currencyCode:aCode];
}

- (RNMoney *)initWithIntegerAmount:(NSInteger)anAmount {
  return [self initWithIntegerAmount:anAmount
                       currencyCode:nil];
}

- (id)init {
  return [self initWithAmount:[NSDecimalNumber zero]];
}

- (id)initWithCoder:(NSCoder *)coder {

  NSDecimalNumber *amount = [coder decodeObjectForKey:
                               kRNMoneyAmountKey];
  NSString *currencyCode = [coder decodeObjectForKey:
                               kRNMoneyCurrencyCodeKey];
  return [self initWithAmount:amount
                 currencyCode:currencyCode];
}

- (void)encodeWithCoder:(NSCoder *)aCoder {
  [aCoder encodeObject:amount_ forKey:kRNMoneyAmountKey];
  [aCoder encodeObject:currencyCode_
                forKey:kRNMoneyCurrencyCodeKey];
}

- (NSString *)localizedStringForLocale:(NSLocale *)aLocale
{
  NSNumberFormatter *formatter = [[NSNumberFormatter alloc]
                                   init];
```

(continued)

```
[formatter setLocale:aLocale];
[formatter setCurrencyCode:self.currencyCode];
[formatter setNumberStyle:NSNumberFormatterCurrencyStyle];
return [formatter stringFromNumber:self.amount];
}

- (NSString *)localizedString {
  return [self localizedStringForLocale:
          [NSLocale currentLocale]];
}

- (NSString *)description {
  return [self localizedString];
}

@end
```

Nib Files and Base Internationalization

iOS 6 adds a new feature called "Base Internationalization." In the Project Info panel, you can select Use Base Internationalization, and Xcode will convert your project to the new system. Prior to Base Internationalization, you needed a copy of all of your nib files for every locale. With Base Internationalization, there is an unlocalized version of your nib and storyboard files, and there is a strings file for every localization. iOS takes care of inserting all of the strings into the nib files for you at runtime, greatly simplifying nib localization. You may still want to create individually localized nib files in some cases.

Some languages require radically different layout. For example, visually "large" languages like Russian and dense languages like Chinese may not fit well in the layout used for English or French. Right-to-left languages may need some special handling as well. Luckily, you can still create per-language nib files while using Base Internationalization.

Localizing Complex Strings

Sentence structure is radically different among languages. This means that you can almost never safely compose a string from parts like this:

```
NSString *intro = @"There was an error deleting";
NSString *num = [NSString stringWithFormat:@"%d", 5];
NSString *tail = @"objects.";
NSString *str = [NSString stringWithFormat:@"%@ %@ %@",
                 intro, num, tail]; // Wrong
```

The problem with this code is that when you translate "There was an error deleting" and "objects" into other languages, you may not be able to glue them together in the same order. Instead, you need to localize the entire string together like this:

```
NSString *format = NSLocalizedString(
          @"There was an error deleting %d objects",
```

```
                    @"Error when deleting objects.");
    NSString *str = [NSString stringWithFormat:format, 5];
```

Some languages have more complex plurals than English. For instance, there may be special word forms for describing two of something versus more than two. Don't assume you can check for greater-than-one and easily determine linguistic plurals. Solving this well can be very difficult, so try to avoid it instead. Don't have special code that tries to add an *s* to the end of plurals because this is almost impossible to translate. A good translator will help you word your messages in ways that translate better in your target languages.

Talk with your localization provider early on to understand its process and how to adjust your development practice to facilitate working with it. Figure 17-1 demonstrates a good approach.

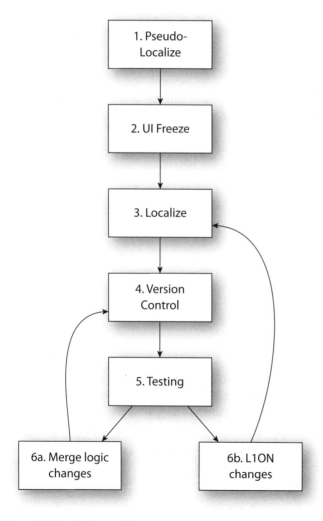

Figure 17-1 Localization workflow

1. **Pseudo-Localize**—During development, it's a good idea to start doing experimental localization to work out any localization bugs early in the process. Pseudo-localization is the process of localizing into a nonsense language. A common nonsense language is one that substitutes all vowels with the letter *x*. For example, "Press here to continue" would become "Prxss hxrx tx cxntxnxx." This kind of "translation" can be done by developers, generally with a simple script, and will make it more obvious where you have used nonlocalized strings. This won't find every problem. In particular, it is not good at discovering strings that are pieced together from other strings, but it can find many simple problems before you pay for real translation services. You will need a language code for this localization. Pick a language that you do not plan to localize your application for. If you're an American English speaker and don't plan to localize for British English, it is particularly useful to use the British English slot for this purpose because you'll still be able to easily read the rest of the iPhone's interface.

2. **UI Freeze**—There should be a clear point in the development cycle at which you freeze the UI. After that point, firmly avoid any changes that affect localizable resources. Many teams ship a monolingual version of their product at this point and then ship a localization update. That's the easiest approach if your market is tolerant of the delay.

3. **Localize**—You will send your resource files to your localizers, and they will send you back localized files. Nib files can be locked in Xcode to prevent changing localizable, nonlocalizable, or all properties. Before sending nib files to a localizer, lock nonlocalizable properties to protect your nib files against changes to connections, class names, and other invisible attributes of the nib file. Figure 17-2 shows the lock option in Interface Builder.

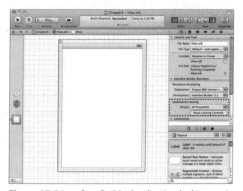

Figure 17-2 Interface Builder localization locking option

4. **Version Control**—As you make changes to your nib files, you will need to keep track of the original files your localizer sent to you. Lock the localizable properties in the nib files (unlocking the nonlocalizable properties). Then put these into a version control system or save them in a separate directory.

5. **Testing**—You'll need to do extensive testing to make sure that everything is correct. Ideally, you will have native speakers of each of your localized languages test all your UI elements to ensure that they make sense and that there aren't any leftover nonlocalized strings. A good localizer can assist in this.

6a. **Merge Logic Changes**—Certain nib file changes do not affect localization. Changes to connections or class names don't change the layout or the resources. These are logic changes rather than localization (L10n) changes. You can merge the localized nib files like this:

```
ibtool --previous-file ${OLD}/en.lproj/MyNib.nib
       --incremental-file ${OLD}/fr.lproj/MyNib.nib
       --strings-file ${NEW}/fr.lproj/Localizeable.strings
       --localize-incremental
       --write ${NEW}/fr.lproj/MyNib.nib
       ${NEW}/en.lproj/MyNib.nib
```

This computes the nonlocalization changes between the old and new English `MyNib.nib`. It then applies these changes to the old French `MyNib.nib` and writes it as the new French nib file. As long as you keep track of the original files you were sent by the localizer, this works quite well for nonlayout changes, and can be scripted fairly easily.

6b. **L10n Changes**—If you make localization changes such as changing the layout of a localized nib file or changing a string, you'll need to start the process over and send the changes to the localizer. You can reuse the previous string translations, which makes things more efficient, but it is still a lot of work, so avoid making these changes late in the development cycle.

Summary

Localization is never an easy subject, but if you work with a good localization partner early and follow the best practices detailed here, you can greatly expand the market for your applications.

Further Reading
Apple Documentation

The following documents are available in the iOS Developer Library at `developer.apple.com` or through the Xcode Documentation and API Reference.

Data Formatting Guide

Internationalization Programming Topics

Locales Programming Guide

WWDC Sessions

WWDC 2012, "Session 244: Internationalization Tips and Tricks"

## Chapter 18

Selling Past the Sale with In App Purchases

The iOS SDK has helped many developers earn enough to make a living, and a variety of business models have helped these developers earn money. The first model is easy to understand (and arguably the most profitable): Make an app and sell it on the App Store. Another model is to release both a free lite version and a pro version of an app. Yet another model, probably pioneered by Web 2.0, is advertisement-based. Developers use a third-party advertiser's SDK (or iAds) to show advertisements, and developers are paid for impressions or click-through. Although all these augment a developer's earnings, the first model, selling apps on the App Store, has been by far the most successful model. In App Purchases augment this model by offering yet another unique way to sell premium contents or features on your iOS app. Some apps take advantage of this and make money only through in app purchases. They fall into a category called *freemium* apps, and they have been quite successful, at least when you look at the top-grossing apps list on U.S. App Store.

In this chapter, I introduce you to Apple's In App Purchase framework, `StoreKit.framework`, and move on to a wrapper framework, the MKStoreKit, and discuss how to use it to integrate a mini–In App Store within your app. Then I provide you with solutions to the problems that developers most commonly face while integrating StoreKit within their app.

In iOS 6, Apple introduced a new feature where you can host your In App purchasable contents with Apple instead of a third-party server or your own server. This should help alleviate receipt validation issues. Later in this chapter, I show you how to set up a purchasable bundle for your app.

Before You Start

The chapter is divided into two major sections, the first half focusing on creating and customizing products on iTunes Connect and the second half focusing on the programming aspects. You'll be using MKStoreKit, and I'll be taking you through its features and showing you how to customize it for every possible business model (that Apple would approve). Throughout the chapter, you'll look at the most common issues and problems faced by developers and solutions for getting around them. Pay special attention to the shaded boxes, which offer helpful tips and important warnings. The information in those boxes explains what can go awry during implementation if something isn't done properly. So, now it's time to start your journey.

In App Purchase Products

Products that can be sold using In App Purchases broadly fall under the following four categories: content, functionality, service, and subscription. The latest SDK provides support for all of them. Apple allows four different product types—namely, consumable, nonconsumable, auto-renewable subscriptions, and nonrenewable subscriptions. Now, match the following categories of products to the needs of your business model:

- **Content**—Products that are categorized as content include digital books, magazines, additional level packs, music, ringtones, and a variety of other data. Content can be either sold as consumable or nonconsumable depending on your business model. Thinking from the user's perspective, content is generally considered nonconsumable. For example, when a user gets a book, it's your responsibility as a developer to remember his purchase and make it available to him for free. The SDK, in most cases, provides this feature for free.

- **Functionality**—Products that are categorized as functionality mainly include locked features. For example, a task manager app can allow users to create a maximum of "n" tasks and to create tasks after that limit, the user has to unlock by paying a fee. Functionality is almost always considered nonconsumable. Apple rejects your app if you submit a functionality as a consumable.

- **Services**—Products that are categorized as services are mostly functionality that cause a recurring expense to the developer. Services are very similar to the functionality category, except that they involve serious computation power and are done on a remote server. A classical example of this is push notifications. A Twitter client, for example, may provide push notifications as a consumable selling, say 1,000 notifications for one dollar.

  Services can also be subscriptions-based if your business model requires it to be so.

- **Subscriptions**—Products that are categorized as subscriptions are mostly content or service. Subscriptions are usually used to provide the said content or service over a period of time as opposed to the time of purchase. Subscriptions are of two types: auto-renewable subscriptions and nonrenewable subscriptions. Auto-renewable subscriptions should be used only for products that provide new content and should not be used for covering your running costs. For example, a Twitter client cannot provide unlimited push notifications as a subscription costing $0.99 every three months using auto-renewable subscriptions, although it's perfectly fine to use nonrenewable subscriptions.

Treating a product as a consumable or a subscription or a nonconsumable is up to the business owner.

> Subscriptions were originally available in iOS 3.0. But they were complicated to use, and the developer bore the burden of renewing and/or restoring them to other devices. As such, adoption was low, and very few developers used them. With iOS 4.3, Apple introduced a new kind of product called, auto-renewable subscriptions, where restoring and renewing subscriptions happens automatically and is taken care of by Apple. From now on, you should almost always use auto-renewable subscriptions. Use the older subscription style only if your business has proper server-side subscription handling in place that customers are already using.

Prohibited Items

In the previous section, I explained about the products that you could sell via In App Purchase. Although Apple is okay with most kinds of business models, you cannot sell certain items via App Store as of this writing.

Arguably the most important point to remember is that you cannot sell physical goods or services through In App Purchases. For example, if you develop a wallpaper app, you can sell digital wallpapers, but you cannot sell printed posters of the same wallpapers through In App Purchase. Similarly, if you're a hotel owner and make an app for booking reservations, you cannot collect reservation fees or booking fees through In App Purchase.

The second kind of item that isn't allowed is intermediate currency that expires. If you're making a music subscription app, you can sell "points" and allow the user to download music for those points, but these points should never "expire". Subscription passes, and pre-purchasable coupons are all allowed as long as they don't expire.

> **Warning:** There might be an app that already sells one of the prohibited items on App Store, but that doesn't automatically entitle you to do the same. You'd be taking the risk of getting rejected by Apple later. If you see an app selling an item that is prohibited by rule, chances are it missed the review process.

Lotteries or sweepstakes are allowed in some cases, if the developer is permitted by law to run a lottery business. Again, you can sell those apps only in countries where you have the legal right to do so. Having a lottery app on the U.S. App Store doesn't automatically entitle you to sell the same app on the U.K. App Store or the Australian App Store. You might need to submit additional documents to the Apple review process along with your app.

> Apple's developer documentation doesn't contain information on what is allowed and what is not allowed. Read the App Store Review Guidelines at `developer.apple.com` and your iOS developer license agreement to understand what is and isn't permitted.

Rethinking Your Business Model

All items that you're planning to sell through In App Purchase (especially content) have to go through Apple's formal review process, which usually take a week and sometimes longer. Remember this when coming up with your business plan.

If you're making an app that provides premium wallpapers for download, you probably won't be able to sell a "wallpaper of the day" through In App Purchases—at least not easily. You can, however, think of different business plans, like offering a free download for any wallpaper of the day if the user has subscribed to a premium membership. Another suggestion is to submit your app's "wallpaper of the day" for at least the next 30 days so that you have full control of releasing it on the correct dates. Ensure that your buffer is longer than the worst-case approval times.

Finally, every product you submit to the App Store needs to be configured on iTunes Connect. This configuration might take anywhere from a couple of minutes to an hour (if complex screenshots are needed). If you're selling digital books or any other digital content, like wallpapers, it might not be feasible (time wise) to configure every product on iTunes Connect. Moreover, there is a limit of 5,000 Stock Keeping Units (SKUs) that you can add to your product via In App Purchases. A recommended alternative in such a case is to make them consumable.

At this point, you've probably decided whether to sell your In App product as a consumable, nonconsumable, or subscription. Now, it's time to move on to the next section, which is perhaps the most important section in App Purchases integration.

Setting Up Products on iTunes Connect

Implementing In App Purchases in your app is 20 percent configuration, 10 percent getting the right business model for your app, and 70 percent implementation. With MKStoreKit, that 70 percent coding reduces to somewhere near zero. However, the addition of new types of products has made configuration confusing, and changes to rules and lack of proper documentation on what is acceptable and what isn't acceptable has made choosing the right business model difficult, so configuration remains the most challenging aspect of the integration.

This section walks you through the steps involved in setting up products on iTunes Connect. I assume that you're already signed up with the iOS developer program and have the necessary credentials to log in to various portals like iOS developer program portal and iTunes Connect.

I refer to the iOS developer program portal and iTunes Connect throughout the next few pages. The following links will be of help. The URLs are pretty easy to remember: iOS developer program portal, `developer.apple.com/devcenter/ios/index.action`; and iTunes Connect, `https://itunesconnect.apple.com/WebObjects/iTunesConnect.woa`.

Step 1: Create a New App ID for Your App

Every app that requires In App Purchases needs an App ID that is unique to the application. The ID cannot include a wild card character (*). The recommended convention for this is reverse DNS notation. Here are a couple of examples:

```
com.mycompany.myapp.levelpack1

org.mycompany.myapp.levelpack2
```

To create a new App ID, log in to the iOS developer program portal and navigate to the iOS provisioning portal, as shown in Figure 18-1.

Click the New App ID button and follow the wizard to create an App ID. Ensure that you use a fully qualified App ID without any wild card characters.

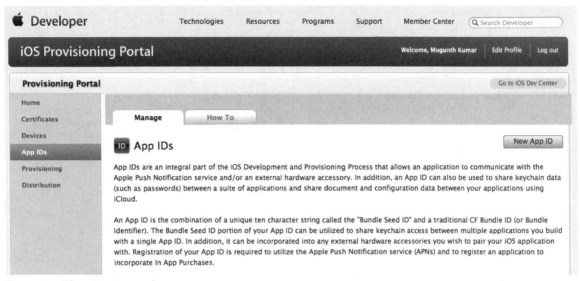

Figure 18-1 iOS provisioning portal

Warning: A wild card character in your App ID will prevent you from adding products for In App Purchases later. If you already have a product on App Store and want to integrate In App Purchases in the next release, but the live app isn't using a unique App ID, it will still work, but your users will not be notified about this update automatically through the App Store; they have to download it again. To push update notifications, App Store relies on the fact that subsequent updates to a product use the same App ID and incrementing version numbers. Although there are workarounds to associate a new App ID to an existing Bundle ID (Apple Technical Note QA1680), I still recommend using a unique product ID for each app you develop. As of this writing, with iOS 6, there are several features that rely on a unique App ID including but not limited to push notifications, Game Center integration, and iCloud integration. If your app might use any one of these features (even in a future release), use a unique App ID from the beginning.

Step 2: Generate Provisioning Profiles

In App Purchases can be tested on iOS Simulator version 5.1 and later; however, I highly recommend using a device to run and test In App Purchases. This means you need to create a provisioning profile for running your app on a device. The second step is to create a provisioning profile. Go to the fifth link from the left side of the navigation pane in the iOS Provisioning Portal. This step is exactly the same as for any other app. Remember to choose the same App ID that you created in the previous step.

Step 3: Create the App's Product Entry

Before you create In App Purchase products, you need to have an app that sells your In App Purchase products. You start by creating an application on iTunes Connect. Open iTunes Connect and click Manage Your Applications. That's the first link in the right column, as shown in Figure 18-2.

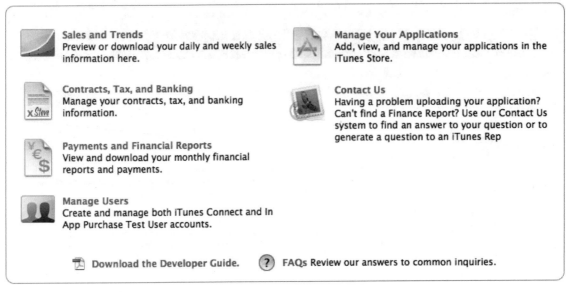

Sales and Trends
Preview or download your daily and weekly sales information here.

Manage Your Applications
Add, view, and manage your applications in the iTunes Store.

Contracts, Tax, and Banking
Manage your contracts, tax, and banking information.

Contact Us
Having a problem uploading your application? Can't find a Finance Report? Use our Contact Us system to find an answer to your question or to generate a question to an iTunes Rep

Payments and Financial Reports
View and download your monthly financial reports and payments.

Manage Users
Create and manage both iTunes Connect and In App Purchase Test User accounts.

Download the Developer Guide.

FAQs Review our answers to common inquiries.

Figure 18-2 iTunes Connect home page

Create a new application from the Manage Your Applications link. It's most important for you to choose your Bundle ID correctly. Choose the App ID you created in Step 1 as your Bundle ID here, as shown in Figure 18-3.

The Bundle ID and App ID are the same. Apple just uses a different name here.

---

 iTunes Connect

Mugunth Kumar M, Mugunth Kumar [Sign Out]

App Information

Enter the following in **UK English**.

App Name   [_____] (?)

SKU Number   [_____] (?)

Bundle ID   [Select   ⬍] (?)
You can register a new Bundle ID here.

Does your app have specific device requirements? Learn more

[Cancel]   [Continue]

Figure 18-3 New application form

Making a mistake in this step means that you will not be able to get detailed information about an In App Product programmatically later. The product identifiers will be returned as Invalid Product IDs, described later in this chapter. This is because your In App Purchase Products are tied to an app using the Bundle ID (App ID).

> An app with a specific Bundle ID cannot sell In App Purchase Products that are meant for a different app that uses a different Bundle ID.

Step 4: Create the In App Purchase Product Entries

Click on the app, and you'll see a screen like the one shown in Figure 18-4.

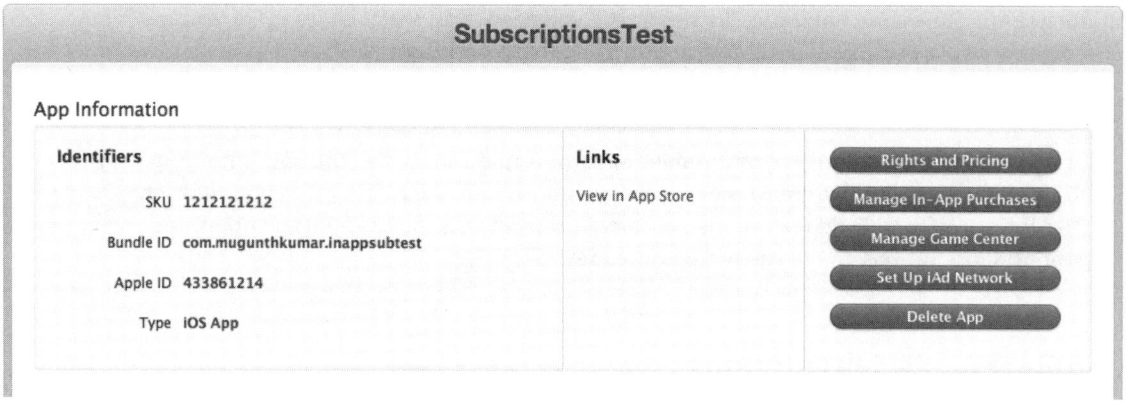

SubscriptionsTest

App Information

Identifiers		Links		
SKU	1212121212	View in App Store	Rights and Pricing	
Bundle ID	com.mugunthkumar.inappsubtest		Manage In-App Purchases	
Apple ID	433861214		Manage Game Center	
Type	iOS App		Set Up iAd Network	
			Delete App	

Figure 18-4 App information

Click the Manage In-App Purchases link to create your first In App Purchase Product.

> If you don't see the Manage In-App Purchases option on your iTunes Connect, be sure your iTunes Connect account has Admin privileges. Then check that your Contract Tax and Banking information is correct. If this is your first app and you haven't yet submitted your tax documents and/or haven't accepted the Paid Applications Contract, you will not see this link on the page. In that case, correct the situation appropriately and wait; Apple normally takes a week or two to approve your documents, depending on your location.

On the first page, you see four different types of products available for you to create. Choose the product type you're creating (discussed earlier in the chapter) and proceed.

> To create the Product ID for your In App Products, I recommend suffixing the product identifier with your Bundle ID. For example, if your Bundle ID is `com.mycompany.myapp`, your In App Product ID would be `com.mycompany.myapp.inapp`. This will ensure that Product IDs across your other apps don't clash with each other.

Consumables, Nonconsumables, Non-Renewing Subscriptions

For all product types except auto-renewable subscriptions, you have to enter a product identifier and choose its price tier. Add a description that shows up when the user buys your product. If your product is multilingual, add descriptions in all supported languages in this page.

You now have to add a screenshot of the product (yes, that's for the product you have not yet created) before you can submit the form. For the time being, upload a 320×480 iPhone screenshot. This screenshot is only for iOS App Store review purposes. In most cases, you need this only for content. For features or other consumables, it's okay to upload a screenshot displaying the In App Store.

> Non-renewing subscriptions are recommended since Apple announced the new auto-renewing subscriptions. Functionally, auto-renewing subscriptions offer everything that non-renewing subscriptions offer and add features like automatic renewal without user intervention, and restoring subscriptions on the customers' other devices.

Auto-Renewable Subscriptions

Auto-renewable subscriptions are slightly different. You create a subscription family and add the duration of the subscription to that family. This allows you to create the same subscription for a magazine with different durations. For instance, you can create a weekly subscription at $5 or a monthly subscription at $20 or a yearly subscription at $300. Other options within a subscription family are similar to consumables.

Step 5: Generating the Shared Secret

For auto-renewable subscriptions, you need to take one more very important step: generate a shared secret.

On the App Information page (refer to Figure 18-4), click the Manage In App Purchases link. On this page, you'll see a link titled View or Generate a Shared Secret. Copy the shared secret safely. You'll need it when you write the real code.

Now that you have the products set up properly on iTunes Connect, it's time to create some user accounts for testing.

Step 6: Creating Test User Accounts

Now create user accounts that you'll use for testing In App Purchases after implementation. You can do this later, after implementation, but doing it now completes all the steps needed for configuring In App Purchases.

To create test user accounts, open the iTunes Connect home page (refer to Figure 18-2) and click Manage Users. You'll see two links, one for creating an iTunes Connect user and another for creating a Test User. Click the second link and create a Test User. This should be fairly simple.

That completes the configuration part of In App Purchasing. If you've done all the steps correctly, you have completed 30 percent of the In App Purchase Integration. The remaining 70 percent is the real code, which you dive into next.

Step 7: Creating Hosted Content

Step 6 is the final step, unless you're planning to host additional downloadable content with Apple, (which I highly recommend). Hosting content is available only for nonconsumables at the moment, and you can choose this option from iTunes Connect when you create your In App purchasable product, as shown in Figure 18-5.

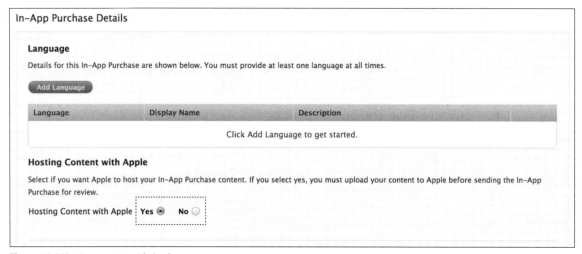

Figure 18-5 Hosting content with Apple

After you do this, you have to create a subproject that signs your hosted content. You can do this in Xcode, as shown in Figure 18-6.

After you create the project, edit the file `ContentInfo.plist` and change the `IAPProduct Identifier` key to your In App Purchase product's product identifier.

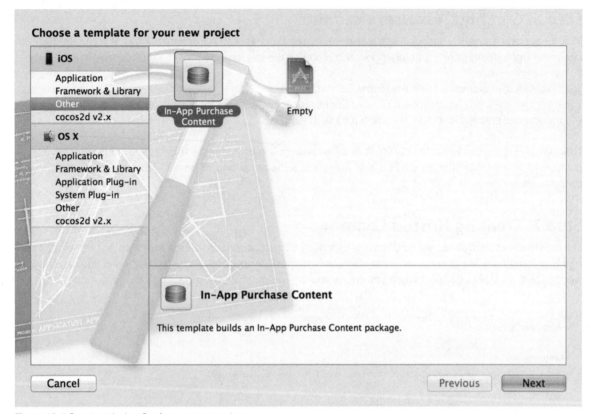

Figure 18-6 Creating a In App Purchase content project

Now, add the assets for this product into this project. You can add them from Finder or another Xcode project by dragging and dropping them. Build and Archive your project. You are all set now. You can submit this archive for App Store review using Xcode's Organizer, much like the way you submit apps.

In App Purchase Implementation

The configuration of In App Purchases was a bit difficult, and so is the programming involved in the implementation. In App Purchase implementation requires you to complete some tedious coding. The following is a comprehensive list of the important tasks that you have to do for implementing In App Purchases in your app.

- First and foremost, your app could be closed (probably by a phone call) while a transaction is in progress. Given that transactions can continue outside of your app, you should have a Store Observer class that initializes at application launch and receives any purchases made while the app is in the background.

- Remember the actual number of consumable items purchased.

- Remember the nonconsumable purchase and allow the user to restore it on his other devices.

- Remembering these purchases should be done securely using iOS keychain.

- Post non-renewing subscriptions to your server and "remember" them there. You should also have the capability to restore them on any other device when the user logs in to your app from another device.

- Consumable contents can be occasionally delivered digitally from your server. In that case, your server should be able to verify the authenticity of the receipt and provide content. This also means that your iOS app should post App Store purchase receipts to your server before it starts downloading content.

- Verify whether auto-renewing subscription receipts are still valid. Even though auto-renewing subscriptions are renewed automatically without manual user intervention, you're still required to check this because a user might have cancelled the subscription. A cancelled subscription remains valid only until the end of the current subscription period. This means that you as a developer need to remember purchases and purchase receipts, and you need to verify the receipts' validity, probably during app launch. If they are no longer valid, stop providing the content.

- To validate auto-renewing subscriptions, post your receipt and the shared secret you generated in Step 5 of the previous section to the App Store, parse the returned JSON (JavaScript Object Notation), and get the subscription's latest purchase date.

- The `StoreKit.framework` doesn't tell you when an auto-renewable subscription ends. Instead, the API returns the actual date of the latest receipt. From this date, you need to calculate the actual date of expiry.

- Display localized product prices and descriptions on your Store View Controller.

Although this might sound complicated, MKStoreKit wraps most of these functionalities. Next, you see how to add MKStoreKit into your app and build it.

Introduction to MKStoreKit

MKStoreKit is an open source framework that makes integrating In App Purchases simpler. As of this writing, the latest version is 4.0, and this is the version used in this chapter.

Why MKStoreKit?

MKStoreKit automatically takes care of the following items in the list in the preceding section.

- When your app is closed while a transaction is in progress, MKStoreKit automatically tracks this, continues to observe StoreKit delegates, and remembers any purchases made outside your app. This takes care of the first item.

- For consumables and nonconsumables, MKStoreKit automatically remembers your purchases in the iOS keychain. Purchases are remembered using your product identifiers as the key, so in most cases, you don't have to customize anything here. The second, third, and fourth items are thus taken care of automatically.

- The MKSKProduct class posts non-renewing subscriptions to your server. So, the fifth item is taken care of, if you do the server customization with the PHP code that comes with MKStoreKit. You learn about this later in the "Customizing MKStoreKit" section.

- For consumables or any content you deliver digitally from your server, you have to post the App Store receipt to your server. The server should verify the receipt and validate it. If the receipt is valid, it should deliver the content to your app. The `MKSKProduct` class takes care of this in tandem with some server configuration, which you learn about later in this chapter. This handles the sixth item.

- For auto-renewing subscriptions, you need to verify the latest receipt when your app is launched and disable access to content when the user has canceled the subscription. The `MKSKSubscriptionProduct` class does this automatically, including parsing the response JSON from the App Store and verifying that the subscriptions are still valid; it notifies you via NSNotificationCenter. You need to tell MKStoreKit how many days a subscription is valid. You learn how to do this later in this chapter. This takes care of the seventh, eighth, and ninth items.

- MKStoreKit has helper methods that return your product descriptions and price formatted in the local currency of the user. This takes care of the last item.

Customizing MKStoreKit to work with your app takes less than 25 lines of code. In some cases, it may be less than 10 lines of code.

Now, before you actually integrate the framework, it's always good to know how it works internally, which the next section explains.

Design of MKStoreKit

MKStoreKit uses blocks instead of delegates to notify you about product purchases. (You learn more about blocks in Chapter 23.) Other notifications like subscription expiry are posted to `NSNotificationCenter`. The framework comprises a main singleton class, the MKStoreManager, and several other support classes listed here:

- **MKStoreManager.h and MKStoreManager.m**—This is the main singleton class that handles most of the implementation. You have to initialize this singleton in the AppDelegate's `applicationDid FinishLaunchingWithOptions:` method.

- **MKSKProduct.h and MKSKProduct.m**—This is an internal class used by MKStoreKit to validate purchases. MKStoreManager uses this class to communicate with your server to check whether the receipt is valid and the actual product can be downloaded. This is used only for Server Product Model where you verify receipts on your server and deliver content digitally.

- **MKSKSubscriptionProduct.h and MKSKSubscriptionProduct.m**—This is another internal class used by MKStoreKit to validate your auto-renewable subscription purchases. MKStoreManager uses this class to communicate with your server to check whether the latest subscription receipts are still valid. If the user has cancelled his subscription, this class notifies MKStoreManager that the subscription is no longer valid, and MKStoreManager posts a notification. You have to observe these notifications (shown below) on your view controller and enable or disable your Subscribe buttons accordingly. If your app doesn't use auto-renewable subscriptions, you don't have to do this.

  ```
  kSubscriptionsPurchasedNotification
  kSubscriptionsInvalidNotification
  ```

■ **MKStoreKitConfigs.h and MKStoreKitConfigs.plist**—These two files in the framework require customization based on your app. You'll learn about customizing them later in this chapter.

■ **Remembering purchases**—MKStoreKit uses iOS keychain to remember a purchase automatically when it's purchased.

Customizing MKStoreKit

Here are two important files in MKStoreKit that you may be required to change:

```
MKStoreKitConfigs.h

MKStoreKitConfigs.plist
```

The `plist` file contains the list of products that you configured on iTunes Connect. You add your products under the corresponding keys in the `plist` depending on the product type. You add your consumables under the consumable key, nonconsumables under the nonconsumable key, and subscriptions under the subscriptions key. Every type of key has its own suboptions. You learn how to configure them later in this chapter.

Initializing MKStoreKit

Before you configure MKStoreKit, initialize it in your AppDelegate's `applicationDidFinishLaunchingWithOptions:` method. This ensures that the StoreObservers are initialized properly to receive transactions completed outside of the app. Just initialize the singleton by calling the following in your AppDelegate:

```
[MKStoreManager sharedManager];
```

> Though singletons initialize automatically when used for the first time, this call is still necessary when the app launches. Initializing MKStoreKit at launch ensures purchases made when the app was closed are received and stored when the app is opened again.

Configuring for Use with Server Product Model

When you sell content in your app and allow users to stream or download the content from your server, you use the server product model. In the server product model, the iOS app makes a purchase and sends the transaction receipt over to the server for verification. The server then verifies the receipt with Apple's receipt verification server and, if the receipt is valid, redirects the request to the requested content.

Server Setup

MKStoreKit comes with server code in PHP ready to verify receipts from the server. Copy the Server Code directory and open it up for access. Copy the public URL for this directory. Let's just assume that it can be accessed at this location:

```
http://api.example.com/servercode
```

Now go back to your iOS source, open the file `MKStoreKitConfigs.h`, and locate these lines:

```
#define SERVER_PRODUCT_MODEL 0
#define OWN_SERVER nil
#define REVIEW_ALLOWED 1
```

Set the `OWN_SERVER` value to `@"http://api.example.com/servercode"`.

You are all set. MKStoreKit will ping the `featureCheck.php` endpoint in this directory to verify receipts and remember the purchase only when the server says receipts are valid. Receipt validation is done by posting the receipt to Apple's receipt validation URL:

```
https://buy.itunes.apple.com/verifyReceipt
```

For sandbox testing, you use

```
https://sandbox.itunes.apple.com/verifyReceipt
```

The server code automatically switches this based on the configuration you defined. The default implementation of `featureCheck.php` returns plain strings—YES or NO—based on whether the receipts are valid or not. You might need to modify it to return in JSONs along with the URL of the content location.

Configuring for Use with Consumables

In a generic sense, a consumable is a product that depletes as it is used. Printer ink is a common example. In App Purchases consumables behave the same way. When a user purchases a consumable product, it's stored on the device and stays there until he uses it up. You're not obliged to restore consumables on other devices. When implementing consumables, you often encounter a business case where bulk purchases are subsidized to the user, just like real-world consumables.

For example, you might have two products in your game, a small box of ammo containing one hundred bullets at $0.99, and a larger box containing a thousand at $5.99. Within your game, both the products are synonymous. However, SKU-wise, they are treated differently, and their cost is different. To implement this model, MKStoreKit allows you to specify names for your products. That way, you can tell MKStoreKit to treat them separately during purchase but treat them the same when consumed. Essentially, this means that buying either of the products increases the count of the same item. To configure MKStoreKit this way, use the suboptions inside the consumable key in `MKStoreKitConfigs.plist`, as illustrated in Figure 18-7.

▼ Consumables	Diction...	(1 item)
▼ com.mugunthkumar.inappsubtest.consumable1	Diction...	(2 items)
Count	Number	500
Name	String	EggBasket

Figure 18-7 Configuring consumables in MKStoreKit

You add the product ID of every consumable within the Consumables dictionary. Every product has a Name and Count. Assume you have two products: `com.myapp.mygame.ammopackSmall` and `com.myapp.mygame.ammopackLarge`. The "Count" key will let you set the count of the virtual consumable for this product purchase. The "Name" key will let you set the real name of the consumable. MKStoreKit normally remembers purchases with a product ID. But for consumables, it uses the "Name" key because multiple SKUs can actually mean the same product within your app.

Now, in the course of your game/app, if the user "consumes" your product, you first check if it's available by calling the methods

```
- (BOOL) canConsumeProduct:(NSString*) productName quantity:(int)
quantity;
- (BOOL) consumeProduct:(NSString*) productName quantity:(int) quantity;
```

`consumeProduct` will properly deduct the quantity of the product consumed from the purchased quantity and store the remaining available quantity in the keychain, all automatically.

Configuring for Use with Auto-Renewable Subscriptions

Auto-renewable subscription configuration is very similar to consumables configuration. The first step is to specify your shared secret. Copy the shared secret you generated earlier in the chapter and paste it here in the file MKStoreKitConfigs.h:

```
#warning Shared Secret Missing Ignore this warning if you don't use auto-
renewable subscriptions
#define kSharedSecret @"<FILL IN YOUR SHARED SECRET HERE>"
```

You can now remove the `#warning` line. Also, if you don't use auto-renewable subscriptions (and thus don't have a shared secret), you can ignore this warning. Just remove the `#warning` line.

Open the `plist` file, and as you would for consumables, instead of specifying the count, specify the duration of the subscription. Now, on your view controllers that display the store, observe the `kSubscriptionsPurchasedNotification` and/or `kSubscriptionsInvalidNotification` and enable or disable your subscribe buttons accordingly.

Making the Purchase

Now that you have configured MKStoreKit, making the real purchase is very simple. It's just a single method call like this:

```
[[MKStoreManager sharedManager] buyFeature:@"com.myapp.myfeature"
        onComplete:^(NSString*) purchasedProduct
{
  // provide the content for the product "purchasedProduct".
}
        onCancelled:^
{
  // optionally display an error
}];
```

After you configure MKStoreKit properly, it takes just a single method call to initiate a purchase. Remembering the purchase is automatically done for you. To check if a product has been purchased previously, you can call this method:

```
[MKStoreManager isProductPurchased:@"com.myapp.feature1"];
```

This returns a `Boolean` that states whether the product is purchased or not. Restoring purchases is done with another one-liner. Read the `MKStoreManger.h` file to see the functionalities exposed by MKStoreKit.

Now that's fewer than 10 lines to get it all running, and maybe another 15 lines of configuration, as I promised at the beginning of the chapter.

Downloading Hosted Content

You set a handler method, `hostedContentDownloadStatusChangedHandler`, to update the UI that shows the download statuses of your user's purchases.

```
[MKStoreManager sharedManager].hostedContentDownloadStatusChangedHandler
= ^(NSArray*
  hostedContents) {

  // Update your UI here. This includes updating your download progress
and
  statuses. The hostedContents is an array of SKDownload objects and each
object
  has an identifier, download state and download progress.
};
```

The best part of hosted content is that you don't have to worry about receipt validation on your server, and you don't have to bother about encrypting your content and getting associated export licenses or depend on a third-party provider like Urban Airship to host your content. The downside is Apple's approval process. Again, if your products change every day, like daily wallpaper, you might have to submit enough products to the queue so that your app has some content to purchase despite Apple's long approval times.

Testing Your Code

Now that you have implemented In App Purchasing, go ahead and test whether your code works. You'll need the credentials of the test user account you created earlier in this chapter for testing. Before you start, open the settings app on the device you'll be using for testing and tap the Store menu. Log out of the App Store and ensure that no user is logged in to the store.

Run the app on a device and initiate a purchase. You're prompted to enter or create a new iTunes account or use an existing account. Choose to use an existing account and enter the test username and password you created previously. App Store will now ask you to confirm the purchase of your In App Purchase Product. Tap on Buy (or subscribe), and your product is now purchased. You have successfully completed the In App Purchase integration in your app.

All this sounds good, but what happens when there is a problem? You see some quick troubleshooting techniques in the next section.

Troubleshooting

Despite all the explanation provided in this chapter, In App Purchases remains one of the most difficult frameworks to troubleshoot.

Invalid Product IDs

The most common problem is that App Store returns your product as invalid. If you have been following the chapter closely, reading every tip and warning, you shouldn't encounter an invalid product ID. However, the problem could happen if you have any of the following issues:

- The product Bundle ID in the `Info.plist` file doesn't match the App ID you created.

- Your contract and tax statements are not yet submitted and/or you have not yet accepted the iOS developer Paid Applications Contract. To correct these issues, go to iTunes Connect and click the Contracts, Tax, and Banking link. It's very important to check this when you submit your first app.

- Jailbroken devices sometimes don't work well with the App Store. An app called AppSync from Cydia seems to be the cause of most problems associated with In App Purchasing. In App Purchases are best tested with a device running an unmodified operating system.

Sometimes, even after you ensure that none of these issues is a problem, the App Store still indicates that your products are invalid. This happens more often to non–U.S.-based developers. Wait several hours before retrying (see "Retrieving Store Information" in the Apple Developer Documentation, 2011). Apple uses distributed servers, and it might take several hours to migrate the products you created from the U.S. servers to other mirror servers near your country.

Cannot Connect to iTunes Store

The other common problem is when you get the message "Cannot connect to iTunes store: Code: -1003." This happens when your firewall blocks iTunes. Test the In App purchases by connecting to a different network or ensure that you have proper Internet connectivity.

You Have Already Purchased This Product, but It's Still Not Downloaded

This error is common when you work with consumables. It happens mostly when you tap on the Buy button too often. The workaround is to disable the Buy button once the purchase is initiated and reenable it after the transaction completes. Follow the interaction pattern similar to the built-in App Store.

If your problem is still not solved, the old school method of deleting the app and redoing all the steps often works.

Summary

In App Purchases, although tricky to implement, offer an innovative and unique way to monetize your apps. Carefully deciding on your business model and implementing In App Purchases can vastly increase the money you make from the App Store. A quick look at the top grossing apps on U.S. App Store shows that at least 25 percent follow the freemium model whereby the app is free but content and features are provided through In App Purchases. This clearly proves that freemium is successful on App Store. With frameworks like MKStoreKit minimizing your coding efforts, why not give it a try?

Further Reading

Apple Documentation

The following documents are available in the iOS Developer Library at `developer.apple.com` or through the Xcode Documentation and API Reference.

App Store Review Guidelines

In App Purchase Programming Guide

Retrieving Store Information

Blogs

MKBlog. "iPhone Tutorial: In App Purchases"
`http://blog.mugunthkumar.com/coding/iphone-tutorial---in-app-purchases`

MKBlog. "MKStoreKit 4.0 — Supporting Auto Renewable Subscriptions"
`http://blog.mugunthkumar.com/coding/mkstorekit-4-0-supporting-auto-renewable-subscriptions`

Other Resources

MugunthKumar / MKStoreKit
`https://github.com/MugunthKumar/MKStoreKit`

Chapter 19

Debugging

The most difficult part in writing software is debugging. Debugging is hard. It's even harder when you're writing software in a low-level language (compared to high-level languages like Java/C#). Instead of a stack trace, you'll often hear iOS developers using buzzwords like dSYM files, symbolication, crash dumps.

This chapter introduces you to LLDB, Apple's Lower Level DeBugger. I'll explain some commonly used terms like dSYM, symbolication, and others that are traditionally different from other programming languages. A lot changed when Apple replaced GDB with LLDB, and nearly everything that I talk about in this chapter is specific to LLDB. If you're still using GDB, it's high time to change to the newer LLDB. I'll show you some of Xcode's features that will help you with debugging and that will unleash the power of the LLDB console. Later in this chapter, you'll find different techniques for collecting crash reports, including a couple of third-party services.

LLDB

LLDB is a next-generation high-performance debugger built using reusable components from LLVM, including the complete LLVM compiler, that includes LLVM's Clang expression parser and the disassembler. What this means to you, the end user/developer, is that LLDB understands the same syntax that your compiler understands, including Objective-C literals and Objective-C's dot notation for properties. A debugger with compiler-level accuracy means that any new feature added to LLVM will automatically be available to LLDB.

Xcode's previous debugger, GDB, didn't really "understand" Objective-C. As such, something as simple as `po self.view.frame` was more complicated to type. You need to type `po [[self view] frame]`. When the compiler was replaced, there was a need to improve the debugger. Because GDB was monolithic, there was no workaround; it had to be rewritten from scratch. LLDB is modular, and providing the debugger with API support and a scripting interface was one of the design goals. In fact, the LLDB command-line debugger links to the LLDB library through this API. In the "Scripting Your Debugger" section later in this chapter I show you how to script LLDB to make your debugging session easier.

Debugging with LLDB

Debugging with LLDB offers little advantage over GDB for most developers. In most cases, you won't even notice the difference, except for a few obvious changes like support for Objective-C properties and literals. However, knowing how LLDB works internally and the subtle differences it brings along will make you a better developer and one who can indeed push the limits. After all, you picked up this book to learn about the advanced concepts that would help you push the limits, right? The next few sections aim at explaining these concepts in detail. With that, it's time to get started.

Debug Information File (dSYM file)

The debug information file (dSYM) stores debugging information about the target. What does it contain? Why do you need a debug information file in the first place? Every programming language you've ever written code for has a compiler that converts your code either to some kind of intermediate language that is understandable by the runtime or to machine code that gets executed natively on the machine's architecture.

A debugger is commonly integrated within your development environment. The development environment often allows you to place breakpoints that stop the app from running and allows you to inspect the values of variables in your code. That is, a debugger effectively freezes the app in real time and allows you to inspect variables and registers. There are at least two important types of debuggers: symbolic debuggers and machine language debuggers. A machine language debugger shows the disassembly when a breakpoint is hit and lets you watch the values of registers and memory addresses. Assembly code programmers generally use this kind of debugger. A symbolic debugger shows the symbol/variable you use in your application when you're debugging through the code. Unlike a machine language debugger, a symbolic debugger allows you to watch symbols in your code instead of registers and memory addresses.

For a symbolic debugger to work, there should be a link or a mapping between the compiled code and the source code you wrote. This is precisely what a debug information file contains. Some languages, such as Java, for example, inject debug information within the byte code. Microsoft Visual Studio, on the other hand, has multiple formats including a standalone PDB file.

> Languages such as PHP, HTML, or Python are different. They usually don't have a compiler, and to some extent, they're not classified as programming languages. PHP and Python are technically scripting languages, whereas HTML is a markup language.

Debuggers use this debug information file to map the compiled code, either intermediate code or the machine code, back to your source code. Think of a debug information file as a map you would refer to if you were to visit an unknown city as a tourist. The debugger is able to stop at the correct location according to the breakpoints you place in your source code by referring to the debug information file.

Xcode's debug information file is called as a dSYM file (because the file's extension is `.dSYM`).

> A dSYM file is technically a package containing a file with your target name in DWARF scheme.

When you create a new project, the default setting is to create a debug file automatically. The Build Options, which is under the Build Settings tab, of your project file (see Figure 19-1) shows this setting.

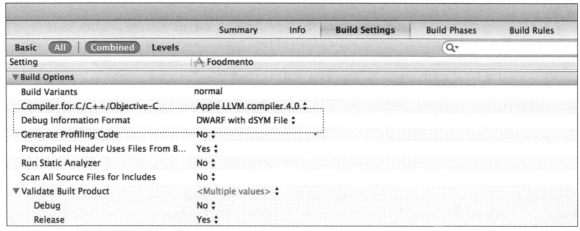

	Summary	Info	**Build Settings**	Build Phases	Build Rules

Basic **All** | **Combined** | Levels   (Q▾

Setting	A Foodmento
▼ **Build Options**	
Build Variants	normal
Compiler for C/C++/Objective-C	Apple LLVM compiler 4.0 ↕
Debug Information Format	DWARF with dSYM File ↕
Generate Profiling Code	No ↕
Precompiled Header Uses Files From B...	Yes ↕
Run Static Analyzer	No ↕
Scan All Source Files for Includes	No ↕
▼ Validate Built Product	<Multiple values> ↕
Debug	No ↕
Release	Yes ↕

Figure 19-1 The Debug Information Format setting in your target settings.

The dSYM file is automatically created every time you build the project. You can also create a dSYM file using the command-line utility `dsymutil`.

Symbolication

Compilers, including the LLVM compiler, convert your source code to assembly code. All assembly codes have a base address, and the variables you define, the stack you use, and the heap you use are all dependent on this base address. This base address could change every time the application runs, especially on iOS 4.3 and above—operating systems that introduced Address Space Layout Randomization. Symbolication is the process of replacing this base address with method names and variable names (collectively known as *symbols*). The base address is the entry point to your application, normally your "main" method unless you are writing a static library. You symbolicate other symbols by calculating their offset from the base address and mapping them to the dSYM file. Don't worry; this symbolication happens (almost transparently) when you're debugging your app in Xcode or when you profile it using Instruments.

> Instruments also needs the debug information file to symbolicate the running target and locates it using Spotlight. If you added the `.dSYM` extension to the exclusion filter in Spotlight, you will not see the variable names (symbol names) in Instruments. More often, this happens when your disk permissions, especially permissions to the Derived Data directory, are corrupted. A quick run by a disk utility should fix that and get you up and running.

Xcode's Symbolication

However, there are times when you want to symbolicate a binary or a crash report (more often the latter) manually. Later in this section, I discuss the various kinds of crash reports that you can use to analyze and fix issues in your app, including a homegrown crash reports collection tool and a couple of third-party crash-reporting services. When you're using the Apple-provided iTunes Connect crash reports, you only see addresses and hex codes scattered throughout. Without proper symbolication, you cannot understand what is going on. Fortunately, Xcode symbolicates crash reports when you just drag a crash report to it. Now, how does Xcode know the corresponding dSYM file? For this "automatic" symbolication to work, you use Xcode's Build and Archive option to build your app for submission to the App Store.

Inside the xcarchive

A quick look at the `xcarchive` bundle created by Xcode reveals the following directories: dSYMs, Products, and an `Info.plist` file. The dSYMs directory contains the list of dSYM files for all the target/static libraries you included in your project. The Products directory contains the list of all executable binaries. The `Info.plist` file is the same `plist` file in your project. The `Info.plist` file plays a very important role in identifying the version of the `target/dsym` in the `xcarchive` bundle. In fact, when you drag a `.crash` file from iTunes Connect to your Xcode, Xcode internally looks up the archives for an `Info.plist` file that matches the crash report and picks up the `.dSYM` file from the dSYMs directory inside that archive. This is the reason why you should *never* delete a submitted archive. When you delete your archives after submission, you'll end up in hot water when you try to symbolicate a crash report.

Committing Your dSYMs to Your Version Control

The other way to store your dSYMs is to commit them to your version control. When you get a crash report from iTunes Connect, check out the dSYM that corresponds to the submitted version and symbolicate your crash reports by matching them with the dSYMs. This way, all developers on a team will have access to the dSYM files, and debugging crashes can be done by anyone on the team without e-mailing the dSYMs around.

> Instead of committing dSYMs for every commit to your develop or feature branch, consider committing them only when you make a release. If you're using Git as explained in Chapter 2, you should be committing dSYMs to the master branch for every release. There are other alternatives, including some third-party services that do server-side symbolication. I discuss them in the "Collecting Crash Reports" section later in this chapter.

Breakpoints

Breakpoints are a way to pause your debugger and inspect your symbols and objects in real time. Some debuggers, including LLDB, allow you to move the instruction pointer and continue debugging from a different location. You can set LLDB breakpoints in your app right from Xcode. You just scroll to the line where you want the breakpoint to be placed and click on Xcode's, Product⇨Debug⇨Add breakpoint at Current line menu, or press Cmd+\.

The Breakpoint Navigator

Breakpoints that you added to your project are listed automatically in your Breakpoint navigator. You can access the Breakpoint navigator using the keyboard shortcut Cmd+6.

The Breakpoint navigator also allows you to add breakpoints for exceptions and symbols.

Exception Breakpoints

An exception breakpoint stops program execution when your code causes an exception to be thrown. Some of the methods in the `Foundation.framework`'s `NSArray`, `NSDictionary`, or `UIKit` classes (like the `UITableView` method) throw exceptions when certain conditions cannot be met. This includes trying to mutate an `NSArray` or trying to access an array element that is beyond the array's bounds. A `UITableView` throws an exception when you declare the number of rows to be "n" and don't provide a cell for every row. In theory, debugging exceptions might look easy, but understanding the source of your exception could be fairly complicated. Your app will just crash with a log that prints out the exception that caused the crash. These `Foundation.framework` methods are used throughout your project, and you won't really understand what happened by looking at the log if you don't have an exception breakpoint set. When you set an exception breakpoint, the debugger pauses the execution of the program just after the exception is thrown, but before it is caught, and you get to see the stack trace of the crashed thread in your Breakpoint navigator.

To make things clear, compare debugging an app with and without the exception breakpoint enabled.

Create an empty application in Xcode (any template should work). In your app delegate, add the following line:

```
NSLog(@"%@", [@[] objectAtIndex:100]);
```

This line creates an empty array, accesses the 100th element and logs it. As this is not legally allowed, Xcode crashes with the following log in Console and takes you to the infamous `main.m`.

```
2012-08-27 15:25:23.040 Test[31224:c07] (null)
libc++abi.dylib: terminate called throwing an exception
(lldb)
```

No one can understand what is happening behind the scenes from that cryptic log message. To debug exceptions like these, you set an exception breakpoint.

You set an exception breakpoint from the Breakpoint navigator. Open the Breakpoint navigator and click the + button on the lower-left corner, choose Add Exception Breakpoint and accept the default settings to set a new exception breakpoint, as shown in Figure 19-2.

Run the same project again. You should see the debugger pausing your application and stopping exactly on the line that raised the exception, as illustrated in Figure 19-3.

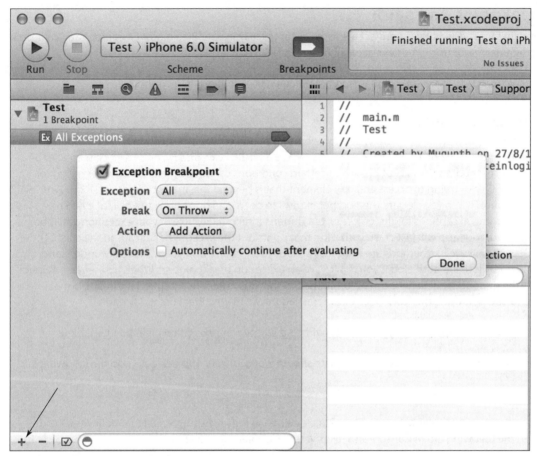

Figure 19-2 Adding an exception breakpoint

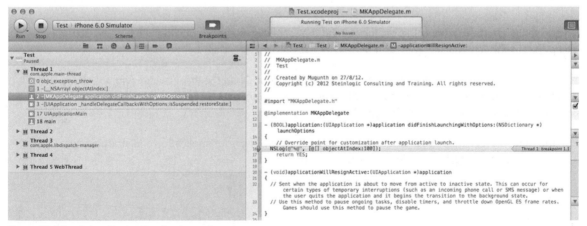

Figure 19-3 Xcode breaking at a breakpoint you just set

Exception breakpoints can help you understand the root cause of the exception. The first thing I do when I start a new project is to set an exception breakpoint. I highly recommend you doing so.

> When you want to run your app quickly without hitting any breakpoints, disable all of them with the keyboard shortcut Cmd-Y.

Symbolic Breakpoints

Symbolic breakpoints are breakpoints that pause your program's execution when a particular symbol gets executed. A symbol can be a method name or a method in a class or any C method (`objc_msgSend`).

You set a symbolic breakpoint from the Breakpoint navigator, much like how you set an exception breakpoint. Just choose Symbolic Breakpoint instead of Exception Breakpoint. Now in the dialog, type the symbol that you are interested in, as shown in Figure 19-4.

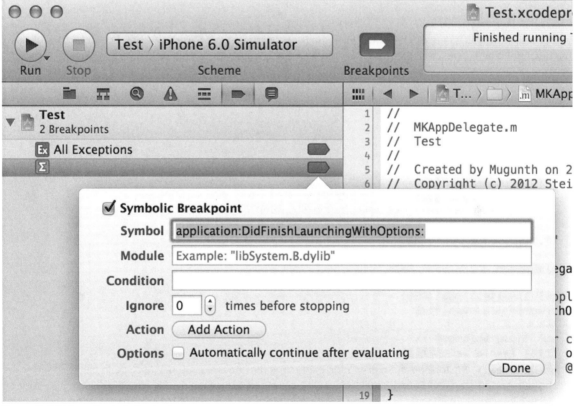

Figure 19-4 Adding a Symbolic Breakpoint

Now type `application:didFinishLaunchingWithOptions:` and press Enter. Build and run your application now. You should see the debugger stopping your application as it starts and showing the stack trace.

Well, the symbol you watched didn't provide any advantage in addition to placing a normal breakpoint in your `application:didFinishLaunchingWithOptions:`. Symbolic breakpoints are often used to watch interesting methods like

```
-[NSException raise]
malloc_error_break
-[NSObject doesNotRecognizeSelector:]
```

In fact, the first exception breakpoint you created in the previous section is synonymous to a symbolic breakpoint on `[NSException raise]`.

The `malloc_error_break` and `[NSObject doesNotRecognizeSelector:]` are symbols that are useful for debugging memory-related crashes. If your app crashes with an `EXC_BAD_ACCESS`, setting a symbolic breakpoint in one or both of these symbols will help you localize the issue.

Editing Breakpoints

Every breakpoint you create can be edited from the Breakpoint navigator. You edit a breakpoint by Ctrl+Clicking the breakpoint and choosing Edit Breakpoint from the menu. You will see a breakpoint-editing sheet, as shown in Figure 19-5.

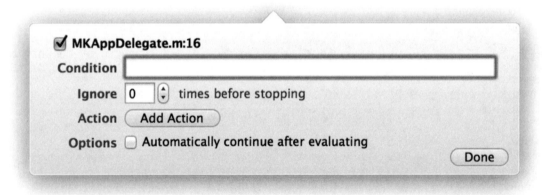

Figure 19-5 Editing breakpoints

Normally, a breakpoint stops program execution every time the line is executed. You can edit a breakpoint to set a condition and create a conditional breakpoint that is executed only when the set condition is reached. Why would this be useful? Imagine you are looping through a large array (n>10000). You're sure that the objects after

5500 are malformed and want to find out what is going wrong. The traditional approach is to write additional code (in your application's code) to check for index values greater than 5500 and remove this code after your debugging session is over.

For example, you could write something like

```
for(int i = 0 ; i < 10000; i ++) {

  if(i>5500) {
    NSLog(@"%@", [self.dataArray objectAtIndex:i]);
  }
}
```

and place a breakpoint in the NSLog. The cleaner way is to add this condition to the breakpoint itself. In Figure 19-5, the sheet has a text field for adding a condition. Set this condition to i>5500 and run your app. Now, instead of breaking for every iteration, your breakpoints stop only when your condition is met.

You can customize your breakpoints to print out a value or play an audio file or execute an action script by adding an action. For example, if the objects that you're iterating are a list of users, and you want to know if a said user is in your list, you can edit your breakpoint to break when the object that you're interested in is reached. Additionally, in the action, you can select from a list of audio clips to play, execute an Apple script and/ or perform a variety of other functions. Click the Action button (refer to Figure 19-5) and choose the custom action, Sound. Now instead of stopping at your breakpoint, Xcode plays the audio clip you selected. If you're a game programmer, you might be interested in capturing an Open GL ES frame when a particular condition occurs, and this option is also available through the Action button.

Sharing Breakpoints

Your breakpoints now have code (or rather code fragments) associated with them that you want to save to your version control. Xcode 4 (and above) allows you to share your breakpoints with your coworkers by committing them to your version control. All you need to do is Ctrl-Click a breakpoint and click Share. Your breakpoints are now saved to the xcshareddata directory inside the project file bundle. Commit this directory to version control to share your breakpoints with all other programmers in your team.

Watchpoints

Breakpoints provide you with the ability to pause the program execution when a given line is executed. Watchpoints provide a way to pause the program execution when the value stored at a variable changes. Watchpoints help to solve issues related to global variables and track which method updates a said global variable. Watchpoints are like breakpoints, but instead of breaking when a code is executed, they break when data is mutated.

In an object-oriented setting, you don't normally use global variables to maintain state, and watchpoints may not be used often. However, you may find it useful to track state changes on a singleton or other global objects like your Core Data persistent store coordinator or API engines like the one we created in Chapter 14. You can set a watchpoint on accessToken in the RESTfulEngine class to know the engine's authentication state changes.

You cannot add a watchpoint without running your application first. Start your app, and open the watch window. The watch window, by default, lists the variables in your local scope. Ctrl+click a variable in the watch window. Now click on the watch <var> menu item to add a watchpoint on that variable. Your watchpoints will now be listed in the Breakpoint navigator.

The LLDB Console

Xcode's debugging console window is a full-featured LLDB debugging console. When your app is paused (at a breakpoint), the debugging console shows the LLDB command prompt. You can type any LLDB debugger command into the console to help you with debugging, including loading external python scripts.

The most frequently used command is po, which stands for print object. When your application is paused in debugger, you can print any variable that is in the current scope. This includes any stack variables, class variables, properties, ivars, and global variables. In short, any variable that is accessible by your application at the breakpoint can be accessed via the debugging console.

Printing Scalar Variables

When you're dealing with scalars like integers or structs (CGRect, CGPoint, etc.), instead of using po, you use p, followed by the type of struct.

Examples include

```
p (int) self.myAge
p (CGPoint) self.view.center
```

Printing Registers

Why do you need to print values on registers? You don't store variables in CPU registers, right? Right, but the registers hold a wealth of information about the program state. This information is dependent on the subroutine calling convention on a given processor architecture. Knowledge of this information reduces your debug cycle time tremendously and makes you a programmer who can push the limits.

Registers in your CPU are used for storing variables that have to be accessed frequently. Compilers optimize frequently used variables like the loop variable, method arguments, and return variables in the registers. When your app crashes for no apparent reason (apps always crash for no apparent reason 'til you find the problem, right?), probing the register for the method name or the selector name that crashed your app will be very useful.

The C99 language standard defines the keyword register that you can use to instruct the compiler to store a variable in the CPU registers. For example, declaring a for loop like for (register int i = 0 ; i < n ; i ++) will store the variable i in the CPU registers. Note that the declaration isn't a guarantee and the compiler is free to store your variable in memory if there are no available free registers.

You can print the registers from the LLDB console using the command `register read`. Now, create an app and add a code snippet that causes a crash.

```
int *a = nil;
NSLog(@"%d", *a);
```

You will create a nil pointer and try accessing the value at the address. Obviously, this is going to throw an `EXC_BAD_ACCESS`. Write the preceding code in your `application:didFinishLaunchingWithOptions:` method (or any method you like) and run the app in your *simulator*. Yes, I repeat, in your *simulator*. When the app crashes, go to the LLDB console and type the command to print the register values.

```
register read
```

Your console should show something like the following.

Register Contents (Simulator)

```
(lldb) register read
General Purpose Registers:
       eax = 0x00000000
       ebx = 0x07408520
       ecx = 0x00001f7e   Test`-[MKAppDelegate
       application:didFinishLaunchingWithOptions:] + 14 at
       MKAppDelegate.m:13
       edx = 0x00003604   @"%d"
       edi = 0x07122070
       esi = 0x0058298d   "application:didFinishLaunchingWithOptions:"
       ebp = 0xbfffde68
       esp = 0xbfffde30
        ss = 0x00000023
    eflags = 0x00010286   UIKit`-[UIApplication _
addAfterCACommitBlockForViewController:] + 23
       eip = 0x00001fca   Test`-[MKAppDelegate
       application:didFinishLaunchingWithOptions:] + 90 at
       MKAppDelegate.m:19
        cs = 0x0000001b
        ds = 0x00000023
        es = 0x00000023
        fs = 0x00000000
        gs = 0x0000000f

(lldb)
```

The equivalent on a device (ARM processor) looks like the following.

Register Contents (Device)

```
(lldb) register read
General Purpose Registers:
        r0 = 0x00000000
```

(continued)

```
        r1  = 0x00000000
        r2  = 0x2fdc676c
        r3  = 0x00000040
        r4  = 0x39958f43    "application:didFinishLaunchingWithOptions:"
        r5  = 0x1ed7f390
        r6  = 0x00000001
        r7  = 0x2fdc67b0
        r8  = 0x3c8de07d
        r9  = 0x0000007f
       r10  = 0x00000058
       r11  = 0x00000004
       r12  = 0x3cdf87f4    (void *)0x33d3eb09: OSSpinLockUnlock$VARIANT$mp +
   1
        sp  = 0x2fdc6794
        lr  = 0x0003a2f3    Test`-[MKAppDelegate
        application:didFinishLaunchingWithOptions:] + 27 at
        MKAppDelegate.m:13
        pc  = 0x0003a2fe    Test`-[MKAppDelegate
        application:didFinishLaunchingWithOptions:] + 38 at
        MKAppDelegate.m:18
      cpsr  = 0x40000030

(lldb)
```

Your output may vary, but pay close attention to the eax, ecx, and esi on the simulator or r0-r4 registers when running on a device. These registers store some of the values that you're interested in. In the Simulator (running on your Mac's Intel processor), the ecx register holds the name of the selector that is called when your app crashed. You print an individual register to console by specifying the register name as shown below

```
register read ecx.
```

You can also specify multiple registers like

```
register read eax ecx.
```

The ecx register on Intel architecture and the r15 register on ARM architecture hold the program counter. Printing the address of the program counter will show the last executed instruction. Similarly, eax (r0 on ARM) holds the receiver address, ecx (r4 on ARM) and holds the selector that was called last (in this case, it's the application:didFinishLaunchingWithOptions: method). The arguments to the methods are stored in registers r1-r3. If your selector has more than three arguments, they are stored on stack, accessible via the stack pointer (r13). sp, lr, and pc are actually aliases to the r13, r14, and r15 registers, respectively. Hence, register read r13 is equivalent to register read sp.

So *sp, *sp+4 , and so on, will contain the address of your fourth and fifth arguments, and so on. On Intel architecture, the arguments start at the address stored in ebp register.

When you download a crash report from iTunes Connect, it normally has the register state, and knowing the register layout on ARM architecture will help you better analyze the crash report. The following is a register state from a crash report.

Register State in a Crash Report

```
Thread 0 crashed with ARM Thread State:
    r0: 0x00000000    r1: 0x00000000    r2: 0x00000001    r3:
0x00000000
    r4: 0x00000006    r5: 0x3f871ce8    r6: 0x00000002    r7:
0x2fdffa68
    r8: 0x0029c740    r9: 0x31d44a4a    r10: 0x3fe339b4    r11:
0x00000000
    ip: 0x00000148    sp: 0x2fdffa5c    lr: 0x36881f5b    pc:
0x3238b32c
  cpsr: 0x00070010
```

Using `otool`, you can print the methods used in your app. Match the address in your program counter using `grep` to see which exact method was being executed when the app crashed.

```
otool -v -arch armv7 -s __TEXT __cstring <your image> | grep 3238b32c
```

Replace `<your image>` with the image of the app that crashed. (You will have either committed this to your repository or stored it in the application archives in Xcode.)

Note that what you learned in this section is highly processor-specific and may change in the future if Apple changes the processor specification (from ARM to something else). However, once you understand the basics, you should be able to apply this knowledge to any new processor that comes along.

Scripting Your Debugger

The LLDB debugger was built from the ground up to support APIs and pluggable interfaces. Python scripting is one of the benefits of these pluggable interfaces. If you're a Python programmer, you'll be pleasantly surprised to learn that LLDB supports importing Python scripts to help you with debugging, which means that you can write a script in Python, import it into your LLDB, and inspect variables in LLDB using your script. If you're not a Python programmer, you can skip this section and probably won't lose anything.

Assume that you want to search for an element in a large array containing, say, 10,000 objects. A simple `po` on the array is going to list all ten thousand objects, which is tedious to manually go through. If you have a script that takes this array as a parameter and searches for the presence of your object, you can import it into LLDB and use it for debugging.

You start the Python shell from the LLDB prompt by typing `script`. The prompt changes from `(lldb)` to `>>>`. Within the script editor, you can access the LLDB frame using the `lldb.frame` Python variable. So `lldb.frame.FindVariable("a")` will get the value of the variable *a* from the current LLDB frame. If you're iterating an array to look for a specific value, you can assign the `lldb.frame.FindVariable("myArray")` to a variable and pass it to your Python script.

The following code illustrates this.

Invoking the Python Script to Search for Occurrence of an Object

```
>>> import mypython_script
>>> array = lldb.frame.FindVariable ("myArray")
>>> yesOrNo = mypython_script.SearchObject (array, "<search element>")
>>> print yesOrNo
```

The preceding code assumes that you wrote a `SearchObject` function in a file `mypython_script`. Explaining the implementation details of the python script is outside the scope of this book.

NSZombieEnabled Flag

A debugging chapter is not complete without mentioning the `NSZombieEnabled` environment variable. The `NSZombieEnabled` variable is used to debug memory-related bugs and to track over release of objects. `NSZombieEnabled` is an environment variable that, when set, swizzles out the default `dealloc` implementation with a zombie implementation that converts an object to a zombie object when the retain count reaches zero. The functionality of the zombie object is to display a log and break into the debugger whenever you send a message to it.

So when you enable NSZombies in your app, instead of crashing, a bad memory access is just an unrecognized message sent to a zombie. A zombie displays the received message and breaks into the debugger allowing you to debug what went wrong.

You enable the `NSZombiesEnabled` environment variable from Xcode's schemes sheet. Click on Product⇨ Edit Scheme to open the sheet and set the Enable Zombie Objects check box, as shown in Figure 19-6.

Zombies were helpful back in the olden days when there was no ARC. With ARC, if you're careful with your ownerships, normally you won't have memory-related crashes.

Different Types of Crashes

Software written using a programming language crashes as opposed to web applications written (or scripted) in a scripting or a markup language. Because a web application runs within the context of a browser, there is little possibility that a web app can corrupt memory or behave in a way that could crash the browser. If you're coming from a high-level language background, terms used by Xcode to denote various crashes may be cryptic to you. This section attempts to shed light on some of these. Crashes are usually signals sent by the operating system to the running program.

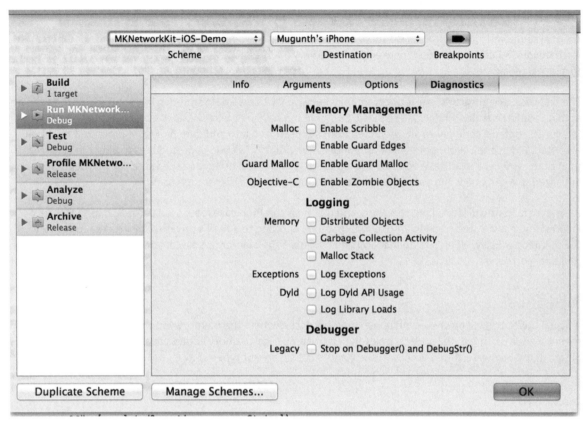

Figure 19-6 Enabling Zombie objects

EXC_BAD_ACCESS

An EXC_BAD_ACCESS occurs whenever you try to access or send a message to a deallocated object. The most common cause of EXC_BAD_ACCESS is when you initialize a variable in one of your initializer methods but use the wrong ownership qualifier, which results in deallocation of your object. For example, you create an NSMutableArray of elements for your UITableViewController in the viewDidLoad method but set the ownership qualifier of the list to unsafe_unretained or assign instead of strong. Now in cellForRowAtIndexPath:, when you try to access the deallocated object, you'll crash with a EXC_BAD_ACCESS. Debugging EXC_BAD_ACCESS is made easy with the NSZombiesEnabled environment variable that you found out about in the last section.

If you don't understand ownership qualifiers, turn back to Chapter 5 to catch up.

SIGSEGV

A signal segment fault (SIGSEGV) is a more serious issue that the operating system raises. It occurs when there is a hardware failure or when you try to access a memory address that cannot be read or when you try to write to a protected address.

The first case, a hardware failure, is uncommon. When you try to read data stored in RAM and the RAM hardware at that location is faulty, you get a SIGSEGV. But more often than not, a SIGSEGV occurs for the latter two reasons. By default, code pages are protected from being written into and data pages are protected from being executed. When one of your pointers in your application points to a code page and tries to alter the value pointed to, you get a SIGSEGV. You also get a SIGSEGV when you try to read the value of a pointer that was initialized to a garbage value pointing to an address that is not a valid memory location.

SIGSEGV faults are more tedious to debug, and the most common case reason a SIGSEGV happens is an incorrect typecast. Avoid hacking around with pointers or trying to a read a private data structure by advancing the pointer manually. When you do that and don't advance the pointer to take care of the memory alignment or padding, you get a SIGSEGV.

SIGBUS

A signal bus (SIGBUS) error is a kind of bad memory access where the memory you tried to access is an invalid memory address. That is, the address pointed to is not a physical memory address at all (it could be the address of one of the hardware chips). Both SIGSEGV and SIGBUS are subtypes of EXC_BAD_ACCESS.

SIGTRAP

SIGTRAP stands for signal trap. This is not really a crash signal. It's sent when the processor executes a trap instruction. The LLDB debugger usually handles this signal and stops at a specified breakpoint. If you get a SIGTRAP for no apparent reason, a clean build will usually fix it.

EXC_ARITHMETIC

Your application receives an EXC_ARITHMETIC when you attempt a division by zero. This should be a straightforward fix.

SIGILL

SIGILL stands for SIGNAL ILLEGAL INSTRUCTION. This happens when you try to execute an illegal instruction on the processor. You execute an illegal instruction when you're trying to pass a function pointer to another function, but for one reason or other, the function pointer is corrupted and is pointing to a deallocated memory or a data segment. Sometimes you get an EXC_BAD_INSTRUCTION instead of SIGILL, and though both are synonymous, EXC_* are machine-independent equivalents of this signal.

SIGABRT

SIGABRT stands for SIGNAL ABORT. This is a more controlled crash where the operating system has detected a condition that is not safe and asks the process to perform cleanup if any is needed. There is no one silver bullet to help you debug the underlying error for this signal. Frameworks like `cocos2d` or `UIKit` often call the C function `abort` (which in turn sends this signal) when certain preconditions are not met or when something really bad happens. When a SIGABRT occurs, the console usually has a wealth of information about what went wrong. Since it's a controlled crash, you can print the backtrace by typing `bt` into the LLDB console.

The following shows a SIGABRT crash on the console.

Printing the Backtrace Often Points the Reason for a SIGABRT

```
(lldb) bt
* thread #1: tid = 0x1c03, 0x97f4ca6a libsystem_kernel.dylib`__pthread_
kill
  + 10, stop reason = signal SIGABRT
    frame #0: 0x97f4ca6a libsystem_kernel.dylib`__pthread_kill + 10
    frame #1: 0x92358acf libsystem_c.dylib`pthread_kill + 101
    frame #2: 0x04a2fa2d libsystem_sim_c.dylib`abort + 140
    frame #3: 0x0000200a Test`-[MKAppDelegate application:didFinishLaunchi
ngWithOptions:] + 58 at MKAppDelegate.m:16
```

Watchdog Timeout

This is normally distinguishable because of the error code `0x8badf00d`. (Programmers, too, have a sense of humor — this is read as *Ate Bad Food*). On iOS, this happens mostly when you block the main thread with a synchronous networking call. For that reason, don't ever do a synchronous networking call.

> For more in-depth information about networking, read Chapter 14 of this book.

Custom Error Handling for Signals

You can override signal handling using the C function `sigaction` and providing a pointer to a signal handler function. The `sigaction` function takes a `sigaction` structure that has a pointer to the custom function that needs to be invoked, as shown in the following code.

Custom Code for Processing Signals

```
void SignalHandler(int sig) {

    // your custom signal handler goes here
}

// Add this code at startup
struct sigaction newSignalAction;
memset(&newSignalAction, 0, sizeof(newSignalAction));
newSignalAction.sa_handler = &SignalHandler;
sigaction(SIGABRT, &newSignalAction, NULL);
```

In this method, a custom C function for SIGABRT signal is added. You can add more handler methods for other signals as well using code similar to this.

A more sophisticated method is to use an open source class written by Matt Gallagher. Matt, author of the popular cocoa blog *Cocoa with Love* made the open source class `UncaughtExceptionHandler` that you can use to handle uncaught exceptions. The default handler shows an error alert. You can easily customize it to save the application state and submit a crash report to your server. A link to Matt's post is provided in the "Further Reading" section at the end of this chapter.

Collecting Crash Reports

When you're developing for iOS, you have multiple ways to collect crash dumps. In this section, I explain the most common way to collect crash reports, namely, iTunes Connect. I also explain about two other third-party services that also offers server-side symbolication in addition to crash reporting.

iTunes Connect

iTunes Connect allows you to download crash reports for your apps. You can log in to iTunes Connect and go to your app's details page to download crash reports. Crash reports from iTunes are not symbolicated, and you should symbolicate them using the exact dSYM that was generated by Xcode when you built the app for submission. You can do this either automatically using Xcode or manually (using the command line utility `symbolicatecrash`).

Collecting Crash Reports

When your app crashes on your customer's device, Apple takes care of uploading the crash dumps to its servers. But do remember that this happens only when the user accepts to submit crash reports to Apple. Although most users submit crash reports, some may opt out. As such, iTunes crash reports may not represent your actual crash count.

Symbolicating iTunes Connect Crash Reports Using Xcode

If you're the sole developer, using Xcode's automatic symbolication makes more sense, and it's easier. There is almost no added advantage of using manual symbolication. As a developer, you only have to use Xcode's Build⇨Archive option to create your product's archive and submit this archive to the App Store. This archive encapsulates both the product and the dSYM. Don't ever delete this archive file, even after your app is approved. For every app (including multiple versions) that you submit, you need to have a matching archive in Xcode.

Symbolicating iTunes Connect Crash Reports Manually

When there's more than one developer, every developer on the team needs to be able to symbolicate a given crash report. When you depend on Xcode's built-in Archive command to store your dSYMs, the only developer able to symbolicate a crash is the one who submits apps to the App Store. To allow other developers to symbolicate, you may have to e-mail archive files around, which is probably not the right way to do it. My recommendation is to commit the archives to your version control for every release build.

From the Organizer, go to the Archives tab and Cmd+click the archive that you just submitted. Reveal it in Finder, copy it to your project directory, and commit your release branch. If you're following my advice from Chapter 2, you've tagged your repository for every release. Now when you get a crash report from Apple, check out the archive file from your version control that corresponds to the crash report. That is, if your crash report is for version 1.1, get the archive from your version control that corresponds to version 1.1. Open the archive file's location in Terminal.

You can now symbolicate your crash report manually using either the following `symbolicatecrash.sh` shell script

```
symbolicatecrash MyApp.crash MyApp.dSYM > symbolicated.txt
```

or using `atos` in interactive mode, as shown here:

```
atos -arch armv7 -o MyApp
```

If your crash report matches your dSYM file, you should see your memory addresses symbolicated in the text file.

Using `atos` in interactive mode will be handy in many cases where you just want to know the address of the thread's stack trace that crashed. `atos` assumes that your dSYM file is located in the same location as your application.

`symbolicatecrash` **is a script that's not in your** `%PATH%` **by default. As such, you might get a command not found error when you type the preceding command on Terminal. As of Xcode 4.3, this script resides in** `/Applications/Xcode.app/Contents/Developer/Platforms/` `iPhoneOS.platform/Developer/Library/PrivateFrameworks/DTDeviceKit.` `framework/Versions/A/Resources/symbolicatecrash`**. Even if this location changes in the near future, you should be able to run a quick search for this file from the command line using the command** `find /Applications/Xcode.app -name symbolicatecrash -type f.`

Third-Party Crash Reporting Services

Whew, that was painful! But as the saying goes, necessity is the mother of invention. A painful crash analysis process leads plenty of third-party developers to make alternative services that, apart from crash log collection and analysis, do symbolication on the server-side.

You can use TestFlight and HockeyApp to collect and symbolicate crash reports. TestFlight is free while HockeyApp is a paid alternative. Both have their own advantages and disadvantages, but an in-depth discussion of them is outside the scope of this chapter. The following section briefly compares them to iTunes Connect.

Advantages of TestFlight or HockeyApp over iTunes Connect

Both TestFlight and HockeyApp provide an SDK that you normally integrate with your app. These SDKs take care of uploading crash reports to their servers. Although Apple uploads crash reports only when the user consents, these SDKs will always upload the crash reports. That means that you get more accurate statistics on the number of times a particular kind of crash occurs. Since crash reports normally don't have personally identifiable information, uploading them without the user's consent is allowed within the rules of the App Store.

Secondly, you can upload a dSYM to TestFlight and symbolicate crash reports for the live versions of your app. For adhoc builds, TestFlight's desktop client automatically uploads dSYMs. Server-side symbolication means that you don't have to know anything about `symbolicatecrash` or `atos` commands or use the Terminal. In fact, both these services upload your dSYM files to their server. You and their SDK collect crash reports from users' devices.

Summary

In this chapter, you read about LLDB and debugging. You learned about breakpoints, watchpoints, and how to edit and share those edited breakpoints. You discovered the power behind the LLDB console and how to use a Python script to speed up your debugging. You also read about the various types of errors, crashes, and signals that normally occur in your iOS app, and you found out how to avoid and overcome them. Finally, you learned about some third-party services that allow you to symbolicate crash reports on the server.

Further Reading

Apple Documentation

The following documents are available in the iOS Developer Library at `developer.apple.com` or through the Xcode Documentation and API Reference.

> *Xcode 4 User Guide*: "Debug Your App"
>
> *Developer Tools Overview*
>
> *LLVM Compiler Overview*

Read the header documentation in the following header files. You can get these files by searching for them using Spotlight.

```
exception_types.h
signal.h
```

WWDC Session

> WWDC 2012: "Session 415: Debugging with LLDB"

Other Resources

Writing an LLVM Compiler Backend
`http://llvm.org/docs/WritingAnLLVMBackend.html`

Apple's "Lazy" DWARF Scheme
`http://wiki.dwarfstd.org/index.php?title=Apple's_%22Lazy%22_DWARF_Scheme`

Hamster Emporium archive. "[objc explain]: So You Crashed in objc_msgSend()"
`http://www.sealiesoftware.com/blog/archive/2008/09/22/objc_explain_So_you_crashed_in_objc_msgSend.html`

Cocoa with Love. "Handling Unhandled Exceptions and Signals"
`http://cocoawithlove.com/2010/05/handling-unhandled-exceptions-and.html`

furbo.org. "Symbolicatifination"
`http://furbo.org/2008/08/08/symbolicatifination/`

LLDB Python Reference
`http://lldb.llvm.org/python-reference.html`

## Chapter 20

Performance Tuning Until It Flies

Performance is one of those things that separates acceptable apps from extraordinary apps. Of course, some performance bugs make an app unusable, but many more just make an app sluggish. Even apps that have good UI responsiveness may be using much more battery life than they should. Every developer needs to spend time periodically evaluating performance and making sure that nothing is being wasted. In this chapter you discover how to best prioritize, measure and improve the performance of your apps, including memory, CPU, drawing, and disk and network performance. Through it all, you learn how to best use one of the most powerful profiling tools available: Instruments. This chapter also provides tips on how to get the most from the powerful math frameworks available in iOS, including Accelerate, GLKit, and Core Image.

The Performance Mindset

Before going down the road of optimizing an application, you need to adopt the right mindset. This mindset can be summarized in a few rules.

Rule 1: The App Exists to Delight the User

I could really stop the rules right here. Everything else is a corollary. As a developer, never forget that your app exists to do more than just "provide value" to the user. It's a competitive market out there. You want an app that the user really loves. Even minor annoyances detract from that goal. Scrolling that stutters, a slow launch, poor response to buttons, all these things accumulate in the user's mind and turn a great app into a "good-enough" app.

Rule 2: The Device Exists for the Benefit of the User

It's very unlikely that your app is why users bought an iDevice. Users likely have many apps, and many reasons for using their devices. Never let your app get in the way of that reality. Your users may really enjoy your app, but that doesn't mean it's okay to drain the battery. It's not okay to fill the disk. You live in a shared environment. Behave accordingly.

Rule 3: Go to Extremes

When your app is working for the user, use every resource available. When it's not doing work for the user, use as few resources as possible. This means that it's okay to use a lot of CPU to maintain a high frame rate while the user is active. But when the user is not using your app directly, it shouldn't be using much of the CPU. As a general rule, it's better for the app to use 100 percent of available resources for a short period and then go completely to sleep than to use 10 percent of available resources all the time. As many have pointed out, a microwave can use more power illuminating its clock than it does heating food, because the clock is on all the time. Look for these low-power, always-on activities and get rid of them.

Rule 4: Perception Is Reality

Much of iOS is devoted to giving the illusion of performance that's unrealistic. For instance, launch images that display the UI give the user the impression that the application is immediately available, even though it hasn't finished launching yet. Similar approaches are acceptable and even encouraged in your apps. If you can display some information while waiting for full data, do so. Keep your application responsive, even if you can't actually let the user do anything yet. Swap images for hard-to-draw views. Don't be afraid to cheat, as long as cheating makes the user experience better.

Rule 5: Focus on the Big Wins

Despite everything I just said, I don't think you must chase every leaked byte or every wasted CPU cycle. Doing so isn't worth the effort. Your time is better spent writing new features, squashing bugs, or even taking a vacation. Cocoa has small leaks, which means that even if your program is perfect, it may still have some leaks you can't remove. Focus on the things that impact the user. A thousand bytes leaked over the run of the program don't matter. A thousand bytes lost every second matter quite a lot.

Welcome to Instruments

Instruments is one of the most powerful performance tools ever developed, and it can be one of the easiest to use. That doesn't mean that using it is *always* easy. Many things in Instruments don't work entirely correctly because of default configurations, but once you know the handful of things to reconfigure, no tool is more adept at finding performance problems quickly.

In this chapter, you will be evaluating a fairly simple piece of code that has several problems. It's an app that reads a file and displays the contents one character at a time in its main view. You can find the full listing in the ZipText project for this chapter on the book's website. You can test each revision by changing the REVISON value in ZipTextView.h. Here is the code you'll be focusing on.

ZipTextView1.m (ZipText)

```
- (id)initWithFrame:(CGRect)frame text:(NSString *)text {
    ... Load long string into self.text ...
    ... Set timer to repeatedly call appendNextCharacter ...
}

- (void)appendNextCharacter {
    for (NSUInteger i = 0; i <= self.index; i++) {
        if (i < self.text.length) {
            UILabel *label = [[UILabel alloc] init];
            label.text = [self.text substringWithRange:NSMakeRange(i,1)];
```

```
      label.opaque = NO;
      [label sizeToFit];
      CGRect frame = label.frame;
      frame.origin = [self originAtIndex:i
                             fontSize:label.font.pointSize];
      label.frame = frame;
      [self addSubview:label];
    }
  }
  self.index++;
}

- (CGPoint)originAtIndex:(NSUInteger)index
              fontSize:(CGFloat)fontSize {
  CGPoint origin;
  if (index == 0) {
    return CGPointZero;
  }
  else {
    origin = [self originAtIndex:index-1 fontSize:fontSize];
    NSString *
    prevCharacter = [self.text
                    substringWithRange:NSMakeRange(index-1,1)];
    CGSize
    prevCharacterSize = [prevCharacter sizeWithFont:
                        [UIFont systemFontOfSize:fontSize]];
    origin.x += prevCharacterSize.width;
    if (origin.x > CGRectGetWidth(self.view.bounds)) {
      origin.x = 0;
      origin.y += prevCharacterSize.height;
    }
    return origin;
  }
}
```

This program starts out well, but quickly slows to a crawl and starts displaying memory warnings. To figure out what is going wrong, start by launching Instruments using Product⇨Profile in Xcode. Here are a few things to note before diving into the details:

▨ By default, Instruments builds in Release mode. This can display radically different performance than Debug mode, but it's usually what you want. If you want to profile in Debug mode, you can modify the mode in the scheme.

▨ Performance in the simulator and on a device can be radically different. Most of the time, it's important to profile only on a device.

▨ Instruments often has a difficult time finding the symbols for the current application. You'll often need to re-symbolicate the document using File⇨Re-Symbolicate Document.

The last bug in Instruments is extremely frustrating and difficult to eliminate completely. Apple says that Instruments is supposed to find its symbols using Spotlight, but this almost never works. I recommend duplicating rdar://10158512 at `bugreport.apple.com`. See `http://openradar.appspot.com/10158512` for the text of this bug report.

In my experience, if you add the following directory to Preferences⇨Search Paths, you can at least press the Symbolicate button, rather than search for the dSYM by hand:

```
~/Library/Developer/Xcode/DerivedData
```

If you have to search by hand, note that the dSYM is in the following directory:

```
~/Library/Developer/Xcode/DerivedData/<app>/Build/Products/Release-
iphoneos
```

In the rest of this chapter, assume that I have symbolicated whenever required.

Finding Memory Problems

Many performance problems can be tracked to memory problems. If you're seeing unexpected memory warnings, it's probably best to track those down first. Use the Allocations template in Instruments. Figure 20-1 shows the result.

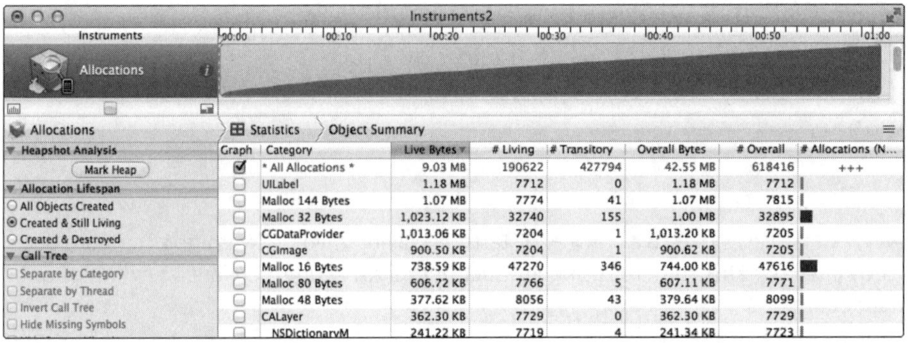

Figure 20-1 Allocations Instrument

Looking at the graph, memory allocations are clearly out of control in this app. Memory use is constantly growing. By sorting the Live Bytes column, you see that the largest memory use is in `UILabel` and that there are over 7,000 of them currently allocated after just one minute. That's very suspicious. Click the arrow beside `UILabel` to get more information and press Cmd-E to show Extended Detail (see Figure 20-2).

Here you see that the `UILabel` is created in `-[ViewController appendNextCharacter]`. If you double-click that stack frame, you will see the code along with hot-spot coloring indicating how much time is spent in each line of code (see Figure 20-3).

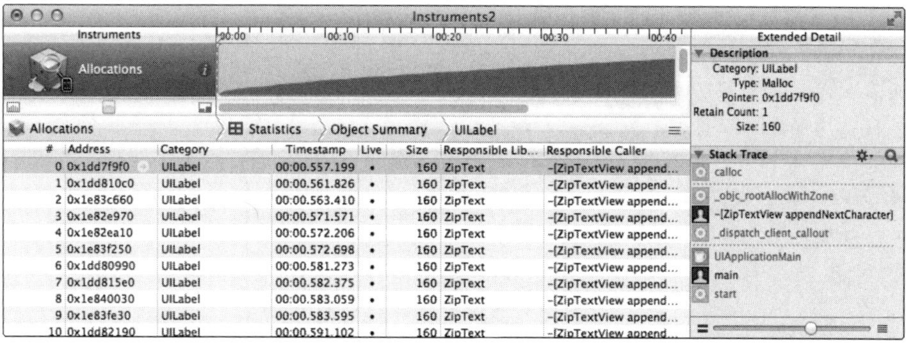

Figure 20-2 Extended Detail for Allocations

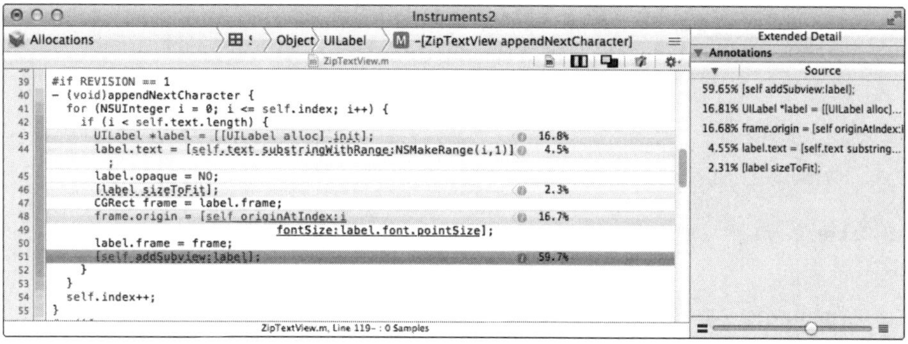

Figure 20-3 Code View

A little investigation here shows the problem. Every time this method is called, it re-creates all of the labels instead of just creating one new one. You just need to remove the loop, as shown here:

ZipTextView2.m (ZipText)

```
for (NSUInteger i = 0; i <= self.index; i++)
```

and replace it with the assignment

```
NSUInteger i = self.index;
```

Now, when you rerun Instruments, memory usage is better. After a minute, you're using less than 2MB, rather than more than 9MB, but steady memory growth over time is still a problem. In this case, it's pretty obvious that the problem is the `UILabel` views, but this is a good opportunity to demonstrate how to use heapshot analysis. Launch ZipText again with the Allocations instrument, and after a few seconds press the Mark Heap button. Let it run few a few seconds and then press Mark Heap again. This shows all the objects that were created between those two points and haven't been destroyed yet. This is a great way to figure out which objects are being leaked when you perform an operation. Figure 20-4 shows that in about six seconds, we've created 148 new `UILabel` objects.

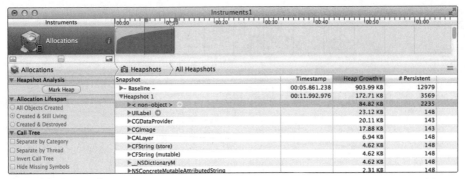

Figure 20-4 Improved memory management

At this point, you should be reevaluating the choice to use a separate `UILabel` for every character. Maybe a single `UILabel`, `UITextView` or even a custom `drawRect:` would have been a better choice here. `ZipTextView3` demonstrates how to implement this with a custom `drawRect:`.

ZipTextView3.m (ZipText)

```
- (void) appendNextCharacter {
    self.index++;
    [self setNeedsDisplay];
}

- (void)drawRect:(CGRect)rect {
    for (NSUInteger i = 0; i <= self.index; i++) {
        if (i < self.text.length) {
            NSString *character = [self.text substringWithRange:
                                     NSMakeRange(i, 1)];
            [character drawAtPoint:[self originAtIndex:i fontSize:kFontSize]
                        withFont:[UIFont systemFontOfSize:kFontSize]];
        }
    }
}
```

One more run, and as you see in Figure 20-5, there's no memory growth at all. But now performance has slowed to a crawl. Before moving on to CPU performance problems, here are a few more useful tips for memory performance investigations:

- Clicking the small *i* beside the Allocations instrument lets you set various options. One of the most useful options is to change the Track Display from Current Bytes to Allocation Density. This graphs how many allocations happen during a sample period. This indicates where you may have excessive memory churn. Allocating memory can be expensive, so you don't want to rapidly create and destroy thousands of objects if you can avoid it.

The Leaks instrument can be useful occasionally, but don't expect too much from it. It only looks for unreferenced memory. It won't detect retain loops, which are more common. Nor will it detect memory you fail to release, like the UILabel views in the previous example. It also can have false positives. For example, it often shows one or two small leaks during program startup. If you see new leaks show up every time you perform an action or regularly at some time interval, then it's definitely worth investigating. Generally, heapshot is a more useful tool in tracking down lost memory problems.

Keep track of whether you're looking at Live Bytes or not. Instruments keeps track of both total and net allocations. Net allocations are allocated memory minus freed memory. If you create and destroy objects regularly, total allocations may be orders of magnitude greater than net allocations. Instruments sometimes refers to net allocations as *live* or *still living*. In the main screen (refer to Figure 20-1), the graph in the final column shows total allocations as a light bar and net allocations as a darker bar.

Remember that the Leaks instrument and the Allocations instrument can tell you only where memory was allocated. This has nothing to do with when memory was leaked. The allocation point probably isn't where your bug is.

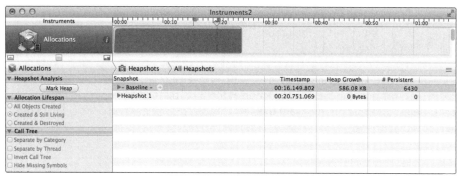

Figure 20-5 Memory footprint of ZipTextView with custom drawing

A Word on Memory Allocation

All objects in Objective-C are allocated on the heap. Calls to +alloc eventually turn into calls to malloc(), and calls to –dealloc eventually turn into calls to free(). The heap is a shared resource across all threads, and modifying the heap requires a lock. This means that one thread allocating and freeing memory can block another thread allocating and releasing memory. Doing this quickly is called *memory churn* and should be avoided, particularly in multithreaded applications (including ones using Grand Central Dispatch). iOS has many tricks to minimize the impact of this. Small blocks of memory can often be allocated without modifying the heap. But as a rule, avoid rapidly creating and freeing memory if you can help it. It can be better to reuse objects than to create new ones rapidly. This can conflict with the advantages of using immutable objects (see Chapter 4), so there is a tradeoff.

As an example, when working with NSMutableArray and NSMutableDictionary, consider using removeAllObjects rather than throwing away the collection and making a new one.

Finding CPU Problems

Although we fixed the memory problems in the example, it's still running much too slowly. After a few paragraphs, it slows to a crawl. Instruments can help with this as well. Profile ZipText again using the Time Profiler Instrument. You will see Time Profiler graph similar to Figure 20-6.

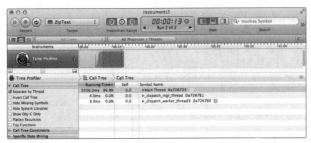

Figure 20-6 Time Profiler analysis of ZipText

What's important about this graph is that it's staying at 100%. Programs should seldom run at 100% CPU for very long. Using Option-Drag, you can select part of the graph. The Call Tree list will update to show you the methods and functions called during those samples.

In the Call Tree list, select Invert Call Tree. You will almost always want an inverted call tree. It is bizarre that it's not the default. The "standard" call tree puts the top-level function at the top. This is always going to be `main()` for the main thread, which is useless information. You want to know the *lowest*-level method or function in the stack. Inverting the call tree puts that on top.

Several other Call Tree options are useful at different times:

- **Top Functions.** This option appeared in Instrumenets at some point, and I've never seen documentation on it, but it's great. It makes a list of the functions (or methods) that are taking the most time. This has become the first place we look when investigating performance problems.

- **Separate by Thread.** There are reasons to turn this on or off. You definitely want to know what's taking the most time on your main thread, since this can block the UI. You also, however, want to know what's taking the most time across all threads. On iPhone and older iPads, there is only one core, so saying that something is being done "in the background" is a bit misleading. The UI thread still has to context-switch to service background threads. Even with the multi-core iPads, "background" threads cost just as much battery life as the UI thread, so you want to keep it under control.

- **Hide System Libraries.** This only shows "your" code. Many people turn this on immediately, but I generally find it useful to leave it turned off most of the time. You generally want to know where the time is being spent, whether or not it's in "your" code. Often the biggest time impacts are in system actions like file writes. Hiding system libraries can hide these problems. That said, sometimes it's easier to understand the call tree if you hide system libraries.

- **Show Obj-C Only.** This is very similar to Hide System Libraries in that it tends to focus on higher-level code. I tend to leave it turned off for the same reasons I leave Hide System Libraries turned off. But again, it can be useful sometimes to help you find your place in the code.

For ZipText, select Invert Call Tree and Hide System Libraries. In Figure 20-7, you see that `originAt Index:fontSize:` is taking most of the time and that it's being called recursively. A little investigation shows that every character recomputes the location of every previous character. No wonder this runs slowly. You can modify `originAtIndex:fontSize:` to cache previous results as shown in `ZipTextView4.m`.

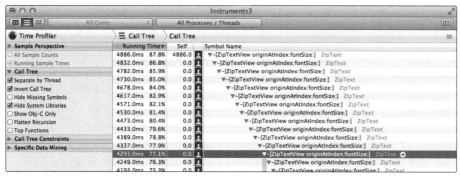

Figure 20-7 ZipText with caching

ZipTextView4.m (ZipText)

```objc
@property (nonatomic) NSMutableArray *locations;
...
- (id)initWithFrame:(CGRect)frame text:(NSString *)text {
  ...
  _locations = [NSMutableArray
                   arrayWithObject:[NSValue
                                      valueWithCGPoint:CGPointZero]];
}


- (CGPoint)originAtIndex:(NSUInteger)index
                fontSize:(CGFloat)fontSize {
  if ([self.locations count] > index) {
    return [self.locations[index] CGPointValue];
  }
  else {
    CGPoint origin = [self originAtIndex:index-1 fontSize:fontSize];
    NSString *
    prevCharacter = [self.text
                      substringWithRange:NSMakeRange(index-1,1)];
    CGSize
    prevCharacterSize = [prevCharacter sizeWithFont:
                          [UIFont systemFontOfSize:fontSize]];
    origin.x += prevCharacterSize.width;
    if (origin.x > CGRectGetWidth(self.bounds)) {
      origin.x = 0;
      origin.y += prevCharacterSize.height;
    }
    self.locations[index] = [NSValue valueWithCGPoint:origin];
```

(continued)

```
        return origin;
    }
}
```

Now things are looking better, but drawing still seems to slow down when it gets about halfway down the screen. You dig into that in the next section.

Here are some other thoughts on hunting down CPU performance issues:

- Operations can take a long time because they are expensive on the CPU or because they block. Blocking operations are a serious problem on the main thread, because the UI can't update. To look for blocking operations, use the Time Profiler instrument, click the *i* button to display the options, and select Record Waiting Threads. In the Sample Perspective, select All Sample Counts. This will let you see what is taking the most time on each thread (particularly the main thread), even if it's blocking.

- By default, the Time Profiler instrument records only User time. It doesn't count Kernel time against you. Most of the time, this is what you want, but if you're making a lot of expensive kernel calls, this can skew the results. Click the *i* button to display the options, and select User & Kernel in Callstacks. You will generally also want to set the Track Display to User and System Libraries, which will separate your code from the system code with a small yellow line.

- Function calls in tight loops are the enemy of performance. An ObjC method call is even worse. The compiler will generally inline functions for you when it makes sense, but just marking a function `inline` doesn't guarantee that it will be inlined. For instance, if you ever take a pointer to the function, that function cannot be inlined anymore. Unfortunately, there's no way to know for certain whether the compiler has inlined a function other than compiling in Release mode (with optimizations) and going and looking at the assembly. The only way to be absolutely certain that something will inline is to make it a macro with `#define`. I don't generally like macros, but if you have a piece of code that absolutely must be inlined for performance reasons, you may want to consider it.

- Always do all performance testing in Release mode with optimizations turned on. It's meaningless to do performance testing on unoptimized Debug code. Optimizations are not linear. One part of your program may be no faster, whereas another may speed up by orders of magnitude. Tight loops are particularly subject to heavy compiler optimization.

- Although Instruments is incredibly powerful, don't neglect an old standby: small test programs that subtract the end time from the start time. For pure computational work, pulling the algorithm into a test program is often the most convenient approach. You can use a standard Single View template and just put the call to your code in `viewDidLoad`.

The Accelerate Framework

The most cited framework for improving performance is Accelerate. In my experience, it's a mixed bag, and you need to approach it as you would any other performance fix—with skepticism and testing. The Accelerate framework is made up of several pieces, including vecLib, vImage, vDSP, BLAS, and LAPACK. Each is distinct, with many similar-but-slightly-different feature sets and data types. This mix makes it difficult to use them together. Accelerate isn't so much a single framework as a collection of mostly preexisting math libraries that have been stuck together.

None of the Accelerate libraries are good at working on small data sets. If you just need to multiply small matrices (particularly 3 × 3 or 4 × 4, which are very common), Accelerate will be much slower than trivial C code. Calling an Accelerate function in a loop will often be much slower than a simple C implementation. If the function offers a *stride* option and you don't need a stride, then writing your own C code may be faster.

> A stride lets you skip through a list by a specific size. For example, a stride of 2 uses every other value. This is very useful when you have interleaved data (such as coordinates or image color information).

On multi-core laptops, the Accelerate libraries can be very fast because they are multithreaded. On mobile devices with one or two cores, this advantage may not outweigh its cost for moderate-sized data sets. The Accelerate libraries are also portable, with a history of focusing on the Altivec (PPC) and SSE3 (Intel) vector processors rather than the NEON processor that's in iOS devices.

For all the Accelerate libraries, you want to gather all your data into one convenient layout so that you can call the Accelerate functions as few times as possible. If at all possible, avoid copying memory, and store your data internally in the format that is fastest for Accelerate, rather than rearrange the data when it's time to make the performance-critical call.

You'll generally need to do significant performance testing to determine the fastest approach. Be sure to do all your testing in Release with optimizations turned on. Always start with a simple C implementation as a baseline to make sure you're getting good payback from Accelerate.

The vImage library is a large collection of image-processing functions and, in my experience, is the most useful portion of Accelerate. In particular, vImage is good at applying matrix operations on high-resolution images very quickly. For most common problems, consider Core Image rather than vImage, but if you need to do the raw math yourself, vImage is quite good at applying matrices and at converting images. Like the other parts of Accelerate, vImage is designed for large data sets. If you're working with a small number of moderate-sized images, Core Image is likely a better choice.

To use vImage effectively, you need to be aware of the difference between an interleaved and a planer format. Most image formats are interleaved. The red data for a pixel is followed by the green data for that pixel, which is then followed by the blue data, which is followed by the alpha data for the same pixel. That's the format you're probably used to from CGBitmapContext with kCGImageAlphaLast. That's RGBA8888. If the alpha comes first, then it's ARGB8888. Alpha generally comes first in vImage.

You can also store color information as 16-bit floats rather than 8-bit integers. If you store the alpha first, then the format is called ARBGFFFF.

Planer formats separate the color information into *planes*. Each plane holds one kind of information. There's an alpha plane, a red plane, a green plane, and blue plane. If the values are 8-bit integers, then this format is called Planar8. If it's 16-bit floats, then it's PlanarF. Although there are planar image formats, most vImage functions work on a single plane at a time, and the functions often don't care what kind of data the plane holds. There's no difference between multiplying red information and multiplying blue information.

It's generally much faster to work on planar formats than interleaved formats. If at all possible, get your data into a planar format as soon as you can, and leave it in that format through the entire transformation process. Generally, you convert back to RGB only when you need to display the information.

vecLib and vDSP are vector libraries with some overlap in their functionality. vecLib tends to be a bit easier to use and is focused on applying a single operation to a large list of numbers. vDSP tends to have much more flexibility (and therefore slower) functions, and it also has very specialized functions relevant to signal processing. If you need those kinds of functions on large tables of numbers, vDSP is likely much faster than your hand-built solutions. The key word here is "large." If you're computing only several hundred or a few thousand numbers, simple C solutions can often be faster. The cost of making the function call overwhelms everything else.

BLAS and LAPACK are C versions of well-established FORTRAN libraries. They're primarily focused on linear algebra. As with other parts of Accelerate, they are best applied for very complex problems. They tend to be slow for solving small systems of equations because of the overhead of the function call. Apple does not provide good documentation for BLAS and LAPACK. If you're interested in these functions, go to `netlib.org` and find the FORTRAN documentation. So far, we've seldom found them to be really helpful in solving our problems.

GLKit

GLKit is great for integrating OpenGL into your applications, but it also has some hidden gems for fast math. In particular, GLKit offers vector-optimized functions for working with 3×3 and 4×4 matrices as well as 2-, 3-, and 4-element vectors. As we mentioned in the previous section on the Accelerate framework, flexibility is often the enemy of performance. Functions that can take matrices of arbitrary sizes can never be as optimized as a hard-coded solution for multiplying a 3×3 matrix. So, for many of these kinds of problems, I recommend GLKit.

Before you get too excited, GLKit's math optimizations are not particularly faster than what you would get by hand-coding a C solution and letting the compiler optimize it. But they're no slower, and they're easy to use and easy to read. Look for `GLKMatrix3`, `GLKMatrix4`, `GLKVector2`, `GLKVector3`, and `GLKVector4`.

Drawing Performance

Set `REVISION` to 4 in `ZipTextView.h` and select the Core Animation instrument. You'll see results like those in Figure 20-8.

The Core Animation instrument displays frames per second. The Time Profiler is configured to show both User and System times in a stacked bar graph. As you see, the frames per second are dropping over time. That suggests we still have a problem.

The Core Animation instrument has several options that can help debug drawing problems. I discuss these shortly. Unfortunately, none of them is particularly helpful in the case of ZipText. The fps output does, however, provide a good metric for testing whether changes are effective. We at least want the fps output to stay constant over time.

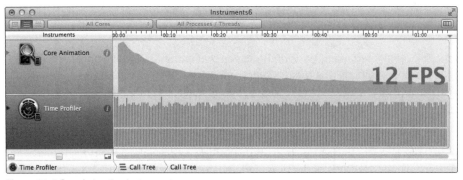

Figure 20-8 Core Animation instrument

Looking at `drawRect:`, it always draws every character, even though most of the characters never change. In fact, only one small part of the view changes during each update. You can improve performance here by drawing only the part that actually needs updating.

In `appendNextCharacter`, calculate the rectangle impacted by the new character:

ZipTextView5.m (ZipText)

```
- (void)appendNextCharacter {
  self.index++;
  if (self.index < self.text.length) {
    CGRect dirtyRect;
    dirtyRect.origin = [self originAtIndex:self.index fontSize:kFontSize];
    dirtyRect.size = CGSizeMake(kFontSize, kFontSize);
    [self setNeedsDisplayInRect:dirtyRect];
  }
}
```

In `drawRect:`, draw only the character that intersects that rectangle:

```
- (void)drawRect:(CGRect)rect {
    ...
    CGPoint origin = [self originAtIndex:i fontSize:kFontSize];
    if (CGRectContainsPoint(rect, origin)) {
      [character drawAtPoint:origin
                   withFont:[UIFont systemFontOfSize:kFontSize]];
    }
    ...
}
```

Instruments now reads a steady frame rate of nearly 60 fps. That's excellent. But don't misread high frame rates. You want high frame rates when you're displaying new data, but it's normal to have low frame rates when nothing is happening. Once your UI is stable, it's normal and ideal for your frame rate to drop to zero.

The Core Animation instrument has several Debug options in the left-hand pane that can be very useful. The following are some of our most-used options:

- **Color Blended Layers.** This applies red to any parts of the screen that require color blending. Parts of the screen that don't require color blending will be green. Color blending generally means nonopaque views. If you have a lot of red on the screen, investigate which views you can make fully opaque.

- **Color Misaligned Images.** As we discuss in Chapter 8, you always want to draw on pixel boundaries when possible. This will indicate areas of the screen that aren't aligned with pixel boundaries.

- **Flash Updated Regions.** This is one of my favorite options. It will flash yellow for any layer that is being updated. For UIs that aren't games, very little of the UI should update from moment to moment. If you turn this option on and see a lot of yellow, then you're probably updating the screen more often than you mean to. The granularity of this is the layer, however. So, if you use `setNeedsDisplayInRect:` to draw only part of a view, the entire view (which is backed by a layer) will flash yellow.

Core Image

For most image-processing needs, Core Image is the best choice. iOS 6 has substantially expanded Core Image. The docs for Core Image tend to be quite good, but here are some tips that may not be obvious:

- If you're down-sampling images (that is, shrinking them), look at `CILanczosScaleTransform`. This provides much better quality than a simple affine transform at little cost.

- `CIColorCube` is generally much faster at color shifting operations than specialized filters like `CISepiaTone`. `CISepiaTone` is focused on accuracy, not speed.

- Whenever you finish with a `CIFilter`, remember to set its input to `nil` if you're not going to destroy the filter right away. They retain a frame buffer, and this can use a lot of memory.

- A `CIContext` can be expensive to create. Reuse it when possible.

- Filtering large images is expensive. See if you can get away with filtering a smaller version and then scaling it up. Often, the quality loss isn't significant. This is particularly noteworthy with blurs.

For an excellent overview of what's new in Core Image for iOS 6, I recommend the Session 511 video from WWDC 2012, "Core Image Techniques."

Optimizing Disk and Network Access

Memory and CPU are the most common performance bottlenecks, but they're not the only resources to be optimized. Disk and network access are also critical, particularly in prolonging battery life.

The I/O Activity instrument is particularly useful for checking whether you're hitting the disk more often than you should be. This instrument is most easily accessed through the System Usage template.

Network access is particularly expensive on iOS devices, and it's worth serious consideration. For instance, creating a new network connection is very expensive. The DNS lookup alone can be incredibly, surprisingly slow. Whenever possible, reuse existing connections. `NSURLConnection` will do this for you automatically if you use HTTP/1.1. This is yet another reason to prefer it versus hand-built network solutions.

Network access is also a serious drain on the battery. It's much better for battery life if you batch all your network activity into short bursts. The cost of network utilization is much more proportional to the length of time than to the amount of data. The device will reduce power on its radios when the network is not in use, so providing long periods of quiet with short bursts of activity is the best design.

The Connections instrument can be useful in tracking down network usage, but keep in mind that it tracks all processes on the device, not just yours.

Summary

You need to strongly consider the limited resources of the platform when designing mobile applications. Things that are acceptable on a desktop device will overwhelm an iPhone, bringing it to a crawl and draining its battery. Continual analysis of performance and improvement should be part of every developer's workflow.

Instruments is one of the best tools available for analyzing your application's performance. Despite being buggy and at times infuriating, it is nevertheless very powerful and a critical tool in tracking down memory and CPU bottlenecks. It's also useful for detecting I/O and network overuse. Spend some time with it and experiment. You'll find it a worthwhile investment.

Accelerate can be useful for very large math operations, but simpler code and higher-level frameworks are often faster for more common operations. Core Image is particularly useful for speeding image processing, and GLKit is useful for small-matrix operations.

In any case, always carefully test the changes you make to improve performance. Many times, the simplest code is actually the best. Find your hotspots with Instruments, and test, test, test.

Further Reading

Apple Documentation

The following documents are available in the iOS Developer Library at `developer.apple.com` or through the Xcode Documentation and API Reference.

Accelerate Framework Reference

Core Image Programming Guide

File-System Performance Guidelines

GLKit Framework Reference

Instruments User Guide

Memory Usage Performance Guidelines

Performance Overview

Performance Starting Point

WWDC 2012, "Session 511: Core Image Techniques"

Other Resources

Bumgarner, Bill. *bbum's weblog-o-mat.* "When Is a Leak Not a Leak?" Good introduction to using heapshot analysis to find lost memory.
`www.friday.com/bbum/2010/10/17`

Netlib. "Netlib Repository at UTK and ORNL." Documentation of LAPACK and BLAS, which are offered by Accelerate.
`www.netlib.org`

Part IV

Pushing the Limits

Chapter 21

Storyboards and Custom Transitions

Prior to iOS 5, interface elements and views were created using Interface Builder (IB) and saved in nib files. Storyboards are a new way to create interfaces, and in addition to creating interface elements, you can now specify the navigation (called *segues*) between those interfaces. This was something you could not do previously without writing code. You can think of storyboards as a graph of all your view controllers connected by segues that dictate the transition between them.

The benefits of storyboards don't stop there. They also make it easy for developers to create static table views without a data source. How many times have you wanted to create a table view that's not bound to a real data source—for example, a table that shows a list of options instead of data. A common use for this is your app's settings page. Storyboards also help co-developers and clients understand the complete workflow of the app.

Storyboards aren't all romantic, and in my opinion, there are some significant drawbacks. Later in this chapter, we show you how to use storyboards without being hurt by those drawbacks.

We begin with how to start using storyboards and how to do things using storyboards that you do with nib files, such as communicating between controllers. In the "Static Tables" section later in this chapter, you find out how to create a static table view without a data source. Finally, you discover the most interesting aspect of storyboards—writing your own custom transition animations. Although these cool transition animations sound complicated, Apple has made it really easy to write them.

Getting Started with Storyboards

You can use storyboards for new projects or add them to an existing project that doesn't have a storyboard yet. For existing projects, you can add storyboards in the same way you add a new file to a project. You find more on how to instantiate view controllers in this storyboard in the upcoming section "Instantiating a Storyboard."

For new projects, storyboards can be created in Xcode 4.5 by using the new project template and selecting to use the default Use Storyboard option, as shown in Figure 21-1.

When you create a new project using storyboards, the `info.plist` key of your app contains a key called `UIMainStoryboardFile`. This key supersedes `NSMainNibFile` that was used prior to iOS 5. You can continue to use `NSMainNibFile` if your app's main window is loaded from a nib file instead of a storyboard. However, you can't use both `UIMainStoryboardFile` and `NSMainNibFile` in the same app. `UIMainStoryboardFile` takes precedence, and your nib file specified in `NSMainNibFile` never gets loaded.

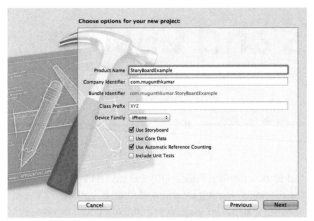

Figure 21-1 New Project Template in Xcode 4.5 showing the Use Storyboard option

> Your application can store the complete storyboard in one file, and IB automatically builds it into separate files optimized for loading. In short, you don't have to be worried about loading time or performance when using storyboards.

Instantiating a Storyboard

When your `UIMainStoryboardFile` is set, the compiler automatically generates code for instantiating it and loads it as your application's startup window. If you're adding storyboards in an existing app, you do so programmatically. The methods for instantiating view controllers within a storyboard are defined in the `UIStoryboard` class.

When you want to display a view controller specified in your storyboard, you load the storyboard using this method:

```
+ storyboardWithName:bundle:
```

Loading View Controllers Within a Storyboard

Loading view controllers within a storyboard is very similar to the nib loading method, and with the `UIStoryboard` object, you can instantiate view controllers using the following method:

```
- instantiateInitialViewController
- instantiateViewControllerWithIdentifier:
```

Segues

Segues are transitions defined in your storyboard file. UIKit provides two default transition styles, Push and Modal. They behave similar to the `pushViewController:animated:completion:` and `presentView Controller:animated:completion` methods you use in iOS 5. In addition to this, you can create custom segues and create new kinds of transitions between view controllers. You look at this later in this chapter in the section "Custom Transitions."

You create segues by connecting certain events on view controllers with other view controllers on your storyboard file. You can drag from a button to a view controller, from a gesture recognizer object to a view controller, and so on. IB creates a segue between them, and you can select the segue and use the attributes inspector panel to modify the transition styles.

The attributes inspector panel also allows you to set a custom class if you select a custom transition style. You can think of a segue as something that connects an action with a transition. Actions that can trigger segues can be button tap events, row selection events on static table views, a recognized gesture, or even audio events. The compiler automatically generates the necessary code to perform a segue when the event to which you connected the segue occurs.

When a segue is about to be performed, a `prepareForSegue:sender:` method is invoked on the source view controller and an object of type `UIStoryboardSegue` is passed to it. You can override this method to pass data to the destination view controller. The next section explains how to do this.

When a view controller performs multiple segues, the same `prepareForSegue:sender:` method gets called for every segue. To identify the performed segue, use the segue identifier to check whether the performed segue is the intended one and pass data accordingly. As a defensive programming practice, I advise you to perform this check even if the view controller performs only one segue. This ensures that later on when you add a new segue, your app will continue to run without crashing.

Passing Data

With iOS 5, when you use storyboards, instantiating view controllers and presenting them to the user are done automatically. You're given a chance to fill in data by overriding the `prepareForSegue:sender:` method. By overriding this method, you can get the pointer to the destination view controller and set the initial values there.

The framework calls the same methods that you used before, such as `viewDidLoad`, `initWithCoder:` and `NSObject`'s `awakeFromNib` method, which means that you can continue writing your view controller's initialization code as with iOS 4.

Returning Data

With storyboards, you communicate data back to the parent view controller exactly as you do with nib files or manually coded user interfaces. Data created/entered by the user on modal forms that you present can be retuned to the parent via delegates or blocks. The only difference is that on your parent view controller, you have to set the delegate in the `prepareForSeque:sender:` method to `self`.

Instantiating Other View Controllers

`UIViewController` has a `storyboard` property that retains a pointer to the storyboard object (`UIStoryBoard`) from which it was instantiated. This property is `nil` if your view controller is created manually or from a nib file. With this back reference, you can instantiate other view controllers defined in your storyboard from any other view controller. You do so by using the view controller's identifier. The following method on `UIStoryBoard` allows you to instantiate a view controller using its identifier:

```
- instantiateViewControllerWithIdentifier:
```

As a result, you can still have view controllers on your storyboard that aren't connected with any other view controllers through segues, and yet these view controllers can be initialized and used.

Performing Segues Manually

While storyboards can automatically trigger segues based on actions, in some cases, you may need to perform segues programmatically. You might do so to deal with actions that cannot be handled by the storyboard file. To perform a segue, you call the `performSegueWithIdentifier:sender:` method of the view controller. When you perform segues manually, you can pass the caller and the context objects in the sender parameter. This sender parameter will be sent to the `prepareForSegue:sender:` method later.

Unwinding Segues

Storyboards originally allowed you to instantiate and navigate to view controllers. iOS 6 introduces methods in `UIViewController` that allows unwinding segues. By unwinding segues, you can implement methods to navigate "back" without creating additional view controllers.

You can add unwinding support to a segue by implementing an `IBAction` method in your view controller that takes a `UIStoryboardSegue` as a parameter. Example

```
- (IBAction)unwindMethod:(UIStoryboardSegue*)sender {
}
```

You can now connect an event in a view controller to its `Exit` object. Xcode automatically enumerates all possible unwind events (any method that is an IBAction and accepts a `UIStoryboardSegue` as a parameter) in a storyboard and allows you to connect to them. This is shown in Figure 21.2.

Building Table Views with Storyboard

One important advantage of storyboards is the capability to create static tables from IB. With storyboards, you can build two types of table views: a static table that doesn't need a special class for providing a data source, and a table view containing a prototype cell (similar to custom table view cells in iOS 4) that binds data from a model.

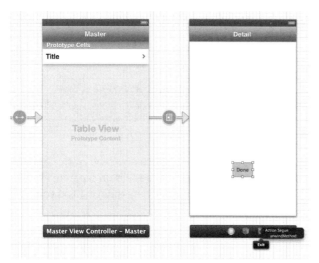

Figure 21.2 Connecting a IBAction for unwinding a segue

Static Tables

You can create static tables in your storyboard by dragging a table, selecting it, and from the attributes inspector, choosing Static Cells, as shown in Figure 21-3.

Figure 21-3 A storyboard illustrating static table view creation

Static cells are great for creating settings pages (or pages whose content doesn't come from a Core Data model or a web service or any such data source) as in Apple's Settings app.

> Static cells can be created only for table views that are from a `UITableViewController`. **You cannot create static cells for table views that are added as a subview of a** `UIViewController` **view.**

Prototype Cells

Prototype cells are similar to custom table view cells, but instead of creating prototype cells on separate nib files and loading them in the data source method `cellForRowAtIndexPath:`, you create them in IB on your storyboard and just set the data on your data source methods.

> You identify all prototype cells using a custom identifier, which ensures proper functioning of the table view cell queuing methods. Xcode will warn you if a prototype cell in your storyboard doesn't have a cell identifier.

Custom Transitions

Another advantage of storyboards is that you can now easily create custom transition effects for your view controllers.

When segues are performed, the compiler generates necessary code to present or push the destination controller based on the transition style you set on your storyboard. You've found that two types of transition styles, Push and Modal, are supported natively by iOS. There's also a third type, Custom. Choose Custom and provide your own subclass of `UIStoryboardSegue` that handles your custom transition effects.

Create a subclass of `UIStoryBoardSegue` and override the `perform` method. In the `perform` method, access the pointer to the source view controller's main view's layer and do your custom transition animation (using Core Animation). Once the animation is complete, push or present your destination view controller (you can get a pointer to this from the segue object). It's as simple as that.

To illustrate, we'll show you how to create a transition where the master view gets pushed down, and the details view appears at the bottom of the screen.

Create a new application using the Master-Details template and open the `MainStoryboard`. Click on the only segue and change the type to Custom. Add a `UIStoryboardSegue` subclass and override the `perform` method and paste the following code.

Custom Transition Using a Storyboard Segue

```
- (void) perform {

  UIViewController *src = (UIViewController *)self.sourceViewController;
  UIViewController *dest = (UIViewController *)self.
destinationViewController;

  CGRect f = src.view.frame;
  CGRect originalSourceRect = src.view.frame;
  f.origin.y = f.size.height;

  [UIView animateWithDuration:0.3 animations:^{
    src.view.frame = f;

  } completion:^(BOOL finished){
    src.view.alpha = 0;
    dest.view.frame = f;
    dest.view.alpha = 0.0f;
    [[src.view superview] addSubview:dest.view];
    [UIView animateWithDuration:0.3 animations:^{

      dest.view.frame = originalSourceRect;
      dest.view.alpha = 1.0f;
    } completion:^(BOOL finished) {

      [dest.view removeFromSuperview];
      src.view.alpha = 1.0f;
      [src.navigationController pushViewController:dest animated:NO];
    }];
  }];
}
```

That's it. You can do all kinds of crazy stuff with the layer pointers of the source and destination view controllers. Justin Mecham has open-sourced a great example of a doorway transition (ported from Ken Matsui) on GitHub. (https://github.com/jsmecham/DoorwaySegue). You can also create your own transition effects by manipulating the layer pointers of the source and destination view controllers. With segues, creating a custom transition effect is far easier.

Another Advantage

When you use storyboards, it becomes easy for co-developers (and clients) to understand the app's flow. Instead of going through multiple nib files and cross-referencing the instantiation code for understanding the flow, co-developers can open the storyboard file and see the complete flow. This alone is a compelling reason to use them.

Disadvantages

Last year, when storyboards were introduced, they had a drawback. They could be used only on apps that are targeting iOS 5+ customers. With the release of iOS 6, you don't need to worry about this issue.

Avoiding the Merge Conflict Hell

Storyboards have another, rather annoying, problem for teams. The default application template in Xcode has one storyboard for the entire app. This means that when two developers work on the UI at the same time, merge conflicts become inevitable, and because storyboards are internally auto-generated XML, these merge conflicts are too complicated to fix. I recommend that you break your storyboards into multiple files, one file for every use case. In most cases, one developer will be working on a use case, and the chances of ending up with a merge conflict are lower.

Examples of use case-based storyboards for a Twitter client are `Login.Storyboard`, `Tweets.Storyboard`, `Settings.Storyboard`, and `Profile.Storyboard`. Instead of breaking the storyboard up into multiple nib files (going back to square one), use multiple storyboards to help solve the merge conflict issue while preserving the elegance of using storyboards. As I mention in the previous paragraph, a developer who is working on the Login use case will probably not be working on Tweets at the same time.

Customizing Your Views Using UIAppearance Protocol

This last section covers an important addition to iOS 5: a method to customize your view appearance through Apple's native classes. Prior to iOS 5, customizing the look and feel of native controls was not natively supported and was often difficult for developers. A common problem that developers face is changing the appearance of all instances of a control. The proper way of doing so was to create the complete control from scratch. But because that was time-consuming, some developers resorted to overriding or swizzling methods like `drawRect:`.

Beginning with iOS 5, Apple provided default support for most `UIKit` controls by formalizing customization using a couple of protocols—namely, `UIAppearance and UIAppearanceContainer`. Any UI control that adheres to the `UIAppearance` protocol can be customized to have a different look and feel. Want more? The `UIAppearance` protocol even allows you to specify a different look and feel based on where the control is contained. That is, you can specify that the appearance of a control (say the `tintColor` of a `UIBarButtonItem`) be different when it's contained within a specific view (`UINavigationBar` or `UIPopoverViewController`). You do so by getting an appearance proxy object for the control's class and using that to customize the appearance. Here's an example.

To customize the tint color of a bar button throughout your application, set the `tintColor` to the `UIBarButtonItem`'s appearance proxy like this:

```
[[UIBarButtonItem appearance] setTintColor:[UIColor redColor]];
```

Note that the `setTintColor` method existed in iOS 4, in `UIBarButtonItem`. But it was applicable only to a particular instance of the control. With the appearance proxy object, you're able to customize the appearance of any object created using the said class.

On similar lines, you can also customize the appearance of a control depending on the contained view by using the following method:

```
[[UIBarButtonItem appearanceWhenContainedIn:[UINavigationBar class], nil]
setTintColor:[UIColor redColor]];
```

The first parameter is a `nil` terminated list of all container classes like `UINavigatorBar`, `UIPopOverController` that conforms to the `UIAppearanceContainer` protocol.

Starting with iOS 5, most UI elements added support to the `UIAppearance` protocol. Additionally, controls like the `UISwitch` in iOS 5 allow you to easily change the color of the "on" gradient to the designer's choice. Now, how do you know when all elements (and which properties of those elements) can be customized through UIKit's appearance proxy? There are two ways. The old way is to open the documentation and read it. The other way is a shortcut that most developers use: reading the header file. When you open the header file for the corresponding UIKit element, any property that is tagged with `UI_APPEARANCE_SELECTOR` supports customization through its appearance proxy. For example, the `UINavigationBar.h` has the property `tintColor` tagged with `UI_APPEARANCE_SELECTOR`:

```
@property(nonatomic,retain) UIColor    *tintColor  UI_APPEARANCE_SELECTOR;
```

This means, you can call

```
[[UINavigationBar appearance] setTintColor:newColor];
```

Although Apple was against UI customization in the beginning (on both Mac and iOS), that is slowly changing, and you can see that Apple's native apps (such as the new Reminders app) have heavily customized skeuomorphic user interfaces. With the `UIAppearance` protocol, you can achieve the same result with far less code.

Summary

In this chapter you learnt about storyboards and methods to implement custom transitions using Storyboards. You also learnt about unwinding segues that are introduced in iOS 6 and the advantages and dis advantages of using storyboards in your app. Lastly, you learn about UI customization using the appearance proxy protocol.

Further Reading

Apple Documentation

The following documents are available in the iOS Developer Library at `developer.apple.com` or through the Xcode Documentation and API Reference.

What's New in iOS 6

TableView Programming Guide

TableViewSuite

UIViewController Programming Guide

WWDC Sessions

The following session video is available at `developer.apple.com`.

WWDC 2011, "Session 309: Introducing Interface Builder Storyboarding"

WWDC 2012, "Session 407: Adopting Storybaords in Your App"

Other Resources

enormego / EGOTableViewPullRefresh
`https://github.com/enormego/EGOTableViewPullRefresh`

jsmecham / DoorwaySegue
`https://github.com/jsmecham/DoorwaySegue`

## Chapter 22

Cocoa's Biggest Trick: Key-Value Coding and Observing

There is no magic in Cocoa. It's just C. But one particular trick borders on "magic," and that's key-value observing (KVO). This chapter explores how and when to use KVO, as well as its nonmagical cousin, key-value coding (KVC).

Key-value coding is a mechanism that allows you to access an object's properties by name rather than by calling explicit accessors. This allows you to determine property bindings at runtime rather than at compile time. For instance, you can request the value of the property named by the string variable `someProperty` using `[object valueForKey:someProperty]`. You can set the value of the property named by `someProperty` using `[object setValue:someValue forKey:someProperty]`. This indirection allows you to determine the specific properties to access at runtime rather than at compile time, allowing more flexible and reusable objects. To get this flexibility, your objects need to name their methods in specific ways. This naming convention is called key-value coding, and this chapter covers the rules for creating indirect getters and setters and how to access items in collections and manage KVC with nonobjects. You also find out how to implement advanced KVC techniques such as Higher Order Messaging and collection operators.

If your objects follow the KVC naming rules, then you can also make use of *key-value observing*. KVO is a mechanism for notifying objects of changes in the properties of other objects. Cocoa has several observer mechanisms including delegation and `NSNotification`, but KVO has lower overhead. The observed object does not have to include any special code to notify observers, and if there are no observers, KVO has no runtime cost. The KVO system adds the notification code only when the class is actually observed. This makes it very attractive for situations where performance is at a premium. In this chapter, you find out how to use KVO with properties and collections, and the trick Cocoa uses to make KVO so transparent.

You can find all the code samples in this chapter in the online files for this chapter in the projects `KVC`, `KVC-Collection`, and `KVO`.

Key-Value Coding

Key-value coding is a standard part of Cocoa that allows your properties to be accessed by name *(key)* rather than by calling an explicit accessor. KVC allows other parts of the system to ask for "the property named `foo`" rather than calling `foo` directly. This permits dynamic access by parts of the system that don't know your keys at compile time. This dynamic access is particularly important for nib file loading and Core Data in iOS. On the Mac, KVC is a fundamental part of the AppleScript interface.

The following code listings demonstrate how KVC works with an example of a cell that can display any object using `valueForKeyPath:`.

KVCTableViewCell.h (KVC)

```
@interface KVCTableViewCell : UITableViewCell
- (id)initWithReuseIdentifier:(NSString*)identifier;

// Object to display.
@property (nonatomic, readwrite, strong) id object;

// Name of property of object to display
@property (nonatomic, readwrite, copy) NSString *property;
@end
```

KVCTableViewCell.m (KVC)

```
- (BOOL)isReady {
  // Only display something if configured
  return (self.object && [self.property length] > 0);
}

- (void)update {
  NSString *text;
  if (self.isReady) {
    // Ask the target for the value of its property that has the
    // name given in self.property. Then convert that into a human
    // readable string
    id value = [self.object valueForKeyPath:self.property];
    text = [value description];
  }
  else {
    text = @"";
  }
  self.textLabel.text = text;
}

- (id)initWithReuseIdentifier:(NSString *)identifier {
  return [self initWithStyle:UITableViewCellStyleDefault
          reuseIdentifier:identifier];
}

- (void)setObject:(id)anObject {
  _object = anObject;
  [self update];
}

- (void)setProperty:(NSString *)aProperty {
  _property = aProperty;
  [self update];
}
```

KVCTableViewController.m (KVC)

```
- (NSInteger)tableView:(UITableView *)tableView
 numberOfRowsInSection:(NSInteger)section {
    return 100;
}

- (UITableViewCell *)tableView:(UITableView *)tableView
        cellForRowAtIndexPath:(NSIndexPath *)indexPath {

    static NSString *CellIdentifier = @"KVCTableViewCell";

    KVCTableViewCell *cell = [tableView
            dequeueReusableCellWithIdentifier:CellIdentifier];

    if (cell == nil) {
      cell = [[KVCTableViewCell alloc]
            initWithReuseIdentifier:CellIdentifier];
      // You want the "intValue" of the row's NSNumber.
      // The property will be the same for every row, so you set it
      // here in the cell construction section.
      cell.property = @"intValue";
    }

    // Each row's object is an NSNumber representing that integer
    // Since each row has a different object (NSNumber), you set
    // it here, in the cell configuration section.
    cell.object = [NSNumber numberWithInt:indexPath.row];

    return cell;
}
```

This example is quite simple, displaying 100 rows of integers, but imagine if `KVCTableViewCell` had animation effects or special selection behaviors. You could apply those to arbitrary objects without the object or the cell needing to know anything about the other. That's the ultimate goal of a good model-view-controller (MVC) design, which is the heart of Cocoa's architecture. (See Chapter 4 for more information on the MVC pattern.)

The `update` method of `KVCTableViewCell` demonstrates `valueForKeyPath:`, which is the main KVC method you use in this example. Here is the important section:

```
id value = [self.object valueForKeyPath:self.property];
text = [value description];
```

In this example, `self.property` is the string `@"intValue"` and `self.target` is an `NSNumber` object representing the row index. So the first line is effectively the same as this code:

```
id value = [NSNumber numberWithInt:[self.object intValue]];
```

The call to `numberWithInt:` is automatically inserted by `valueForKeyPath:`, which automatically converts number types (`int`, `float`, and so on) into `NSNumber` objects and all other nonobject types (structs, pointers) into `NSValue` objects.

While this example utilizes an `NSNumber`, the key take-away is that `target` could be any object and `property` could be the name of any property of `target`.

Setting Values with KVC

KVC can also modify writable properties using `setValue:forKey:`. For example, the following two lines are roughly identical:

```
cell.property = @"intValue";
[cell setValue:@"intValue" forKey:@"property"];
```

Both of these will call `setProperty:` as long as `property` is an object. See the later section "KVC and Nonobjects" for a discussion on how to handle `nil` and nonobject properties.

Methods that modify properties are generally called *mutators* in the Apple documentation.

Traversing Properties with Key Paths

You may have noticed that KVC methods have `key` and `keyPath` versions. For instance, there are `valueForKey:` and `valueForKeyPath:`. The difference between a key and a key path is that a key path can have nested relationships, separated by a period. The `valueForKeyPath:` method traverses the relationships. For instance, the following two lines are roughly identical:

```
[[self department] name];
[self valueForKeyPath:@"department.name"];
```

On the other hand, `valueForKey:@"department.name"` would try to retrieve the property `department.name`, which in most cases would throw an exception.

The `keyPath` version is more flexible, whereas the `key` version is slightly faster. If the key is passed to me, I generally use `valueForKeyPath:` to provide the most flexibility to my caller. If the key is hard-coded, I generally use `valueForKey:`.

KVC and Collections

Object properties can be one-to-one or one-to-many. One-to-many properties are either ordered (arrays) or unordered (sets).

Immutable ordered (`NSArray`) and unordered (`NSSet`) collection properties can be fetched normally using `valueForKey:`. If you have an `NSArray` property called `items`, then `valueForKey:@"items"` returns it as you'd expect. But there are more flexible ways of managing this.

For this example, you create a table of multiples of two. The data model object keeps track of only the number of rows, not the actual results, but it provides the results as though it were an `NSArray`. This project is available as `KVC-Collection` in the sample code. Here is how to create it:

1. Create a new iPhone project in Xcode using the Model-Detail Application template with storyboard and automatic reference counting.

2. Select `MainStoryboard.storyboard` and remove the Master View Controller and the Detail View Controller.

3. Drag a view controller from the library and set its class to `KVCTableViewController`. Click-drag from the navigation controller to your new view controller and set the relationship to `rootViewController`.

4. Add labels and buttons as shown in Figure 22-1. The hash marks (###) are separate labels from the titles.

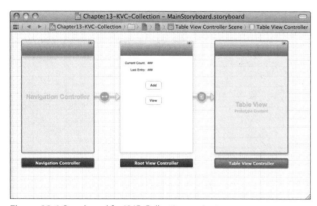

Figure 22-1 Storyboard for KVC-Collection project

5. Delete the `MasterViewController` and `DetailViewController` source files.

6. Add a new source file using the `UIViewController` template and name it `RootViewController`. Do not use a XIB for its user interface.

7. In the storyboard, select the root view controller and show the assistant editor. Click-drag from the labels and buttons to `RootViewController.h` to create the outlets shown in Figure 22-2. Click-drag from the View button to the table view controller and select the push segue.

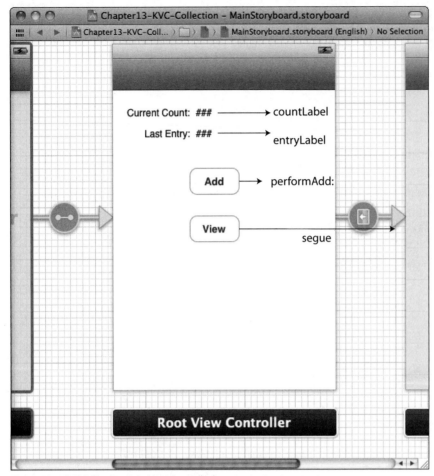

Figure 22-2 KVC-Collection Root View Controller layout

The following code implements the example project and demonstrates KVC access to a collection. When you press the Add button, the number of items stored in `DataModel` is incremented. When you press the View button, a table view is constructed to display the information in the `DataModel` using KVC proxy collections in `RootViewController` and KVC collection accessors in `KVCTableViewController`. Following the code, we explain how both of these access mechanisms work.

RootViewController.h (KVC-Collection)

```
@interface RootViewController : UIViewController
@property (nonatomic, strong) IBOutlet UILabel *countLabel;
@property (nonatomic, strong) IBOutlet UILabel *entryLabel;
- (IBAction)performAdd;
@end
```

RootViewController.m (KVC-Collection)

```objc
- (void)refresh {
  DataModel *model = [DataModel sharedModel];

  // There is no property called "items" in DataModel. KVC will
  // automatically create a proxy for you.
  NSArray *items = [model valueForKey:@"items"];
  NSUInteger count = [items count];
  self.countLabel.text = [NSString stringWithFormat:@"%d",
                            count];

  if (count > 0) {
    self.entryLabel.text = [[items objectAtIndex:(count-1)]
                            description];
  } else {
    self.entryLabel.text = @"";
  }
}

- (void)viewWillAppear:(BOOL)animated {
  [self refresh];
  [super viewWillAppear:animated];
}

- (IBAction)performAdd {
  [[DataModel sharedModel] addItem];
  [self refresh];
}
```

KVCTableViewController.m (KVC-Collection)

```objc
- (NSInteger)tableView:(UITableView *)tableView
 numberOfRowsInSection:(NSInteger)section {

  // countOfItems is a KVC method, but you can call it directly
  // rather than creating an "items" proxy.
  return [[DataModel sharedModel] countOfItems];
}

- (UITableViewCell *)tableView:(UITableView *)tableView
        cellForRowAtIndexPath:(NSIndexPath *)indexPath {
  static NSString *CellIdentifier = @"Cell";

  UITableViewCell *cell = [tableView
        dequeueReusableCellWithIdentifier:CellIdentifier];

  if (cell == nil) {
    cell = [[UITableViewCell alloc]
```

(continued)

```
            initWithStyle:UITableViewCellStyleDefault
            reuseIdentifier:CellIdentifier];
  }

  DataModel *model = [DataModel sharedModel];
  id object = [model objectInItemsAtIndex:indexPath.row];
  cell.textLabel.text = [object description];

  return cell;
}
```

DataModel.h (KVC-Collection)

```
@interface DataModel : NSObject

+ (DataModel*)sharedModel;

- (void)addItem;
- (NSUInteger)countOfItems;
- (id)objectInItemsAtIndex:(NSUInteger)index;
@end
```

DataModel.m

```
@interface DataModel ()
@property (nonatomic, readwrite, assign) NSUInteger count;
@end

@implementation DataModel

+ (DataModel*)sharedModel {
  static DataModel *sharedModel;
  static dispatch_once_t onceToken;
  dispatch_once(&onceToken, ^{ sharedModel = [DataModel new]; });
  return sharedModel;
}

- (NSUInteger)countOfItems {
  return self.count;
}

- (id)objectInItemsAtIndex:(NSUInteger)index {
  return [NSNumber numberWithInt:index * 2];
}

- (void)addItem {
  self.count++;
}
@end
```

Note how `RootViewController` accesses the array of items from `DataModel`:

```
NSArray *items = [model valueForKey:@"items"];
```

Normally, you'd expect this to call `[DataModel items]`, but there is no such method. `DataModel` doesn't ever create an array. So where does this `NSArray` come from?

`DataModel` implements `countOfItems` and `objectInItemsAtIndex:`. These are very specially named methods. When `valueForKey:` looks for `items`, it searches for the following methods:

- `getItems` or `items` or `isItems`—These are searched in order, and the first one found is used to return the value.

- `countOfItems` and either `objectInItemsAtIndex:` or `itemsAtIndexes`—This is the combination you use in this example. KVC generates a proxy array that's discussed shortly.

- `countOfItems` and `enumeratorOfItems` and `memberOfItems`—This combination causes KVC to return a proxy set.

- An instance variable named `_items`, `_isItems`, `items` or `isItems`—KVC will directly access the ivar. You generally want to avoid this behavior. Direct instance variable access breaks encapsulation and makes the code more fragile. You can prevent this behavior by overriding `+accessInstanceVariablesDirectly` and returning `NO`.

In this example, `valueForKey:` automatically generates and returns a proxy `NSKeyValueArray`. This is a subclass of `NSArray`, and you can use it like any other array, but calls to `count`, `objectAtIndex:` and related methods are forwarded to the appropriate KVC methods. The proxy caches its requests, making it very efficient. See the *Key-Value Coding Programming Guide* in the iOS Developer Library for the full set of methods you can implement for this form.

In this example, the property is `items`, so KVC looks for `countOfItems`. If the property were to have been `boxes`, KVC would look for `countOfBoxes`. KVC requires that you name your methods in a standard way so that it can construct these method names. This is why getters must begin with a lowercase letter.

For mutable collection properties, there are two options. You can use the mutator (property-changing) methods such as the following (again, see the *Key-Value Coding Programming Guide* for the full list):

```
- (void)insertObject:(id)object
    inChildrenAtIndex:(NSUInteger)index;
- (void)removeObject:(id)object
    inChildrenAtIndex:(NSUInteger)index;
```

Or you can return a special proxy object by calling `mutableArrayValueForKey:` or `mutableSetValueForKey:`. Modifying this object automatically calls the appropriate KVC methods on your object.

KVC and Dictionaries

Dictionaries are just a special kind of nested relationship. For most keys, calling `valueForKey:` is the same as calling `objectForKey:` (the exception is if the key begins with @, which is used to refer to the `NSDictionary` itself, if needed). This is a convenient way to handle nested dictionaries, because you can use `valueForKeyPath:` to access arbitrary layers.

KVC and Nonobjects

Not every method returns an object, but `valueForKey:` always returns an `id`. Nonobject return values are wrapped in an `NSValue` or `NSNumber`. These two classes can handle just about anything from numbers and Booleans to pointers and structures. While `valueForKey:` will automatically wrap scalar values into objects, you cannot pass nonobjects to `setValue:forKey:`. You must wrap scalars in `NSValue` or `NSNumber`.

Setting a nonobject property to `nil` presents a special case. Whether doing so is permissible or not depends on the situation, so KVC does not guess. If you call `setValue:forKey:` with a value of `nil`, the key will be passed to `setNilValueForKey:`. You need to override this method to do the right thing if you want to handle setting `nil` for a nonobject property. Its default behavior is to throw an exception.

Higher-Order Messaging with KVC

`valueForKey:` is filled with useful special cases, such as the fact that it's overridden for collections like `NSArray` and `NSSet`. Rather than operating on the collection, `valueForKey:` is passed to each member of the collection. The results are added to the returned collection. This allows you to easily construct collections from other collections such as

```
NSArray *array = [NSArray arrayWithObjects:@"foo",
                     @"bar", @"baz", nil];
NSArray *capitals =
            [array valueForKey:@"capitalizedString"];
```

This passes the method `capitalizedString` to each item in the `NSArray` and returns a new `NSArray` with the results. Passing messages (`capitalizedString`) as parameters is called Higher Order Messaging. Multiple messages can be passed using key paths:

```
NSArray *array = [NSArray arrayWithObjects:@"foo",
                     @"bar", @"baz", nil];
NSArray *capitalLengths =
        [array valueForKeyPath:@"capitalizedString.length"];
```

The preceding code calls `capitalizedString` on each element of `array`, then calls `length`, and wraps the return into an `NSNumber` object. The results are collected into a new array called `capitalLengths`.

You looked at more flexible approaches using trampolines in Chapter 4, but KVC provides a very easy solution for many problems, as long as you don't need to pass parameters.

Collection Operators

KVC provides a few complex functions as well. It can, for instance, sum or average a list of numbers automatically. Consider this:

```
NSArray *array = [NSArray arrayWithObjects:@"foo",
                  @"bar", @"baz", nil];
NSUInteger totalLength =
    [[array valueForKeyPath:@"@sum.length"] intValue];
```

@sum is an operator that sums the indicated property (length). Note that this can be hundreds of times slower than the equivalent loop:

```
NSArray *array = [NSArray arrayWithObjects:@"foo",
                  @"bar", @"baz", nil];
NSUInteger totalLength = 0;
for (NSString *string in array) {
  totalLength += [string length];
}
```

The performance issues are generally significant when dealing with arrays of thousands or tens of thousands of elements. Beyond @sum, you can find many other operators in the *Key-Value Coding Programming Guide* in the iOS Developer Library. The operations are particularly valuable when working with Core Data and can be faster than the equivalent loop because they can be optimized into database queries. You cannot create your own operations, however.

Key-Value Observing

Key-value observing is a mechanism for transparently notifying observers of changes in object properties. At the beginning of the "Key-Value Coding" section, you built a table view cell that could display any object. In that example, the data was static. If you changed the data, the cell wouldn't update. You improve that now. You can make the cell automatically update whenever its object changes. You need a changeable object, so use the current date and time. You use key-value observing to get a callback every time a property you care about changes.

KVO has a lot of similarities to NSNotificationCenter. You start observing using addObserver:for KeyPath:options:context:. To stop observing, you use removeObserver:forKeyPath:conte xt:. The callback is always observeValueForKeyPath:ofObject:change:context:. Here are the modifications required to create 1,000 rows of date cells that automatically update every second.

KVCTableViewCell.m (KVO)

```
- (void)removeObservation {
  if (self.isReady) {
    [self.object removeObserver:self
                 forKeyPath:self.property];
```

(continued)

```objc
    }
  }

- (void)addObservation {
  if (self.isReady) {
    [self.object addObserver:self forKeyPath:self.property
                      options:0
                      context:(__bridge void*)self];
  }
}

- (void)observeValueForKeyPath:(NSString *)keyPath
                      ofObject:(id)object
                        change:(NSDictionary *)change
                       context:(void *)context {

  if ((__bridge id)context == self) {
    // Our notification, not our superclass's
      [self update];
  }
  else {
    [super observeValueForKeyPath:keyPath ofObject:object
                           change:change context:context];
  }
}

- (void)dealloc {
  if (_object && [_property length] > 0) {
    [_object removeObserver:self
                 forKeyPath:_property
                    context:(__bridge void*)self];
  }
}

- (void)setObject:(id)anObject {
  [self removeObservation];
  _object = anObject;
  [self addObservation];
  [self update];
}

- (void)setProperty:(NSString *)aProperty {
  [self removeObservation];
  _property = aProperty;
  [self addObservation];
  [self update];
}
```

KVCTableViewController.m (KVO)

```objc
#import "RNTimer.h"
@interface KVCTableViewController ()
@property (readwrite, retain) RNTimer *timer;
```

```
@property (readwrite, retain) NSDate *now;
@end

@implementation KVCTableViewController

- (void)updateNow {
  self.now = [NSDate date];
}

- (void)viewDidLoad {
  [self updateNow];

  __weak id weakSelf = self;
  self.timer =
      [RNTimer repeatingTimerWithTimeInterval:1
                                        block:^{
                                          [weakSelf updateNow];
                                        }];
}
- (void)viewDidUnload {
  self.timer = nil;
  self.now = nil;
}

...

- (UITableViewCell *)tableView:(UITableView *)tableView
          cellForRowAtIndexPath:(NSIndexPath *)indexPath {

  static NSString *CellIdentifier = @"KVCTableViewCell";

  id cell = [tableView
    dequeueReusableCellWithIdentifier:CellIdentifier];

  if (cell == nil) {
    cell = [[[KVCTableViewCell alloc]
      initWithReuseIdentifier:CellIdentifier] autorelease];
    [cell setProperty:@"now"];
    [cell setObject:self];
  }

  return cell;
}
@end
```

In KVCTableViewCell, you observe the requested property on your target in addObservation. When you register for KVO, you pass self as the context pointer (after casting to void* for ARC) so that in the callback you can determine if this was your observation. Because there's only one KVO callback for a class, you may be receiving a callback for a property your superclass registered for. If so, you need to pass it along to super. Unfortunately, you can't always pass to super because NSObject will throw an exception. So you use a unique context to identify your observations. There's more about this in section "KVO Tradeoffs," later in this chapter.

In `RootViewController`, you create a property `now` and ask the cell to observe it. Once a second, you update it. Observers are notified and the cells update. This is all quite efficient because, at any given time, there's only about one screen of cells because of cell reuse.

The real power of KVO is seen in `[KVCTableViewController updateNow]`:

```
- (void)updateNow {
    self.now = [NSDate date];
}
```

The only thing you have to do is update your data. You don't have to worry that someone might be observing you, and if no one is observing you, then you don't pay any overhead as you would for `NSNotificationCenter`. The incredible simplicity on the part of the model class is the real benefit of KVO. As long as you use accessors to modify your ivars, all the observation mechanism is handled automatically, with no cost when you don't need it. All the complexity is moved into the observer rather than the observed. It's no wonder that KVO is becoming very popular in low-level Apple frameworks.

KVO and Collections

Observing collections often causes confusion. The thing to remember is that observing a collection is not the same as observing the objects in it. If a collection contains Adam, Bob, and Carol, then adding Denise changes the collection. Changes to Adam do not change the collection. If you want to observe changes to the objects in a collection, you must observe those objects, not the collection. Generally, that's done by observing the collection and then observing objects as they're added, and stopping when they're removed.

How Is KVO Implemented?

Key-value observing notifications rely on two `NSObject` methods: `willChangeValueForKey:` and `didChangeValueForKey:`. Before an observed property change is made, something must call `willChangeValueForKey:`. This records the old value. After the change is made, something must call `didChangeValueForKey:`, which calls `observeValueForKeyPath:ofObject:change:context:`. You can do this by hand, but that's fairly uncommon. Generally, you do so only if you're trying to control when the callbacks are made. Most of the time, the change notifications are called automatically.

There is very little magic in Objective-C. Even message dispatching, which can seem mysterious at first, is actually pretty straightforward. (Message dispatching is covered in Chapter 28.) However, KVO borders on magic. Somehow when you call `setNow:`, there is an extra call to `willChangeValueForKey:`, `didChangeValueForKey:`, and `observeValueForKeyPath:ofObject:change:context:` in the middle. You might think that this is because you synthesized `setNow:`, and occasionally you'll see people write code like this:

```
- (void)setNow:(NSDate *)aDate {
    [self willChangeValueForKey:@"now"]; // Unnecessary
    now_ = aDate;
    [self didChangeValueForKey:@"now"]; // Unnecessary
}
```

This is redundant, and you don't want to do it because then the KVO methods will be called twice. KVO always calls `willChangeValueForKey:` before an accessor and `didChangeValueForKey:` afterward. How? The answer is class swizzling. Swizzling is discussed further in Chapter 28, but when you first call `addObserver:forKeyPath:options:context:` on an object, the framework creates a new KVO subclass of the class and converts the observed object to that new subclass. In that special KVO subclass, Cocoa creates setters for the observed properties that work effectively like this:

```
- (void)setNow:(NSDate *)aDate {
    [self willChangeValueForKey:@"now"];
    [super setValue:aDate forKey:@"now"];
    [self didChangeValueForKey:@"now"];
}
```

This subclassing and method injection is done at runtime, not compile time. That's why it's so important that you name things correctly. KVO can figure this out only if you use the KVC naming convention.

It's difficult to detect the KVO class swizzling. It overrides `class` to return the original class. But occasionally you'll see references to NSKVONotifying_MYClass instead of MYClass.

KVO Tradeoffs

KVC is powerful technology, but other than possibly being slower than direct method calls, it's generally a good thing. The one major downside is that you lose compile-time checks of your property names. Always code following KVC naming conventions, whether or not you use KVC directly. Doing so will save you a lot of grief when you want to instantiate objects from a nib file, which requires KVC. It also makes your code readable by other Objective-C developers, who expect certain names to mean certain things. For the most part, this means naming your getters and setters `property` and `setProperty:`, respectively.

KVO, on the other hand, is a mixed bag. It can be useful, and it can cause trouble. It's implemented in a highly magical way, and some of its usage is quite awkward. There's only one callback method, which means you often wind up with a lot of unrelated code in that method. The callback method passes you a change dictionary that is somewhat tedious to use.

Because `removeObsever:forKeyPath:context:` will crash if you're not an observer for that key path, you must keep track of exactly which properties you're observing. KVO has no equivalent to `NSNotificationCenter removeObserver:`, which conveniently cleans up all observations you might have.

KVO creates subtle code-path surprises. When you call `postNotification:`, you know that some other code may run. You can search your code for the notification name and generally find all of the things that might happen. It can be surprising that just setting one of your own properties can cause code in another part of the program to execute. It can be difficult to search the code to discover this interaction. KVO bugs in general are difficult to solve because so much of the activity "just happens," without any visible code causing it.

So KVO's greatest strength is also its greatest danger. It can sometimes dramatically reduce the amount of common code you write. In particular, it can get rid of the frequent problem of hand-building all your setters just so you can call an `updateSelf` method. In this way, KVO can reduce bugs because of incorrectly cut-and-pasted code. But it can also inject really confusing bugs, and with the introduction of Automatic Reference Counting, handwritten setters are even easier to write correctly.

> Several third parties have attempted to improve KVO. In particular, KVO is an obvious candidate for a block-based interface. In the "Further Reading" section at the end of this chapter, you'll find links to some examples by very accomplished developers. Still, we're very nervous about using a third-party solution here. KVO is complicated and magical. Blocks are complicated and magical. Combining the two without a large testing group seems extremely dangerous. Personally, we'll wait for Apple to improve this interface, but if you're interested in other options see the "Further Reading" section.

Our recommendation is to use KVO sparingly, simply, and only in places where doing so is a real benefit. For situations in which you need a very large number of observations (a few hundred or more), its performance scales much better than `NSNotification`. It gives you the advantages of `NSNotification` without modifying the observed class. And it sometimes requires less code, although you need to include all the special-case code you may need to work around subtle KVO problems. In the `KVCTableViewCell` example, hand-coding `setProperty:` and `setTarget:` makes the file about 15 lines shorter than the equivalent KVO solution.

Avoid KVO in situations where you have complex interdependencies or a complicated class hierarchy. Simple solutions with delegates and `NSNotification` are often better than excessively clever solutions using KVO.

On the other hand, Apple is clearly moving toward KVO in performance-critical frameworks. It's the primary way to deal with `CALayer` and `NSOperation`. You can expect to see it more often in new low-level classes. It has the advantage of zero-overhead observation. If there are no observers of a given instance, then KVO costs nothing because there is no KVO code. Delegate methods and `NSNotification` still have to do work even if there are no observers. For low-level, performance-critical objects, KVO is an obvious win. Use it wisely. And hope Apple improves the API.

Summary

In this chapter you have learned two of the most powerful techniques in Objective-C, KVC and KVO. These techniques provide a level of runtime flexibility that is difficult to achieve in other languages. Writing your code to conform to KVC is a critical part of a Cocoa program, whether you call `valueForKey:` directly or not. KVO can be challenging to use well, but is a powerful tool when you need high-performance observations. As a Cocoa developer, you need to keep KVC and KVO in mind when designing your classes. Following a few simple naming rules will make all the difference.

Further Reading

Apple Documentation

The following documents are available in the iOS Developer Library at `developer.apple.com` or through the Xcode Documentation and API Reference.

Key-Value Coding Programming Guide

Key-Value Observing Programming Guide

NSKeyValueCoding Protocol Reference

NSKeyValueObserving Protocol Reference

Other Resources

Matuschak, Andy. "KVO+Blocks: Block Callbacks for Cocoa Observers." This is a very promising category for adding block support to KVO. In our opinion, it hasn't gotten the kind of testing time it needs to trust it in a complex project, but we definitely like the approach. It's the foundation of several other wrappers.
`http://blog.andymatuschak.org/post/156229939/kvo-blocks-block-callbacks-for-cocoa-observers`

Waldowski, Zachary. *BlocksKit*. An extensive set of blocks-based enhancements, including a KVO observations wrapper. Although we haven't used it extensively enough to recommend it, this package has the most users and active development. It is the one we'd most likely use. Its KVO wrapper is based on Andy Matuschak's code, among others.
`https://github.com/zwaldowski/BlocksKit`

Wight, Jonathan. "Key-Value Observing Done Right (again)." Another interesting example, based on Andy Matuschak's `MAKVONotificationCenter`.
`http://toxicsoftware.com/kvoblocks/`

Chapter 23
Think Different: Blocks and Functional Programming

Programming paradigms fall into three major categories: procedural programming, object-oriented programming, and functional programming. Most modern programming languages—such as Java, C#, or Objective-C—don't fall purely into a specific paradigm. They are usually inclined to one while including some methodologies from the others. Objective-C is primarily object-oriented, yet it borrows some functional aspects using *blocks*. This chapter is about the functional programming aspects in Objective-C.

The functional programming (FP) paradigm is easy to explain theoretically, but its highly abstract nature makes it hard to understand and realize its power. The easiest way to really comprehend and appreciate FP is to implement a practical system. Later in this chapter, you dirty your hands with a dash of FP by writing an FP equivalent of a commonly used Cocoa method, and then go deeper by writing a category addition that makes UIAlertView to use blocks. Finally, you learn about some of the block-based methods added to the Cocoa framework from iOS 4 onward. Now, it's time to get started.

What Is a Block?

Simply defined, a *block* is an ad hoc piece of code. Just as you would with a primitive data type like an integer or double, you declare a block and start using it. You can pass blocks as parameters, "copy" them for use later, and do pretty much anything that you would normally do to a primitive data type. Veteran C programmers have been using function pointers to do pretty much what I just explained. What sets a block apart from a function pointer (which is different enough to be called a programming paradigm by itself) is that it can be declared within the lexical scope of another method, and that it can "capture the state" of the variables in that scope. This means that a block has contextual information without the programmer needing to do anything, unlike a function pointer.

> A block differs from a function pointer by the way it is written and by its inherent nature to capture the state of local variables and optionally modify them.

Now, that paragraph doesn't really help you to understand the concept. It is analogous to a chapter about object-oriented programming that says: An object is an "entity" that encapsulates code with data. You will be hesitant to use blocks and the functional programming paradigm if you don't comprehend it conceptually. So, before I talk about another abstract point, let's get this one straight by thinking outside the box. To do that, I'll show you a practical, nonprogramming-related example.

Why Use Functional Programming?

To understand why you should use a functional programming paradigm, it might help to relate it to something outside of the programming world. I'm going to digress a bit here, but I promise it will take fewer than a couple of minutes of your reading time.

The Human Brain Versus the Microprocessor

A microprocessor stores data in an addressable fashion and executes instructions to process them. The human brain, on the other hand, thinks and stores data associatively. Here's an example: On a microprocessor, you store a number, the value 10, and access it by its address 0x8BADF00D. When you ask the microprocessor about the location of the number, it returns the address. The human brain works differently. The answer to the question "Where is my iPhone?" would probably be similar to "It's in the living room on top of the DVD player." The human answer won't even remotely be like "It's 5.04 meters northeast at an angle 27.23 degrees from the main entrance to the living room." Humans associate the location of the iPhone with the location of the DVD player and the location of the DVD player with the location of the living room. That's a huge difference in the thought processes between how a microprocessor works and how a human brain works. This difference in thought process is akin to the difference between the procedural and functional programming paradigm.

Procedural Versus Functional Paradigm

Procedural programming is all about feeding your microprocessor with instructions. While that has worked very well for decades, some real-world programming tasks are easier to think about and express as abstract concepts than as a bunch of instructions. Let me give you an example. (No, not that clichéd Factorial example; read on, this is going to be Objective-C.)

A "Functional" UIAlertView

For this example, you want to show a `UIAlertView` and take an action when the user taps on the Affirmative button. The procedural way of doing this is to create a `UIAlertView` object, own its delegate and implement the callbacks, show the `UIAlertView`, and release it.

UIAlertView Example (The Procedural Way)

```
-(IBAction) buttonTapped:(id) sender  {
  UIAlertView *alert = [[UIAlertView alloc] initWithTitle:@"Send email"
  message:@"Are you sure you want to send it now?"
  delegate:self
  cancelButtonTitle:@"Cancel"
  otherButtonTitles:@"Send", nil];

  [alert show];
  [alert release];
}

-(void)alertView:(UIAlertView*) alertView didDismissWithButtonIndex:(NSInt
```

```
eger) buttonIndex  {

  if(buttonIndex != [alertView cancelButtonIndex])  {
    [self sendTheMail];
  }
}

-(void) sendTheMail  {
  // write actual code for sending the mail
}
```

Implementation (and syntax) aside, now look at how you call a `UIAlertView` that adheres to the functional paradigm.

UIAlertView Example (The Functional Way)

```
[UIAlertView showAlertViewWithTitle:@"Send email"
message:@"Are you sure you want to send it now?"
cancelButtonTitle:@"Cancel"
otherButtonTitles:[NSArray arrayWithObjects:@"Send", nil] onCompletion:^{
  // write actual code for sending the mail
}
onCancel:^{
  // write code for handling other cases
}];
```

That's much cleaner.

■ You don't have to implement a delegate.

■ You don't have to allocate or release objects.

■ You don't have to explicitly show the alert.

Instead, you specify your intent and things happen behind the scenes automatically for you.

In the functional paradigm example, your code reads like English. You declaratively say what your title is, and what the message you need to show is. When the user closes the alert, you declaratively state the code to execute for both cases, closing by accepting the alert and closing by cancelling the alert instead of implementing a delegate.

The point is that instead of instructing, you specify your intent, and the code becomes much cleaner to read and follow.

You will appreciate this code when you show multiple alerts in a single view controller and use tags to differentiate the alert views on the callback delegate. If you understand this, you have grasped the essence of the functional programming paradigm.

You learn about implementing this method as a category addition to `UIAlertView` along with syntax and other associated matters in the "Block-Based UIAlertView" section later in this chapter.

Because microprocessors still execute instructions one by one, you may wonder why you should bother about functional programming. That's a valid question. The functional paradigm is almost always for writing software that's easy for co-developers to read and understand. In nearly every case, an equivalent imperative logic is less expensive to execute, but with Moore's law, and the rate at which microprocessor speed increases, code clarity is often more important than writing efficient code. Again, that doesn't always mean you should adhere to a functional paradigm. Not every real-world problem can be represented in a functional way.

I hope I've persuaded you to learn about and use the functional paradigm, so now it's time to dive into the technical bits of it.

Declaring a Block

You declare a block using the ∧ (caret) character.

```
int (^MyBlock) (int parameter1, double parameter2);
```

This syntax is for a block called `MyBlock` that takes an `int` and `double` as parameters and returns an integer. In most applications, you would `typedef` a block like this:

```
typedef int (^MyBlock) (int parameter1, double parameter2);
```

You can now declare instances of your block as you would declare any other data type.

```
MyBlock firstBlock, secondBlock;
```

You assign a block like this:

```
firstBlock = ^(int parameter1, double parameter2)   {
// your block code here
NSLog(@"%d, %f", parameter1, parameter2);
}
```

Because the block takes in an `integer` and a `double` as parameters, you can invoke them like this:

```
firstBlock(5, 2.3);
```

Invoking this block in this case will `NSLog` the numbers to the console. That's pretty much all for the syntax.

Scope of Variables

You learned earlier in this chapter that compared to a function pointer, a block has access to the variables in its lexical scope. Optionally, a block can modify a variable in the lexical scope even if the variable goes out of scope. To allow a block to access variables within a lexical scope, the Objective-C runtime allocates blocks on the stack.

When you're passing a block as a parameter to other methods, you copy the block to the heap. This is because blocks, when used as parameters, will be invoked after the stack in which they were originally created is deallocated. You can use the normal Objective-C copy message to copy a block. This also copies/retains variables in the block's lexical scope. This is a very important concept to understand when you use blocks in your code. In the next section, you read scenarios about when a variable is copied and when it is retained.

Warning: A block cannot be retained. Sending a retain message is a no-op and doesn't increase the retain count. Normally, you wouldn't send a retain message explicitly. But you should keep an eye on the storage semantics you use when you declare a block as a property. Even the omniscient LLVM compiler doesn't warn you when you use retain as a storage type for a block property.

Stack Versus Heap

A block is a different kind of Objective-C object. Unlike traditional objects, blocks are not created on the heap. This is primarily for two reasons. The first is performance: A stack allocation is almost always faster than heap. The second reason is the necessity to access other local variables.

Now, a stack gets destroyed when the scope of the function ends. If your block is passed to a method that needs it even after its scope is destroyed, you should copy your block. This is where Objective-C runtime performs some magic behind the scenes. When a block is copied, it's moved from stack to heap. Along with the block, the local variables defined in its scope are copied when you reference it within your block. All NSObject subclasses that are referenced are retained instead of copied (because they are already on heap, and retain is less time-consuming than a copy). The Objective-C runtime gives a const reference to every local variable to a block. This also means that a block cannot modify the contextual data by default, and code like the following will result in a compilation error.

```
int statusCode   = -1;

Myblock b = ^{
  statusCode = 4;
};
```

But I previously said that blocks could "optionally" modify the local variables. To allow modification, you declare variables with a __block modifier. So the declaration of statusCode is now __

```
block int statusCode = -1;
```

The reason for this additional modifier is to instruct the compiler to copy the __block variables when the block is copied. Copying is a more time-consuming operation than either retain or passing by a const reference, and the implementers decided to leave this in the hands of the developer.

__block **variables are copied instead of being retained.**

Now, coming back to the previous example of a blocks-based `UIAlertView`, your block has all contextual information available without declaring any additional data structures. Your `UIAlertView` `onDismiss` or `onCancel` block methods can access the local variables without the developer managing them (through context parameters).

However, it comes with a minor catch that you should be wary of: the retain cycle.

The Retain Cycle Problem

```
__block TWTweetComposeViewController *controller =
  [[TWTweetComposeViewController alloc] init];
        [controller setInitialText:@"Test Tweet"];

      controller.completionHandler =
       ^(TWTweetComposeViewControllerResult result)   {

          controller = nil; // retain cycle issue
          [self dismissModalViewControllerAnimated:YES];
      };
```

If you attempt to capture a variable within a block as shown in the preceding code, you end up with a retain cycle that never gets deallocated. With the new LLVM compiler, you normally don't have to worry much about this because it's clever enough to point out the issue.

Implementing a Block

Now that you know the workings of a block, let's implement the `UIAlertView` block-based example I showed you in the previous section.

Block-Based UIAlertView

You want the syntax for this example to be like the following:

```
[UIAlertView showAlertViewWithTitle:(NSString*)title
message:(NSString*) message
cancelButtonTitle: (NSString*) cancelTitle
otherButtonTitles: (NSArrat*) otherButtons
onCompletion:^{
  // write actual code for sending the mail
}
onCancel:^{
  // write code for handling other cases

}];
```

Follow these steps to implement the `UIAlertView` block-based example:

1. Add a Category class on `UIAlertView`. You can use the Category template provided in Xcode 4.5. Call the Category class `UIAlertView` (Blocks).

2. `typedef` your Dismiss and Cancel blocks:

```
typedef void (^DismissBlock)(int buttonIndex);
typedef void (^CancelBlock)();
```

3. Add the method definition to the header file:

```
+ (UIAlertView*) showAlertViewWithTitle:(NSString*) title
                           message:(NSString*) message
                 cancelButtonTitle:(NSString*) cancelButtonTitle
                 otherButtonTitles:(NSArray*) otherButtons
                         onDismiss:(DismissBlock) dismissed
                          onCancel:(CancelBlock) cancelled;
```

4. Declare static storage for the blocks on the implementation:

```
static DismissBlock _dismissBlock;
static CancelBlock _cancelBlock;
```

5. Implement your block-based method:

```
+ (UIAlertView*) showAlertViewWithTitle:(NSString*) title
                           message:(NSString*) message
                 cancelButtonTitle:(NSString*) cancelButtonTitle
                 otherButtonTitles:(NSArray*) otherButtons
                         onDismiss:(DismissBlock) dismissed
                          onCancel:(CancelBlock) cancelled {

_cancelBlock  = [cancelled copy];

_dismissBlock  = [dismissed copy];

  UIAlertView *alert = [[UIAlertView alloc] initWithTitle:title
                                        message:message
                                        delegate:[self self]
                              cancelButtonTitle:cancelButtonTitle
                              otherButtonTitles:nil];

  for(NSString *buttonTitle in otherButtons)
    [alert addButtonWithTitle:buttonTitle];

  [alert show];
  return alert;
}
```

Note that you have copied the block parameters passed to the method to the static storage. That's because you have to invoke those block methods later in the `UIAlertViewDelegate`.

6. Handle the `UIAlertViewDelegate`:

```
+ (void)alertView:(UIAlertView*) alertView
   didDismissWithButtonIndex:(NSInteger) buttonIndex {

   if(buttonIndex == [alertView cancelButtonIndex])  {
     _cancelBlock();
   }
   else  {
     _dismissBlock(buttonIndex - 1);// cancel button is button 0
   }

}
```

The delegate method is nothing fancy. Just handle it and call the appropriate block method.

With about 50 lines of code, you implemented a block-based `UIAlertView`. This should help write clean, readable code on your view controllers. You can implement similar methods for `UIActionSheet` as well.

You can download the complete source code for the block-based UIAlertView from the book's website.

This should give you a complete understanding of how to use blocks in a much more sophisticated example. That completes the example; in the next section, you learn about Cocoa functions that take blocks as parameters.

Blocks and Concurrency

By now, you know how to use blocks and should be comfortable with the syntax. Now I'll talk about one other important benefit you get from blocks: concurrency.

Managing concurrency has always been the hardest part of programming. Blocks are an excellent use case here because they can be used for creating units of programming or tasks that can be executed independently. Blocks can be used with dispatch queues in *Grand Central Dispatch* (GCD) or `NSOperationQueue` without needing to create threads explicitly. Using an `NSOperationQueue` is something you have already done in the RESTEngine implementation.

In the next section, you briefly go through a feature of iOS and OS X called GCD and discover how blocks and dispatch queues in GCD work together to make concurrency implementation easier. Later on, I'll compare GCD with `NSOperationQueue` and provide suggestions on when to use what.

Dispatch Queues in GCD

Grand Central Dispatch is a very powerful feature that allows you to write concurrent code easily. It shifts the burden of managing multiple threads and thread synchronization to the operating system (iOS or OS X). When you use GCD, you create units that can be executed independently of each other and let the operating system handle the queuing and synchronization for you. The GCD implementation on iOS (and OS X) consists of a set of C language extensions, APIs, and a runtime engine. GCD automatically ensures that your independent units are executed on multiple processors if available (like iPad 2). As a developer, the only thing you have to focus on is designing your heavy worker processes so that they work independently of each other (as opposed to threads with shared synchronized data). GCD also provides context pointers to share data across your blocks. You learn more about GCD in Chapter 13.

GCD provides three types of dispatch queues—serial, concurrent, and main—to which you can enqueue your task. The serial queue executes one task at a time in first-in-first-out (FIFO) order, and the concurrent queue executes them in parallel, also in FIFO order. The main dispatch queue executes operations on the main thread, which is usually used to synchronize execution across threads executing in different serial/concurrent queues. Now, you'll see how to create a dispatch queue and submit tasks to it.

Creating a dispatch queue is as easy as one C function call:

```
myQueue = dispatch_queue_create("com.mycompany.myapp.myfirstqueue", NULL);
```

To dispatch tasks asynchronously to this queue, use the `dispatch_async` method. That method takes your block as the second parameter. It essentially queues your block to the queue specified in the first parameter. This is yet another few lines of code:

```
dispatch_async(myQueue, ^(void) {
    [self doHeavyWork];
});
```

That's it. Without explicitly using a thread and in less than ten lines of code, you have implemented GCD in your app! Designing your blocks in an independent way already solved most of the complexities involved around synchronization. For example, in the preceding code, the `doHeavyWork` method is designed to work independently with its own data.

Imagine that the `doHeavyWork` method is an image manipulation method. To design it so that it runs independently, slice the image (vertically or horizontally), pass each slice to a block, and send this to a dispatch queue instead of using the complete image data on a shared synchronized variable. That is, if you have a 3200 × 2000-pixels image and you want to apply a filter on it, slice it to 10 different images of 320 × 2000 pixels each and process them independently. After processing is done, stitch them back together on the main dispatch queue and notify the relevant observers.

NSOperationQueue Versus GCD Dispatch Queue

You already know that iOS provides another queuing mechanism called NSOperationQueue. This also takes blocks as parameters and queues them just like a dispatch queue. Now, you might have a question: When should I use GCD and when should I use NSOperationQueue?

Here are some similarities and differences between `NSOperationQueue` and GCD:

- `NSOperationQueue` is built using GCD and is a higher-level abstraction of it.

- GCD supports only FIFO queues, whereas operations queued to an `NSOperationQueue` can be reordered (reprioritized).

- Setting dependencies between operations is possible with `NSOperationQueue` but not with GCD. If one of your operations needs data that is generated by the other, you can set the operation to be dependent on the other operation, and `NSOperationQueue` automatically executes them in the correct order. With GCD, there is no built-in support to set dependencies.

- `NSOperationQueue` is KVO-compliant. This means you can observe the state of the tasks. Does that mean you should always use `NSOperationQueue` instead of GCD? The answer is no. `NSOperationQueue` is slower than GCD in terms of execution speeds. If you profile your code using Instruments and you think you need more performance, use GCD. Usually in lower-level code, you may not have task dependencies or a necessity to observe state using KVO. As always, follow Donald Knuth's quote, "We should forget about small efficiencies, say about 97% of the time: premature optimization is the root of all evil," and use the lower-level GCD only if it improves performance gains when profiled with Instruments.

Block-Based Cocoa Methods

With iOS 4 and the introduction of blocks, many of the built-in Cocoa framework methods have block-based equivalents. Covering every single block-based method is impossible in a single chapter and merits a complete book of its own. But Apple follows a pattern. In this section, I briefly explain some of the methods that take block parameters and give some hints and tips on when to look out for a block-based equivalent method in the framework.

UIView Animations Using Blocks

Prior to iOS 4, view-based animations were usually done using `UIView`'s class methods, `beginAnimations` and `commitAnimations`. You write the code you want to be animated within these two statements, and the animation is performed after the call to `commitAnimations`.

Code to animate the alpha value of a view will look something like the following.

Animation in iOS 3 (Without Blocks)

```
 [UIView beginAnimations:@"com.mycompany.myapp.animation1"
  context:&myContext];
[UIView setAnimationCurve:UIViewAnimationCurveEaseInOut];
[UIView setAnimationDuration:1.0f];
[UIView setAnimationDelay:1.0f]; // start after 1 second
[UIView setAnimationDidStopSelector:@selector(animationDidStop:finish
ed:)];
   self.imageView.alpha = 0.0;
 [UIView commitAnimations];
```

Starting with iOS 4, UIView has several equivalent block-based animation methods. One method that you'll quite commonly use is `animateWithDuration:delay:options:animations:completion:`. The previous code snippet can be expressed using this method as follows.

Animation in iOS 4 and Above (with Blocks)

```
[UIView animateWithDuration:1.0 delay:1.0
  options:UIViewAnimationCurveEaseInOut
                    animations:^ {
                        self.imageView.alpha = 0.0;
                    }
                    completion:^(BOOL finished) {
                        [self. imageView removeFromSuperView];
                    }];
```

Note that the blocks version also removes the `imageView` after animation is complete, something that is done on the callback method `animationDidStop:animated:` in the iOS 3 version (not illustrated). Other block-based animation methods are permutations of this method, omitting parameters like delay, options, and completion block. A huge advantage of this method is that you don't have to maintain context you normally set with the methods `setAnimationWillStartSelector:` and `setAnimationDidStopSelector:`. Because a block is aware of the local context, you don't even need a context parameter.

Presenting and Dismissing View Controllers

iOS 5 introduced a new method for presenting and dismissing view controllers that takes a block parameter that must be called when the presenting (or dismissing) animation is completed. This block parameter is called after `viewDidDisappear` is called. Prior to iOS 5, you might have done some cleanup code in `viewDidDisappear`. With this method in iOS 5, you can easily do that in the completion block. The methods can be invoked like this.

Presenting

```
[self presentViewController:myViewController animated:YES completion:^  {
  //Add code that should be executed after view is presented
        }];
```

Dismissing

```
[self dismissViewControllerAnimated:YES completion:^  {
  //Add code that should be executed after view is presented
}];
```

TweetComposer Versus In App E-mail/SMS

In iOS 5, Apple added native support for Twitter, and apps that need to send out a tweet could just instantiate a `TWTweetComposerViewController`, prepopulated with the text to be tweeted, and present it to the user. The implementation is very similar to how you normally send an in app e-mail or SMS. However, the `TWTweetComposeViewController` reports completion by a block parameter instead of a delegate. Your completion handler will look something like the following code.

TwTweetComposeViewController Completion Handler

```
controller.completionHandler = ^(TWTweetComposeViewControllerResult
result)
  {
            [self dismissModalViewControllerAnimated:YES];
            switch (result) {
                case TWTweetComposeViewControllerResultCancelled:
                    break;
                case TWTweetComposeViewControllerResultDone:
                    break;
            }
        };
```

Dictionary Enumeration Using NSDictionary enumerateWithBlock

Dictionary enumeration using block-based methods is sure to make your code cleaner. You no longer have to deal with `keyEnumerator` or `objectForKey` methods. With block-based equivalents, it's much easier, as shown here:

```
[dictionary enumerateKeysAndObjectsUsingBlock:^(id key, id val, BOOL
  *stop)  {

      //NSLog(@"%@, %@", key, val);
}];
```

Looking for Block-Based Methods

The Cocoa framework follows a pattern when using blocks. Here are a few patterns that can help you in searching for equivalent block-based methods in the Cocoa framework:

- Check whether the current method has a context parameter for a Cocoa method. If it does, then the chances are that there will be a block-based equivalent.

- Look for delegates with one or two optional methods. You might find a `completionHandler` for classes that were previously notifying results via delegates.

- Enumeration, sorting, and filtering methods mostly have block equivalents. Examples include, `NSArray`, `NSDictionary`, `NSString`, `NSAttributedString`, `NSFileManager`, and several others.

Once you get used to the Cocoa framework design pattern and functional programming paradigm, you should be able to intuitively guess whether a method might have an equivalent block-based method.

Supported Platforms

Blocks are supported from iOS 4 and Snow Leopard. This means that when you use blocks, you should raise your minimum deployment target to iOS 4.0 for iOS projects and Mac OS X 10.6 for Mac projects. In 2012–2013, you shouldn't be worrying about this as almost all devices run iOS 4 and above so nothing should stop you from using blocks. When you are writing an iOS app that has a Mac counterpart and you are sharing the code across platforms, remember that Leopard and prior operating systems don't support blocks natively. But there is a third-party open-source block-based runtime called PLBlocks (see the "Further Reading" section at the end of this chapter) that allows support for blocks on those operating systems. For Mac projects, my recommendation is to start using Apple's equivalent block-based Cocoa methods using conditional compilation and then remove it altogether when Snow Leopard/Lion/Mountain Lion usage is high enough, which I predict isn't more than a year away.

Summary

This chapter discussed the functional paradigm, a very powerful paradigm that can make your code more readable (as in case of the `UIAlertView` example) and help you write less code. Functional programming will be the next big programming paradigm change after object-oriented programming, and you'll be seeing more and more Cocoa methods using and accepting blocks.

Apple usually makes older technologies obsolete faster than its competitors. As much as this is true for the products it makes, it's also true for the Apple API. The `TWTweetComposeViewController` class is a perfect example of this. While it is similar to the `MFMailComposeViewController` class, handling responses from `TWTweetComposeViewController` is via a `completionHandler` block, unlike `MFMailComposeViewController` that uses a `MFMailComposeViewControllerDelegate`. Note that `TWTweetComposeViewController` doesn't even support delegates.

On similar lines, for maybe another couple of years, there will be block-based additions to existing methods, but newer classes will have only block-based parameters, and you'll eventually be forced to use them. A paradigm change like this is tough. But the sooner you get accustomed to the changes, the better.

Further Reading

Apple Documentation

The following documents are available in the iOS Developer Library at `developer.apple.com` or through the Xcode Documentation and API Reference.

iOS Programming Guide. "Blocks Programming Topics"

Concurrency Programming Guide. "Migrating Away from Threads—Apple Developer"

Blogs

mikeash.com. "Friday Q&A 2009-08-14: Practical Blocks"
`http://www.mikeash.com/pyblog/friday-qa-2009-08-14-practical-blocks.html`

Cocoa with Love. "How blocks are implemented (and the consequences)"
`http://cocoawithlove.com/2009/10/how-blocks-are-implemented-and.html`

JCMultimedia. *How to make iPhone Apps and Influence People.* "Is it worth supporting iOS 3 devices?"
`http://blog.jcmultimedia.com.au/2011/03/is-it-worth-supporting-ios-3-in-2011.html`

"Plblocks: Block-capable Toolchain: Runtime for Mac OS X 10.5 and iPhone OS 2.2+"
`http://code.google.com/p/plblocks/`

Eschatology. "When to use NSOperation vs. GCD"
`http://eschatologist.net/blog/?p=232`

Other Resources

MugunthKumar / UIKitCategoryAdditions
`https://github.com/MugunthKumar/UIKitCategoryAdditions`

## Chapter 24

Going Offline

The iPhone can connect to the Internet from nearly anywhere. Most iOS apps use this capability, which makes these apps one of the best Internet-powered devices ever made. However, because it's constantly on the move, connectivity, reception, or both can be poor. This poses a problem for iOS developers, who need to ensure that their apps' perceived response time remains more or less constant, as though the complete content were available locally. You do this by caching your data locally. Caching data means saving it temporarily so that it can be accessed faster than making a round trip to the server. That's easier said than done, and most apps don't get this right.

This chapter shows you the caching techniques you can use to solve the problem of slow performance caused by poor or unavailable connectivity. In Chapter 14, you found that Internet-connected apps fall into two major categories. The first category of apps behaves like a front end to an online web service. The iHotelApp in Chapter 14 is one such app. In this chapter, you add caching support to the iHotelApp that you developed in Chapter 14. The second category of apps synchronizes user-generated content with a remote server and optionally downloads the most recent "n" items from the server. iOS 5 introduced a new cloud platform specifically targeting the second category of apps. Chapter 25 introduces you to iCloud where you'll also discover ways to sync users' data across all their devices through iCloud and other similar competing services, such as Parse.

Reasons for Going Offline

The main reason why your app might need to work offline is to improve the perceived performance of the app. You go offline by caching your app's content. You can use two kinds of caching to make your app work offline. The first is *on-demand caching,* where the app caches request responses as and when they're made, much as your web browser does. The second is *precaching,* where you cache your contents completely (or a recent "n" items) for offline access.

Web service apps like the one developed in Chapter 14 use on-demand caching techniques to improve the perceived performance of the app rather than to provide offline access. Offline access just happens to be an added advantage. Twitter and foursquare are great examples of this. The data that these apps bring in often becomes stale quickly. How often are you interested in a tweet that was posted a couple of days ago or in knowing where a friend was last week? Generally, the relevance of a tweet or a check-in is important only for a couple of hours, but loses some or all of its importance after 24 hours. Nevertheless, most Twitter clients cache tweets, and the official foursquare client shows you the last state of the app when you open it without an active Internet connection.

You can even try this on your favorite Twitter client, Twitter for iPhone, Tweetbot, or whatever you prefer: Open a friend's profile and view his timeline. The app fetches the timeline and populates the page. While it loads the timeline, you see a loading spinner. Now go to a different page and come back again and open the timeline. You will see that it's loaded instantly. The app still refreshes the content in the background (based on when you previously opened it), but instead of showing a rather uninteresting spinner, it shows previously cached content, thereby making it appear fast. Without this caching, users will see the spinner for every single page, which slowly frustrates them. Whether the Internet connection is fast or slow, it's your responsibility as an iOS developer to mitigate this effect and provide the perception that the app is loading fast. This goes a long way toward improving your customers' satisfaction and thereby boosting your app's ratings on the App Store.

The other kind of caching gives more importance to the data being cached and the ability to edit the cached items on the fly without connecting to the server. Examples include apps such as Google Reader clients, read-later apps such as Instapaper, and so on.

Strategies for Caching

The two caching techniques discussed in the previous section—on-demand caching and precaching—are quite different when it comes to design and implementation. With on-demand caching, you store the content fetched from the web service locally on the file system (in some format), and then for every request, you check for the presence of this data in the cache and perform a fetch from the server only if the data isn't available (or is stale). Hence, your cache layer will behave more or less like cache memory on your processor. The speed of fetching the data is more important than the data itself. On the other hand, when you precache, you save content locally for future access. With precaching, a loss of data or a cache-miss is not acceptable. For example, consider a scenario where the user has downloaded articles to read while on the subway, only to find that they're no longer present on her device.

Apps like Twitter, Facebook, and foursquare fall into the on-demand category, whereas apps like Instapaper and Google Reader clients fall into the precaching category.

To implement precaching, you'll probably use a background thread that accesses data and stores it locally in a meaningful representation so that local cache can be edited without reconnecting to the server. Editing can be either "marking items as read" or "favoriting items" or a similar operation on the item. By *meaningful representation,* I mean you save your contents in a way that allows you to make these kinds of modifications locally without talking to the server and then are able to send the changes back once you're connected again. This capability is in contrast to apps like foursquare where you cannot become a Mayor of a place without an Internet connection, though you can see the list of your Mayorships without an Internet connection (if it is cached). Core Data (or any structured storage) is one way to do this.

On-demand caching works like your browser cache. It allows you to view content that you've viewed/visited before. You can implement on-demand caching by caching your data models (to create a data model cache) when you open a view controller on-demand rather than on a background thread. You can also implement on-demand caching when a URL request returns a successful (200 OK) response (to create a URL cache). There are advantages and disadvantages of both the methods, and I show you the pros and cons of each of them in the "Data Model Cache" and "URL Cache" sections later in this chapter.

A quick-and-dirty way of deciding whether to go for on-demand caching or precaching is to determine whether you might ever post-process any data after downloading it. Post-processing could be in the form of either user-generated edits or updates made to downloaded data, such as rewriting image links in an HTML page to point to locally cached images. If your app requires any of the above-mentioned post-processing, you must implement precaching.

Storing the Cache

Third-party apps can save information only to the application's sandbox. Because cache data is not user-created, it should be saved to the `NSCachesDirectory` instead of the `NSDocumentsDirectory`. A good practice is to create a self-contained directory for all your cached data. In this example, you'll create in the `Library/caches` folder a directory named `MyAppCache`. You can create this directory using the following code:

```
NSArray *paths = NSSearchPathForDirectoriesInDomains(NSCachesDirectory,
   NSUserDomainMask, YES);
NSString *cachesDirectory = [paths objectAtIndex:0];
cachesDirectory = [cachesDirectory
   stringByAppendingPathComponent:@"MyAppCache"];
```

The reason for storing the cache in the caches folder is that iCloud (and iTunes) backups exclude this directory. If you create large cache files in the Documents directory, they get uploaded to iCloud during backup and use up the limited space (about 5GB at the time of this writing) fairly quickly. You don't want to do that—you want to be a good citizen on your user's iPhone, right? `NSCachesDirectory` is meant for that.

Precaching is implemented using a higher-level database such as raw SQLite or an object serialization framework such as Core Data. You need to carefully choose the technology based on your requirements. I offer suggestions on when to use URL cache or data model cache and when to use Core Data in the "Which Caching Technique Should You Use?" section later in this chapter. Now, I'll show you the implementation-level details of a data model cache.

Implementing Data Model Caching

You implement data model caching using the `NSKeyedArchiver` class. In order to archive your model objects using `NSKeyedArchiver`, the model class must conform to the `NSCoding` protocol.

NSCoding Protocol Methods

```
- (void)encodeWithCoder:(NSCoder *)aCoder;
- (id)initWithCoder:(NSCoder *)aDecoder;
```

When your models conform to `NSCoding`, archiving them is as easy as calling one of the following methods:

```
[NSKeyedArchiver archiveRootObject:objectForArchiving
toFile:archiveFilePath];
```

```
[NSKeyedArchiver archivedDataWithRootObject:objectForArchiving];
```

The first method creates an archive file specified at the path `archiveFilePath`. The second method returns an `NSData` object. `NSData` is usually faster because there's no file-access overhead, but it is stored in your application's memory and will soon use up memory if it's not checked periodically. Periodic caching to flash memory on the iPhone is also not advisable because cache memory, unlike hard drives, comes with limited read/write cycles. You need to balance both in the best possible way. You will learn in detail about caching using archives later in the "Data Model Cache" section later in this chapter.

The `NSKeyedUnarchiver` class is used to unarchive your models from a file (or a NSData pointer). You can use either one of the following class methods, depending on from where you have to unarchive.

```
[NSKeyedUnarchiver unarchiveObjectWithData:data];

[NSKeyedUnarchiver unarchiveObjectWithFile:archiveFilePath];
```

These four methods come in handy when converting to and from serialized data.

Use of any of the `NSKeyedArchiver`/`NSKeyedUnarchiver` methods requires that your models implement the `NSCoding` protocol. However, doing so is so easy that you can automate implementing the `NSCoding` protocol using tools such as Accessorizer. (See the "Further Reading" section at the end of this chapter for a link to Accessorizer at the Mac App Store.)

The next section explains strategies that can be used for precaching. You learned previously that precaching requires that you use a more structured data format. I introduce both Core Data and SQLite here.

Core Data

Core Data, as Marcus Zarra says, is more of an object serialization framework than just a database API:

> It's a common misconception that Core Data is a database API for Cocoa. . . . It's an object
> framework that can be persisted to disk (Zarra, 2009).

For a good, in-depth explanation of Core Data, read *Core Data: Apple's API for Persisting Data on Mac OS X* by Marcus S. Zarra (Pragmatic Bookshelf, 2009. ISBN 9781934356326).

To store data in Core Data, first create a Core Data model file and create your Entities and Relationships; then write methods to save and retrieve data. Using Core Data, you get true offline access for your app, such as Apple's built-in Mail and Calendar apps. When you implement precaching, you must periodically delete data that's no longer needed (stale); otherwise, your cache will start growing and hurt the app's performance. Synchronizing local changes is done by keeping track of changesets and sending them back to the server. There are many algorithms for changeset tracking, and the one I recommend is the one used by the Git version control system (I don't cover syncing your cache with a remote server because it's beyond the scope of this book).

Using Core Data for On-Demand Caching

Although technically you can use Core Data for on-demand caching, I advise against using it as such. The benefit Core Data offers is individual access to the models' properties without unarchiving the complete data. However, the complexity of implementing Core Data in your app defeats the benefits; moreover, for on-demand cache implementation, you probably wouldn't require individual access to the models' properties.

Raw SQLite

SQLite can be embedded into your app by linking against the libsqlite3 libraries, but it has significant drawbacks. All sqlite3 libraries and Object Relational Mapping (ORM) mechanisms are almost always going to be slower than Core Data. In addition, while sqlite3 is thread-safe, the binary bundled with iOS is not. So unless you ship a custom-built sqlite3 library (compiled with the thread-safe flag), it becomes your responsibility to ensure that data access to and from the sqlite3 database is thread-safe. Because Core Data has so much more to offer and has thread-safety built in, I suggest avoiding native SQLite as much as possible on iOS.

> The only exception for using Raw SQLite over using Core Data in your iOS app is when you have application-specific data in the resource bundle that is shared by all other third-party platforms your app supports—for example, a location database for an app that runs on iPhone, Android, BlackBerry, and, say, Windows Phone. But again, that's not caching, either.

Which Caching Technique Should You Use?

Of the different techniques available to save data locally, three of them stand out: URL cache, data model cache (using `NSKeyedArchiver)`, and Core Data.

If you're developing an app that needs to cache data to improve perceived performance, you should implement on-demand caching (using a data model cache or URL cache). On the other hand, if you need your data to be available offline and in a more meaningful way so that editing offline data is possible, use a higher-level serialization like Core Data.

Data Model Cache Versus URL Cache

On-demand caching can be implemented using either a data model cache or a URL cache. Both have advantages and disadvantages, and choosing which to use depends on the server implementation. A URL cache is implemented like a browser's cache or a proxy server's cache. It works best when your server is correctly designed and conforms to the HTTP 1.1 caching specifications. If your server is a SOAP server (or servers implemented like RPC servers, other than a RESTful server), you need to use data model caching. If your server adheres to the HTTP 1.1 caching specification, use URL caching. Data model caching allows the client (iOS app) to have control over cache invalidation, whereas when you implement a URL cache, the server dictates invalidation through HTTP 1.1 cache control headers. Although some programmers find this approach counter-intuitive and complicated to implement, especially on the server, it's probably the right way to do caching. As a matter of fact, `MKNetworkKit` provides native support for the HTTP 1.1 caching standard.

Cache Versioning and Invalidation

When you cache data, you need to decide whether to support version migration. If you're using an on-demand caching technique, version migration might be necessary if you use a data model cache. But the easiest way is to delete the cache when the user downloads the new version because old data is not important. On the other hand, if you have implemented precaching, chances are that you have cached multiple megabytes of data, and it only makes sense to migrate them to the new version. With Core Data, data migration across versions is easy (at least compared to raw sqlite).

> When you're using URL cache-based on-demand caching, URL responses are stored against the URLs as raw data. Versioning never becomes a problem. A change in version is either reflected by a URL change or invalidated from the server through cache-control headers.

In the following sections, I show you how to implement the two different types of on-demand caching: data model caching (using `AppCache`) and URL caching (using `MKNetworkKit`). You can download the complete source code located in this chapter's files on the book's website.

Data Model Cache

In this section you add on-demand caching to the iHotelApp from Chapter 14 by implementing a data model cache. On-demand caching is done as and when the view disappears from the hierarchy (technically, in your `viewWillDisappear:` method). The basic construct of the view controller that supports caching is shown in Figure 24-1. You can get the complete code for AppCache Architecture from the downloaded source code for this chapter. From this point on, I'm assuming that you've downloaded the code and have it available to use.

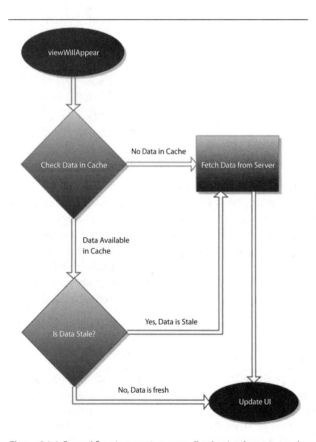

Figure 24-1 Control flow in your view controller that implements on-demand caching

In your `viewWillAppear` method, check your cache for the data necessary to display this view. If it's available, fetch it and update your user interface with cached data. Now, check whether your data from the cache is old. Your business rules should dictate what's new and what's old. If you decide that the content is old, show the data on the UI, and in the background, fetch data from the server and update the UI again. If the data is not available in the cache, fetch the data from the server while showing a loading spinner. After fetching data, update your UI.

The preceding flowchart assumes that what you show on the UI are models that can be archived. Implement the `NSCoding` protocol in the `MenuItem` model in the iHotelApp. The `NSKeyedArchiver` mandates that this protocol be implemented, as illustrated by the following code snippets.

NSCoding encodeWithCoder Method for the MenuItem Class (MenuItem.m)

```
- (void)encodeWithCoder:(NSCoder *)encoder
{
    [encoder encodeObject:self.itemId forKey:@"ItemId"];
    [encoder encodeObject:self.image forKey:@"Image"];
    [encoder encodeObject:self.name forKey:@"Name"];
    [encoder encodeObject:self.spicyLevel forKey:@"SpicyLevel"];
    [encoder encodeObject:self.rating forKey:@"Rating"];
    [encoder encodeObject:self.itemDescription forKey:@"ItemDescription"];
    [encoder encodeObject:self.waitingTime forKey:@"WaitingTime"];
    [encoder encodeObject:self.reviewCount forKey:@"ReviewCount"];
}
```

initWithCoder Method for the MenuItem Class (MenuItem.m)

```
- (id)initWithCoder:(NSCoder *)decoder
{
    if ((self = [super init])) {
        self.itemId = [decoder decodeObjectForKey:@"ItemId"];
        self.image = [decoder decodeObjectForKey:@"Image"];
        self.name = [decoder decodeObjectForKey:@"Name"];
        self.spicyLevel = [decoder decodeObjectForKey:@"SpicyLevel"];
        self.rating = [decoder decodeObjectForKey:@"Rating"];
        self.itemDescription = [decoder
          decodeObjectForKey:@"ItemDescription"];
        self.waitingTime = [decoder decodeObjectForKey:@"WaitingTime"];
        self.reviewCount = [decoder decodeObjectForKey:@"ReviewCount"];
    }
    return self;
}
```

As mentioned previously, you can generate the `NSCoding` protocol implementation using Accessorizer.

Based on the cache flow flowchart you saw in Figure 24-1, you have to implement the actual caching logic in the `viewWillAppear:` method. The following code added to `viewWillAppear:` implements that.

Code Snippet in the viewWillAppear: of Your View Controller That Restores Your Model Objects from Cache

```
NSArray *paths = NSSearchPathForDirectoriesInDomains(NSCachesDirectory,
    NSUserDomainMask, YES);
NSString *cachesDirectory = [paths objectAtIndex:0];
NSString *archivePath = [cachesDirectory
    stringByAppendingPathComponent:@"AppCache/MenuItems.archive"];

NSMutableArray *cachedItems = [NSKeyedUnarchiver
    unarchiveObjectWithFile:archivePath];

if(cachedItems == nil)
  self.menuItems = [AppDelegate.engine localMenuItems];
else
  self.menuItems = cachedItems;

NSTimeInterval stalenessLevel = [[[[NSFileManager defaultManager]
    attributesOfItemAtPath:archivePath error:nil]
fileModificationDate] timeIntervalSinceNow];

if(stalenessLevel > THRESHOLD)
  self.menuItems = [AppDelegate.engine localMenuItems];

[self updateUI];
```

The logical flow of the caching mechanism is as follows:

1. The view controller checks for previously cached items in the archive file `MenuItems.archive` and unarchives it.

2. If the `MenuItems.archive` is not present, the view controller makes a call to fetch data from server.

3. If `MenuItems.archive` is present, the view controller checks the archive file modification date to determine how stale this cached data is. If it's old (as determined by your business requirements), fetch the data again from the server. Otherwise, display the cached data.

Next, the following code added to the `viewDidDisappear` method saves (as `NSKeyedArchiver` archives) your models to the `Library/Caches` directory.

Code Snippet in the viewWillDisappear: of Your View Controller That Caches Your Models

```
NSArray *paths = NSSearchPathForDirectoriesInDomains(NSCachesDirectory,
    NSUserDomainMask, YES);
NSString *cachesDirectory = [paths objectAtIndex:0];
NSString *archivePath = [cachesDirectory stringByAppendingPathComponent:@"
AppCache/MenuItems.archive"];

[NSKeyedArchiver archiveRootObject:self.menuItems toFile:archivePath];
```

As the view disappears, you save the contents of the menuItems array to an archive file. Take care not to cache this if you didn't fetch from the server in viewWillAppear:.

So, just by adding fewer than ten lines in the view controller (and a bunch of Accessorizer-generated lines in the model), you add caching support to your app.

Refactoring

When you have multiple view controllers, the preceding code will probably get duplicated. You can avoid that duplication by abstracting out the common code and moving it to a new class called AppCache. This AppCache is the heart of the application that handles caching. By abstracting out common code to AppCache, you can avoid duplicated code in viewWillAppear: and viewWillDisappear:.

Refactor this code so that your view controller's viewWillAppear/viewWillDisappear block looks like the following code. The lines in bold show the changes made while refactoring, and I explain them following the code.

Refactored Code Snippet in the viewWillAppear: of Your View Controller That Caches Your Models Using the AppCache Class. (MenuItemsViewController.m)

```
-(void) viewWillAppear:(BOOL)animated {

  self.menuItems = [AppCache getCachedMenuItems];
  [self.tableView reloadData];

  if([AppCache isMenuItemsStale] || !self.menuItems) {

    [AppDelegate.engine fetchMenuItemsOnSucceeded:^(NSMutableArray
     *listOfModelBaseObjects) {

      self.menuItems = listOfModelBaseObjects;
      [self.tableView reloadData];
    } onError:^(NSError *engineError) {
      [UIAlertView showWithError:engineError];
    }];
  }

  [super viewWillAppear:animated];
}

-(void) viewWillDisappear:(BOOL)animated {

  [AppCache cacheMenuItems:self.menuItems];
  [super viewWillDisappear:animated];
}
```

The AppCache class abstracts the knowledge of staleness from the view controller. It also abstracts exactly where the cache is stored. Later in this chapter, you modify this AppCache to introduce another layer of cache where the content is stored in memory.

Because the `AppCache` class abstracts out exactly where the cache is stored, you don't have to worry about copying and pasting code that gets the application's cache directory. In case your app is like the iHotelApp example, you also can easily add security to the cached data by creating subdirectories for every user. The helper method in `AppCache` then currently returns the cache directory that can be modified to return the correct subdirectory for the currently logged-in user. This way, data cached by user 1 will not be visible to user 2, who logs in later.

The complete code listing is available from the source code download for this chapter on the book's website.

Cache Versioning

The `AppCache` class you wrote in the last section, abstracted out on-demand caching from your view controllers. When the view appears and disappears, caching happens behind the scenes. However, when you update the app, you might change your model classes, which means that any previously archived data will no longer be restored on your new models. As you learned earlier, in on-demand caching, your data is not that important, and you can delete it when you update the app. I'll show you a code snippet that can be used to delete the cache directory when version upgrades are performed.

Invalidating the Cache

First, save the application's current version somewhere. `NSUserDefaults` is a candidate. To detect version updates, every time the app launches, check whether the previously saved version is older than the app's current version. If it is, delete the cache folder and resave the new version to `NSUserDefaults`. Following is the code for doing so. Add it to your `AppCache init` method.

Code Snippet in the AppCache Initialize Method That Handles Cache Versioning (AppCache.m)

```
+(void) initialize
{
  NSString *cacheDirectory = [AppCache cacheDirectory];
  if(![[NSFileManager defaultManager] fileExistsAtPath:cacheDirectory])
  {
    [[NSFileManager defaultManager]  createDirectoryAtPath:cacheDirectory
    withIntermediateDirectories:YES
    attributes:nil
    error:nil];
  }

  double lastSavedCacheVersion = [[NSUserDefaults standardUserDefaults]
   doubleForKey:@"CACHE_VERSION"];
  double currentAppVersion = [[AppCache appVersion] doubleValue];

  if( lastSavedCacheVersion == 0.0f || lastSavedCacheVersion <
    currentAppVersion)
  {
    [AppCache clearCache];
    // assigning current version to preference
    [[NSUserDefaults standardUserDefaults] setDouble:currentAppVersion
```

```
      forKey:@"CACHE_VERSION"];
    [[NSUserDefaults standardUserDefaults] synchronize];
  }
}
```

Note that this code depends on a helper method that gets the application's current version. You can read the version from your app's `Info.plist` file using this block of code:

Code to Get the Current App Version from the Info.plist File (AppCache.m)

```
+(NSString*) appVersion
{
  CFStringRef versStr =
  (CFStringRef)CFBundleGetValueForInfoDictionaryKey
   (CFBundleGetMainBundle(), kCFBundleVersionKey);
  NSString *version = [NSString stringWithUTF8String:CFStringGetCStringPtr
                        (versStr,kCFStringEncodingMacRoman)];

  return version;
}
```

The preceding code calls a method to clear the cache directory. The following snippet illustrates that.

Code Snippet That Clears All Cached Files from the Cache Directory (AppCache.m)

```
+(void) clearCache
{
  NSArray *cachedItems = [[NSFileManager defaultManager]
                            contentsOfDirectoryAtPath:[AppCache
                            cacheDirectory] error:nil];

  for(NSString *path in cachedItems)
    [[NSFileManager defaultManager] removeItemAtPath:path error:nil];
}
```

Again, the cache invalidation and versioning issue is also abstracted out of the view controllers using the `AppCache` architecture. Now, go ahead and create an in-memory cache for the `AppCache` class. An in-memory cache improves the performance of caching drastically, but at the expense of memory. However, because on iOS, only one app runs in the foreground, this shouldn't be a problem.

Creating an In-Memory Cache

Every iOS device shipped so far has included flash memory, and this flash memory has one little problem: It has limited read-write cycles before it wears out. Although this limit is generally very high compared to the device's life span, it's still important to avoid writing to and reading from flash memory too often. In the previous example, you were caching directly to disk when the view was hidden and reading directly from disk whenever the view was shown. This behavior could tax the flash memory on users' devices. To avoid this problem, you can introduce another cache layer, which uses the device RAM instead of flash (`NSMutableDictionary`).

In the "Implementing Data Model Caching" section, you read about two methods for creating archives: one for saving them to a file and one for saving them as `NSData` objects. You will use the second method, which gives you a `NSData` pointer that you can store in a `NSMutableDictionary` rather than as flat files in the file system. The other advantage you get by introducing an in-memory cache layer is slightly higher performance when you archive and unarchive contents. Although this sounds complicated, it isn't really. In this section, you look at how to add a transparent in-memory cache to the `AppCache` class. (In-memory cache is transparent in the sense that the calling code—the ViewController—doesn't even know about its presence and doesn't need any code changes.) You also design a Least Recently Used (LRU) algorithm to save the cached data back to disk.

The following list outlines the steps you follow to create the in-memory cache. These steps are explained in more detail in the following sections.

1. Add variables to hold your cached data in memory.

2. Limit the size of the in-memory cache and write the least recently used items to a file and remove it from in-memory cache. RAM is limited, and when you hit the limit, you'll get a memory warning. Failing to release memory when you receive this warning will crash your app. You obviously don't want that to happen, right? So you set a maximum threshold for the memory cache. When anything is added to the cache after it's full, the last used object (least recently used) should be saved to file (flash memory).

3. Handle memory warnings and write the in-memory cache to flash memory (as files).

4. Write all in-memory cache to flash memory (files) when the app is closed or quit or when it enters the background.

Designing the In-Memory Cache for AppCache

You start designing the `AppCache` class by adding the variables to hold the cache data. Add an `NSMutableDictionary` for storing your cache data, an `NSMutableArray` to keep track of recently used items, in chronological order, and an integer that limits the maximum size of this cache, as shown in the following code.

Variables in AppCache

```
static NSMutableDictionary *memoryCache;
static NSMutableArray *recentlyAccessedKeys;
static int kCacheMemoryLimit;
```

Now you have to make changes to the `cacheMenuItems:` and `getCachedMenuItems` methods in `AppCache` to save the model objects transparently to this in-memory cache.

```
+(void) cacheMenuItems:(NSMutableArray*) menuItems
{
  [self cacheData:[NSKeyedArchiver archivedDataWithRootObject:menuItems]
        toFile:@"MenuItems.archive"];
}

+(NSMutableArray*) getCachedMenuItems
```

```
{
   return [NSKeyedUnarchiver unarchiveObjectWithData:[self
dataForFile:@"MenuItems.archive"]];
}
```

Instead of writing directly to the file, the preceding code calls a helper method, `cacheData:toFile:`. This
method will save the `NSData` from the `NSKeyedArchiver` to the in-memory cache. It also checks and
removes the least recently accessed data and saves it to file when the prefixed memory limit for the number
of in-memory items is reached. The implementation for this is shown in the following code.

Helper Method That Transparently Caches Data to In-Memory Cache (AppCache.m)

```
+(void) cacheData:(NSData*) data toFile:(NSString*) fileName
{
   [memoryCache setObject:data forKey:fileName];
   if([recentlyAccessedKeys containsObject:fileName])
   {
     [recentlyAccessedKeys removeObject:fileName];
   }

   [recentlyAccessedKeys insertObject:fileName atIndex:0];

   if([recentlyAccessedKeys count] > kCacheMemoryLimit)
   {
     NSString *leastRecentlyUsedDataFilename = [recentlyAccessedKeys
                                         lastObject];
     NSData *leastRecentlyUsedCacheData =
       [memoryCache objectForKey:leastRecentlyUsedDataFilename];
     NSString *archivePath = [[AppCache cacheDirectory]
                             stringByAppendingPathComponent:fileName];
     [leastRecentlyUsedCacheData writeToFile:archivePath atomically:YES];

     [recentlyAccessedKeys removeLastObject];
     [memoryCache removeObjectForKey:leastRecentlyUsedDataFilename];
   }
}
```

Similarly to the preceding code, which caches data (`cacheData:toFile:`), in the following code, you need
to write a method that checks the in-memory cache and returns this data, instead of directly reading from a file.
The method should access the file only if it isn't present in the in-memory cache.

Helper Method That Transparently Retrieves the Cached Data from In-Memory Cache (AppCache.m)

```
+(NSData*) dataForFile:(NSString*) fileName
{
   NSData *data = [memoryCache objectForKey:fileName];
   if(data) return data; // data is present in memory cache

   NSString *archivePath = [[AppCache cacheDirectory]
```

(continued)

```
                                  stringByAppendingPathComponent:fileName];
    data = [NSData dataWithContentsOfFile:archivePath];

    if(data)
      [self cacheData:data toFile:fileName]; // put the recently accessed
                                             data to memory cache

    return data;
}
```

This method also saves the data read from flash memory back to in-memory cache, which is just as a Least Recently Used caching algorithm works.

Handling Memory Warnings

For the most part, the `AppCache` is now complete, and you've added a transparent in-memory cache without modifying the calling code. However, you need to do one more important thing. Because you're retaining data used by views in `AppCache`, the memory consumption of your app continues to grow, and the chances of receiving a memory warning become very high. To avoid this situation, you handle the memory warning notifications in `AppCache`. In the static initialize method, add a notification observer to `UIApplicationDidReceiveMemoryWarningNotification`:

```
[[NSNotificationCenter defaultCenter] addObserver:self
  selector:@selector(saveMemoryCacheToDisk:)
  name:UIApplicationDidReceiveMemoryWarningNotification object:nil];
```

Now write a method to save the in-memory cache items to files:

```
+(void) saveMemoryCacheToDisk:(NSNotification *)notification
{
  for(NSString *filename in [memoryCache allKeys])
  {
    NSString *archivePath = [[AppCache cacheDirectory]
                              stringByAppendingPathComponent:filename];
    NSData *cacheData = [memoryCache objectForKey:filename];
    [cacheData writeToFile:archivePath atomically:YES];
  }

  [memoryCache removeAllObjects];
}
```

This method ensures that your `AppCache` doesn't eat up the available system memory, and is faster than writing directly to files from your view controller.

Handling Termination and Entering Background Notifications

You also need to ensure that your in-memory cache is saved when the app quits or enters the background. This gives an added advantage to your on-demand caching: offline access.

Now, you add the third and final step, which is to watch for the app's resigning active or closing notifications and handle memory warnings as you did in the previous section. No extra methods are needed; just add observers in the initialize method for `UIApplicationDidEnterBackgroundNotification` and `UIApplicationWillTerminateNotification`. This is to ensure that your in-memory cache is saved to the file system.

Observing Notifications and Saving In-Memory Cache to Disk (AppCache.m)

```
[[NSNotificationCenter defaultCenter] addObserver:self
   selector:@selector(saveMemoryCacheToDisk:)
name: UIApplicationDidEnterBackgroundNotification object:nil];

[[NSNotificationCenter defaultCenter] addObserver:self
   selector:@selector(saveMemoryCacheToDisk:)
name: UIApplicationWillTerminateNotification object:nil];
```

Remember to call `removeObserver` in `dealloc` as well. For the complete AppCache code, download the code sample from the book's website.

Whew! That was a bit of overload. But it's not finished yet. I told you that on-demand caching can be implemented using a data model cache or a URL cache. The `AppCache` implementation you learned about earlier is a data model cache. I'll now show you how to implement a URL cache. But don't worry; implementing a URL cache is much simpler on the client side. Most of the heavy lifting and cache invalidation is done remotely on the server, and the server dictates cache invalidation through cache control headers.

Creating a URL Cache

You implement a URL cache by caching the responses for every URL request made by the app. This cache is very similar to the `AppCache` you implemented in the previous section. The difference is that the key and value you stored in the cache will differ. A data model cache uses a filename as the key and the archives of the data models as values. A URL cache uses the URL as the key and the response data as value. Most of the implementation will be very similar to the `AppCache` you wrote earlier, except for caching invalidation.

A URL cache works like a proxy server in the way it handles caching. As a matter of fact, `MKNetworkKit` handles the HTTP 1.1 caching standard transparently for you. But you still need to understand how it works under the hood.

Earlier, I told you that the server dictates caching invalidation for a URL cache. The HTTP 1.1 (RFC 2616 Section 13) specification explains the different cache control headers that a server might send. The RFC specifies two models, an expiration model and a validation model.

Expiration Model

The expiration model allows the server to set an expiry date after which the resource (your image or response) is assumed to be stale. Intermediate proxy servers or browsers are expected to invalidate or *expire* the resource after this set amount of time.

Validation Model

The second model is a validation model where the server usually sends a checksum (Etag). All subsequent requests that get fulfilled from cache should be *revalidated* with the server using this checksum. If the checksum matches, the server returns an HTTP 304 Not Modified status.

Example

Here's an example of both the expiration model and the validation model, along with tips on when to use which model on your server. Although this information may be beyond the usual scope of iOS developers, understanding how caching works will help you become a better developer. If you've ever configured your server, you've already written something like this to your `nginx.conf` file.

nginx Configuration Setting for Specifying an Expiration Header

```
location ~ \.(jpg|gif|png|ico|jpeg|css|swf)$ {
            expires 7d;
    }
```

This setting tells `nginx` to emit a cache control header (`Expires` or `Cache-Control: max-age=n`) that instructs the intermediate proxy servers to cache files ending with jpg and others for 7 days. On API servers, the sysadmin probably does this, and on all your image requests you see the cache control header. Caching your images on your device by respecting these headers is probably the right way to go, rather than invalidating all images in your local cache every few days.

While static responses, images, and thumbnails use the expiration model, dynamic responses mostly use the validation model for controlling cache invalidation. This requires computing the checksum of the response objects and sending it along with the Etag header. The client (your iOS app) is expected to send the Etag on the `IF-NONE-MATCH` header for every subsequent request (revalidates its cache). The server, after computing the checksum, checks whether the checksum matches the `IF-NONE-MATCH` header. If yes, the server sends a 304 Not Modifed, and the client serves the cached content. If the checksum is different, the content has indeed changed. The server can be programmed either to return the complete content or a change set explaining what was exactly changed.

Let me give you an example to make this clear. Assume that you're writing an API that returns the list of outlets in the iHotel App. The endpoint */outlets* return you the list of outlets. On a subsequent request, the iOS client sends the Etag. The server fetches the list of outlets again (probably from the database) and computes the checksum of the response. If the checksum matches, there were no new outlets added, and the server should send a 304 Not Modified. If the checksum is different, a couple of things could have happened. A new outlet may have been added or an existing outlet may have been closed. The server may choose to send you the complete list of outlets again or to send you a change set that contains the list of new outlets and the list of closed outlets. The latter is more efficient (though it's slightly more complicated in terms of implementation) if you don't foresee frequent changes to outlets.

The World Wide Web increased from a dozen computers in 1990 to billions of computers worldwide within a decade. The one driving factor was the HTTP protocol being built with scalability and caching, and the fact that every major browser adhered to these HTTP standards. With the computing world becoming more and more *app-centric,* your web service application needs to reimplement what browsers did to help the Web grow. Adhering to some of these simple standards would go a long way in improving the performance of your application and thereby pushing the limits of what it can do.

Caching Images with a URL Cache

A URL cache like the preceding one is transparent to the data it caches. This means that it's immaterial to a URL whether you cache images or music or videos or URL responses. Just remember that when you use a URL cache for caching images, the URL of the image becomes the key and the image data becomes the cached object. The advantage you get here is superior performance because a URL cache doesn't post-process responses (like converting them to JPEGs or PNGs). This fact alone is a good reason to use a URL caching for caching your images.

Summary

In this chapter, you read about the various types of caching and learnt the pros and cons of different methods of caching. Caching goes a long way in improving the performance of an app, and yet a majority of the apps on App Store don't implement it properly. The techniques you discovered (for both iOS and your API server) in this chapter can help you push the limits and take your app to the next level.

For information about synchronizing your local data with a remote server, turn to Chapter 25.

Further Reading

Apple Documentation

The following documents are available in the iOS Developer Library at `developer.apple.com` or through the Xcode Documentation and API Reference.

Archives and Serializations Programming Guide

iCloud

Books

The Pragmatic Bookshelf | *Core Data*
`http://pragprog.com/titles/mzcd/core-data`

Other Resources

Callahan, Kevin, Accessorizer. *Mac App Store.* 2011
`http://itunes.apple.com/gb/app/accessorizer/id402866670?mt=12`

Chapter 25
Data in the Cloud

With iOS 5 (and Mac OS X Lion), Apple introduced a new and very important feature, or service, called iCloud. iCloud stores content and continuously pushes it to all devices associated with a specific Apple account. To the end user, the integration is seamless and happens automatically. However, to give your customers that kind of user experience, you need to do some challenging development.

Prior to iCloud, developers implemented Dropbox Sync to synchronize data across user devices. Although this approach handled the problem of syncing data in the iOS 4 era, it's limited to flat files, and more important, users needed to have a Dropbox account and needed to sign in with their Dropbox credentials on your app. With iCloud, no such explicit signing-in is necessary. You get automatic access to the iCloud data store you declare for your app without extra action on the part of the user. Moreover, iCloud has support for key-value data storage and excellent support for Core Data-managed apps.

Recently, many companies have begun rolling out backend as a service (BaaS). Parse, StackMob, Kinvey, CloudRec, and Applicasa are just a few of these services. Two of them, StackMob and Parse, are becoming increasingly popular among developers, which means that their longevity as companies will be much higher than others.

In this chapter, I briefly explain iCloud, Parse, and StackMob. Parse and StackMob have important advantages (and a differentiator) over iCloud. First, they have a REST API, which you can use to access your data from a web-based application. Second, they have an Android, a Windows Phone 7, and an HTML 5 SDK. If your product requires an Android client or an HTML 5 client, using a third-party service rather than iCloud may be beneficial.

Having said all that, it's time to get started.

iCloud

iCloud allows you to store two kinds of data on iCloud: traditional files (document file storage) and key-value data. If your app currently uses `NSUserDefaults` to store app-specific settings, consider using iCloud key-value data storage to sync those settings across the users' devices.

The preferred way to store documents on iCloud is to design your persistent models as subclasses of `UIDocument` or `UIManagedDocument`. `UIDocument` is for storing document-based content, whereas `UIManagedDocument` is for storing database-related content. The next sections discuss the differences between them.

UIDocument

Using `UIDocument` is not an iCloud prerequisite; however, if you manage individual files, you also need to manage file presenters and coordinators to support iCloud's locking mechanism. `UIDocument` has built-in support for this mechanism. Additionally, it helps to resolve version conflicts created by updates on other devices.

Another important feature provided by `UIDocument` is its excellent support for managing *file packages*. (A file package is a directory of files that appears to the user as a single file. On the Mac, applications in your `/Applications` folder are file packages.) File packages can store individual components in your app as separate files, so a change to one of those components means that only the changed file needs to be sent to iCloud. For example, in a drawing app, you may consider saving the actual drawing in one file and save sets of custom brushes and fonts in separate files within the same file package. You can still manage the entire file package using just one `UIDocument` subclass. Now, when a user opens a large drawing and edits a custom brush, only the file representing the custom brush needs to be sent, which is usually much smaller than the drawing.

UIManagedDocument

`UIManagedDocument` is a concrete subclass of `UIDocument` that integrates with Core Data. You initialize it with the URL of your persistent store, and the document object does the rest. It then creates a Core Data stack based on your Core Data model file (.xcdatamodeld file). If you're currently using Core Data to store your user content, you should adopt `UIManagedDocument` in your app to support iCloud data syncing. This is the easiest way to migrate apps that currently use Core Data.

Key-Value Data Storage

The second type of data you can store is key-value data. If your app currently stores app-specific settings in `NSUserDefaults`, consider using iCloud's key-value data storage to sync these settings across users' devices. The total size of data you store here cannot exceed 64K and no key can be more than 4K. In most cases, you'll be saving a Boolean or an integer or a string, which usually is less than 4K. If you're storing serialized models, however, ensure that they're less than 4K, or use `UIDocument` as discussed in the previous section.

Understanding the iCloud Data Store

To start using iCloud in your app, you first need to configure your App ID for iCloud usage. After doing so, generate your provisioning profile and request entitlements in your app. Depending on the data storage requirements in your app, you need to request either one or both of the following entitlements.

iCloud Entitlements Key

```
com.apple.developer.ubiquity-container-identifiers
com.apple.developer.ubiquity-kvstore-identifier
```

The first one is for document storage, and the second one is for key-value storage. Once this is done, you can start moving documents from your application's sandbox to iCloud storage. The entitlements ensure that data generated by your app is sandboxed and will not be accessible to other apps.

Sharing Data Within Apps (or App Suites)

On iCloud, you identify and request a data store container by its unique ID. This is normally your application's App ID, but not necessarily. You can use the same iCloud data store container for two of your apps and share data within them. This feature is very powerful, especially if you're developing lite and pro versions of an app. You can now use the same ubiquitous container identifier, which means that data created by a user using the lite version becomes automatically available if he purchases your pro version, making the migration process easier. You can also use this technique to share data among a suite of apps you build.

The iCloud data store resides on the user's iPhone until it's moved to iCloud by the iCloud daemon running on the iOS device. Once you move the document to iCloud storage, you can safely delete the original copy because all subsequent edits will happen directly to the file stored on iCloud. As a developer, you don't have to worry about uploading to a remote source or be bothered about network disconnection. The iCloud daemon automatically takes care of this syncing. You should, however, handle conflict resolutions. The default conflict resolution strategy iCloud follows is to choose the last modified document. Although this might be okay in some cases, you should evaluate this strategy on a case-by-case basis for your app.

Storing Data Within Your iCloud Container

Starting from iOS 5 and later, the Settings app on every device has an iCloud section that allows users to see how much data they've used on iCloud for backup and how much data has been used by apps syncing with iCloud. When you store your files in the iCloud container, they appear as a big blob of data to the user. When you store them in the Documents directory, users can see the individual files and their sizes. Files in the Documents directory can be deleted one by one, whereas data stored outside this directory appears as a big blob and must all be deleted at the same time. To avoid confusion and to play nice, always store any user-generated files in the Documents directory and store miscellaneous metadata that you don't want the users to see outside this directory.

A Word About iCloud Backup

With versions of iOS prior to iOS 5, when the device syncs with iTunes, the contents of your app's Documents folder are automatically backed up. With iOS 5 and later devices, the same contents are backed up to the iCloud, with one exception: Documents you store on iCloud manually are not included in this backup (because they're already on iCloud). Do note that this automatic backup is different than iCloud syncing. Backed-up documents are treated as opaque data that can be used only to completely restore an iOS device. Individual file access is not possible programmatically or by the user.

Disadvantages

Although iCloud offers seamless syncing of user data and easy conflict management, no REST API is available for accessing the data stored by your application. A REST API is important if you want to provide a web frontend to your product. Any software as a service that offers syncing across devices is looking at the possibility of rolling out a web-based application in the near future, and iCloud's lack of a REST API makes that almost impossible. Although Apple's own apps (Mail, Contacts, Calendar, Reminders, and Notes) have an impressive web-based frontend, as of iOS 6, there's no mention of a REST API for iCloud.

The lack of a web interface for looking at your data makes debugging harder as well. With iCloud, debugging becomes harder, and the only way you will know that the data has been backed up to iCloud is when the second device successfully downloads and displays the data. A longer debugging cycle exponentially increases the product's development time.

Third-Party Cloud Offerings

Several third-party offerings help developers move and sync their data on the cloud without worrying about maintaining a server. Parse and StackMob seem to be the choices these days because of their reliability over the last couple of years and the plethora of features they offer, including a REST API, a JavaScript API, and SDKs for competing platforms like Android and Windows Phone.

In addition, almost all third-party offerings have good support for debugging, meaning that data synced from a mobile device is readily available for viewing on a web application. This is something that iCloud lacks, which makes debugging difficult. With iCloud, the only way to know that something has been successfully stored in iCloud is when another device can read (and display) that data.

As I mention earlier, I discuss both Parse and StackMob in this chapter. Both turn out to be reliable alternatives to Apple's iCloud. I also discuss the disadvantages of these providers later in this chapter (in the "Disadvantages of Backend as a Service" section) to help you make an educated decision about which to use.

Parse

Parse is a third-party service that allows users to create a complete back-end stack on the cloud. Although Parse has a REST API, when you want to interact with it from your iOS app, you can use the iOS SDK instead. The iOS SDK offers the same set of features offered by the REST API, and there's no reason not to use it.

Getting Started with Parse

Getting started with Parse is almost as easy as getting started with iCloud, minus the provisioning- and entitlement certificates-generation nightmare. Instead, when you use Parse (or most other equivalent competing services), after logging in to your Parse account, you create an application and use the Application ID and Client Key to authenticate your app with Parse's backend.

The Application ID/Client Key combo is almost synonymous to the ubiquitous container identifiers you use for iCloud. After you create your app on Parse, you should see a screen similar to the one shown in Figure 25-1.

After you create the application, download the iOS SDK and drag `Parse.Framework` to your application. Add the dependent frameworks and import the header file, `Parse/Parse.h` in your `AppDelegate`.

Copy the Application ID and the Client Key you created for your application and add the following line to the `AppDelegate`:

```
[Parse setApplicationId:@"<Your Application ID>" clientKey:@"<Your Client
    Key>"];
```

That completes the setup.

Parse Top Level Objects

The complete Parse framework relies on the following objects: `PFObject`, `PFUser`, `PFRole`, and `PFACL`. `PFObject` is like a "God object" that you use throughout your application. It's almost synonymous with `UIManagedObject` in iCloud. One important difference, however, is that when you use Parse, there's almost no support for model versioning, and in fact, there's no versioned structure for your models. `PFObject` is like a tableless nosql-style object where you store elements using a key-value pair. In fact, from your app, you create a `PFObject` for every object that you want to store on the cloud. The other two objects, `PFRole` and `PFUser`, are subclasses of `PFObject`. You'll be using `PFRole` and `PFUser` in your app, if you provide a user account and want to store associated data with a user and want to control access to the data based on the logged in user. Although implementing roles is not hard using iCloud's `UIManagedObject`, Parse provides it free and ready for use right out of the box.

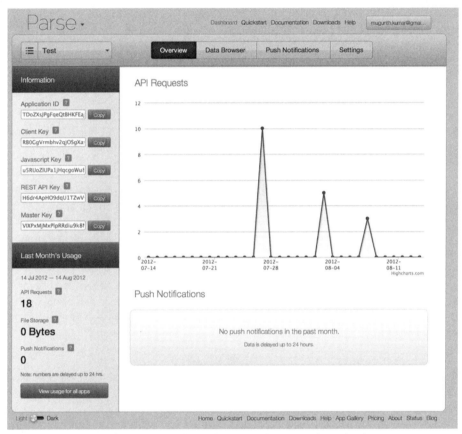

Figure 25-1 Screenshot of Parse showing the Application ID/Client Key and your data browser

Code

Now that I've introduced you to the basics, it's time to dirty your hands with some code. The main classes you'll be using are `PFObject`, `PFUser`, `PFACL`, and `PFRole`.

Storing Objects

Almost everything that you save to cloud using Parse is done through the `PFObject`. You create a `PFObject` with a specific class name, add the necessary data as a key-value pair (think `NSMutableDictionary`), and call the `save` method. Here's an example.

Creating and Saving an Object to the Cloud Using Parse

```
PFObject *object = [PFObject objectWithClassName:@"UserProfile"];
[object setValue:@"Mugunth" forKey:@"userName"];
[object setValue:@"mugunthkumar" forKey:@"twitter"];
[object setValue:@"imk.sg" forKey:@"skype"];
[object setValue:@"blog.mugunthkumar.com" forKey:@"blog"];

[object saveInBackgroundWithBlock:^(BOOL succeeded, NSError *error) {

  if(!succeeded) {
    NSLog(@"%@", error);
  }
}];
```

With very few changes, you can link the same `UserProfile` object with a user account (`PFUser`). This means only that the logged in user can read/modify the object. Again, Parse allows full access control to developers using the access control object `PFACL`. Now, take a look at these two objects in action.

```
[PFUser logInWithUsernameInBackground:@"mugunth" password:@"magic"
 block:^(PFUser *user, NSError *error) {

  if(!user) {
    NSLog(@"%@", error);
    return;
  }

  PFACL *acl = [PFACL ACLWithUser:user];
  PFObject *object = [PFObject objectWithClassName:@"UserProfile"];
  [object setACL:acl];

  [object setValue:@"Mugunth" forKey:@"userName"];
  [object setValue:@"mugunthkumar" forKey:@"twitter"];
  [object setValue:@"imk.sg" forKey:@"skype"];
  [object setValue:@"blog.mugunthkumar.com" forKey:@"blog"];

  [object saveInBackgroundWithBlock:^(BOOL succeeded, NSError *error) {

    if(!succeeded) {
      NSLog(@"%@", error);
    }
  }];
}];
```

The code remains the same except for the initial block that logs a user in and creates an access control list object. When you run the preceding code, you should see the objects when you open the Data Browser on Parse, as shown in Figure 25-2.

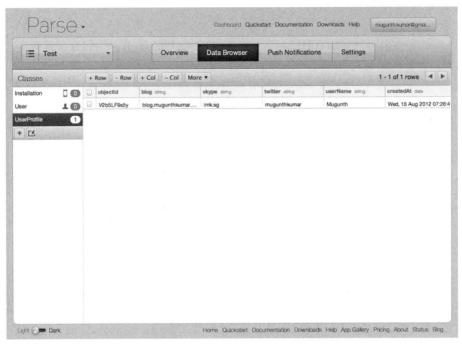

Figure 25-2 Data browser showing your objects

Retrieving Objects

Retrieving objects from the cloud can't be easier than this. You use the `PFQuery` object to get objects from the cloud.

Retrieving Objects from Parse Using PFQuery

```
PFQuery *query = [PFQuery queryWithClassName:@"UserProfile"];
[query getFirstObjectInBackgroundWithBlock:^(PFObject *object, NSError
 *error) {

  NSLog(@"%@", object);
}];
```

When you execute the following code, you'll see the saved user profile object. You can even filter objects by applying a filter to the `PFQuery` object.

```
[query whereKey:@"twitter" equalTo:@"mugunthkumar"];
```

Offline Mode

What makes Parse powerful is the ability it gives you to work and save objects offline. The Parse framework automatically takes care of synchronizing changes to the cloud when the device comes back online.

All you need to do is call the `saveEventually` method instead of `saveInBackgroundWithBlock:`.

With all these features, the most heavy-duty task is writing code to convert objects that must be saved to the cloud to a `PFObject`. You can do this manually, by writing a method `saveToParse` in all the models. Another much cleaner and elegant way is to piggyback on Objective-C's runtime methods. The MKFoundation library has a runtime method that converts your objects to `PFObjects`, piggybacking on Objective-C's powerful runtime methods.

> `MKFoundation` **is a library available on GitHub that allows you to convert any object in your application to a** `PFObject`. `MKFoundation` **uses Objective-C runtime methods to inspect your objects and creates a** `PFObject`. **With small changes the** `PFObject`, **such as changing the user/role, you can instantly bring Parse back-end support to your application.**
>
> **For more in-depth information about Objective-C runtime, read Chapter 28 of this book.**

StackMob

StackMob is an alternative to Parse. Although slightly more complicated to use and understand than Parse, the additional complexities come from the fact that it has many more features that might be useful to your application.

Setting Up StackMob

The procedures for setting up StackMob are similar to those for any other third-party framework. Unlike Parse, Stackmob provides the source code for the iOS SDK. You download the sample code and add the source code files directly into your app. As of this writing, the StackMob code base is not ARC-ready, so you have to add `-fno-objc-arc` to all of the StackMob files.

Again, instead of following the complicated provisioning and certificate entitlements approach, StackMob, like Parse, generates a bunch of API access keys.

Logging In and Uploading and Retrieving Data

Most of StackMob APIs work similar to those in Parse, including logging in and saving and retrieving objects from the cloud. Instead of a PFObject, you create a normal `NSDictionary` and upload it to StackMob.

A quick look at the code will make things clear.

Storing Data to StackMob

```
NSMutableDictionary *object = [NSMutableDictionary dictionary];

    [object setValue:@"Mugunth" forKey:@"username"];
    [object setValue:@"mugunthkumar" forKey:@"twitter"];
    [object setValue:@"imk.sg" forKey:@"skype"];
    [object setValue:@"blog.mugunthkumar.com" forKey:@"blog"];

//schema name "blogentry" must be 3-25 alphanumeric characters, lowercase
[[StackMob stackmob] post:@"UserProfile"
        withArguments:object andCallback:^(BOOL success, id result){
            if(success){
                //action after successful call
            } else {
                //action after a call results in error
            }
        }];
```

The preceding code block explains how to save a set of data to the cloud. After you successfully execute this block of code, you'll see something like Figure 25-3 in your StackMob account.

Figure 25-3 StackMob's Object Datastore browser

The other methods, such as deleting or retrieving data, also operate in a very similar manner to Parse, so I won't explain them in detail here. However, one thing does set StackMob apart from the competition, which I discuss in the next section.

StackMob Custom Code

The real power and distinguishing feature of StackMob is the capability of supporting custom code/scripts in Java/Scala/Closure. You can write your own custom server-side code and upload it to StackMob and call those APIs from your Mobile App. Why write custom server code when you can do all that on the client? That's the essence of a backend as a service, right? Although you can write queries, filters, and so on on iOS, you'll still find that it's necessary to write back-end code for analytics and data mining. Running complex analytics-related queries from a user's device is slow. Some startups work on products that depend heavily on data analytics. Without a real back-end server and access to the raw database, it becomes almost impossible to do data analytics from your server. Although a REST API can help, you're still limited by available Internet bandwidth, whereas custom code running on StackMob servers aren't.

Disadvantages of Using a Backend as a Service

Both StackMob and Parse are easy for beginners to start with, and both offer out-of-box support for Apple's Push Notification Services, In App Purchases, friction-free user registration, and Facebook/Twitter support.

The main disadvantage is the dependence of your complete product on a third-party service that, as of 2012, is not yet established (compared to companies like Amazon or Rackspace that provide Infrastructure as a Service). Service disruptions in these back-end providers could potentially bring down your app just as an AWS outage brought down Instagram, Quora, foursquare, and several others at the same time. Although Apple's iCloud is not immune to this (no company is), think about the risks before considering a third-party solution to replace your backend. There is no silver bullet.

Summary

In this chapter, you read about the different methods available to developers for storing and syncing user data in the cloud. I explained the pros and cons of using iCloud and the pros and cons of using a third-party backend as a service. Moving your data to the cloud should now be easier than you had thought. Knowing about alternatives to iCloud will help you make an educated decision about how to move ahead with your product.

Further Reading

Apple Documentation

The following documents are available in the iOS Developer Library at `developer.apple.com` or through the Xcode Documentation and API Reference.

iCloud

UIManagedObject Class Reference

WWDC Sessions

WWDC 2012, "Session 209: iCloud Storage Overview"

WWDC 2012, "Session 227: iCloud with Core Data"

WWDC 2012, "Session 237: Advanced iCloud Document Storage"

WWDC 2012, "Session 218: Using iCloud with UIDocument"

WWDC 2012, "Session 224: "Using iCloud with NSDocument"

Other Resources

Parse. "iOS Developer Guide"
`https://parse.com/docs/ios_guide`

StackMob. "StackMob Docs"
`http://www.stackmob.com/devcenter/`

MugunthKumar / MKFoundation
`http://github.com/MugunthKumar/MKFoundation`

Chapter 26
Fancy Text Layout

There are several options for displaying text in iOS. There are simple controls like `UILabel` and full rendering engines like `UIWebView`. For complete control, including laying out columns or drawing text along a path, you can implement your own layout engine with Core Text.

iOS 6 brings rich text to all of UIKit. Finally, you can display and edit bold and underline without using a web view or Core Text, which greatly simplifies many programs. To get the most out of iOS rich text, you need to understand `NSAttributedString` as explained in this chapter. After you understand attributed strings, handling rich text is mostly just a matter of requesting it from your text views.

After you've mastered attributed strings, you can take full control of text layout using Core Text. In this chapter, you learn about the major parts of Core Text, how to use them to lay out text any way you like, and the limitations that you still need to work around.

The Normal Stuff: Fields, Views, and Labels

You're probably already familiar with `UILabel`, `UITextField`, and `UITextView`, so I discuss them only briefly here. They're the basic controls you use for day-to-day text layout.

`UILabel` is a lightweight, static text control. It's very common for developers to have problems with dynamic labels (labels with contents set programmatically) they create in Interface Builder. Here are a few tips for using `UILabel`:

- `UILabel` can display multiline text. Just set its `numberOfLines` property in code or the Lines property in Interface Builder. This sets the maximum number of lines. If you want the number of lines to be unbounded, set it to zero.

- By default, Interface Builder turns on `adjustsFontSizeToFitWidth`. This can be surprising if you assign text that's wider than the label. Rather than truncating or overflowing the label, the text shrinks. Generally, it's a good idea to make your dynamic labels very wide to avoid truncating or resizing.

- Unlike other views, user interaction is disabled by default for `UILabel`. If you attach a `UIGestureRecognizer` to a `UILabel`, you must remember to set `userInteractionEnabled` to `YES`. Don't confuse this with the `enabled` property, which controls only the appearance of the label.

- `UILabel` is not a `UIControl` and doesn't have a `contentVerticalAlignment` property. Text is always vertically centered. If you want to adjust the vertical location of the text, you need to resize the label with `sizeToFit` and then adjust the label's `frame.origin`.

UITextField provides simple, single-line text entry. It includes an optional "clear" button and optional overlay views on the left and right (leftView and rightView). The overlay views can be used to provide hints about the field's purpose. For instance, a search icon on the left is a very space-efficient way of indicating a search field. Remember that the overlay views are UIView objects. You can use a UIButton here or any other kind of interactive view. Just make sure they're large enough for the user to touch easily. You generally should not move these views with setFrame:. Override the methods leftViewRectForBounds: or rightViewRectForBounds:. UITextField then lays them out appropriately. The text rectangle (textRectForBounds:) is clipped automatically, so it doesn't overlap these rectangles.

A common problem with UITextField is detecting when the user presses the Return key. In many cases, you will want to automatically process the data and dismiss the keyboard when the user presses Return. To do so, you need to implement a UITextFieldDelegate method, as shown here.

ViewController.m (AutoReturn)

```
- (BOOL)textField:(UITextField *)textField
        shouldChangeCharactersInRange:(NSRange)range
        replacementString:(NSString *)string {
  if ([string isEqualToString:@"\n"]) {
    self.outputLabel.text = [textField text];
    [textField resignFirstResponder];
    return NO;
  }
  return YES;
}
```

Whenever the user presses the Return key, a newline character (\n) is sent to this method. When that happens, you apply whatever processing you want (in this case, setting the text of another label) and then call resignFirstResponder to dismiss the keyboard. You can use a similar technique with UITextView.

UITextView is easily confused with UITextField, but it serves a somewhat different function. It's intended for multiline, scrolling text, editing, or viewing. It's a type of UIScrollView, not a UIControl. You can apply a font to the text, but the same font is used for the entire view. UITextView cannot display rich text from an NSAttributedString. There is no Apple-provided view that can do that. Generally, the choice between UITextField and UITextView is obvious based on how much text the user is going to type.

Rich Text in UIKit

One of the best new features in iOS 6 is the addition of rich text to UIKit. You can now display all kinds of useful information that used to require complex custom views or web views. Although there are still some reasons to use web views for rich text (see the later section "Web Views for Rich Text"), whenever possible use attributed strings. Before I discuss attributed strings, you need to understand the basics of the three most common rich text attributes: bold, italic, and underline.

Understanding Bold, Italic, and Underline

Bold, italic, and underline are generally presented to the user as simple attributes, but they're quite different from each other. In typography, you do not "bold" a font by drawing it with thicker lines. Instead, the font designer provides a heavier weight version of the font, called a *variation*. In iOS the `Helvetica` font has a variation called `Helvetica-Bold`. Although these fonts are related, they're completely different `UIFont` objects.

Italic is similar, but there are two related typefaces that are commonly treated as "italic." True italic type is based on calligraphy and uses different shapes (*glyphs*) than the regular (or *roman*) font. Some fonts do not have a true italic variation and merely slant the roman type to the right. This is called *oblique*. When users request italic, they generally mean either italic or oblique. Text with both bold and italic requires yet another font variation, such as `Helvetica-BoldOblique`.

Unlike bold and italic, underline is a decoration like color or shadow. You do not change font when you add decorations. See Figure 26-1 for examples of bold, italic, and underline. Note carefully the significant difference between the glyphs of the roman font and the italic variation, whereas the underlined glyphs are identical to the roman font. Also note that underline is not as simple as drawing a line under the text. Proper underline includes breaks for descenders like *p*. All of these small details greatly improve the appearance and legibility of text in iOS.

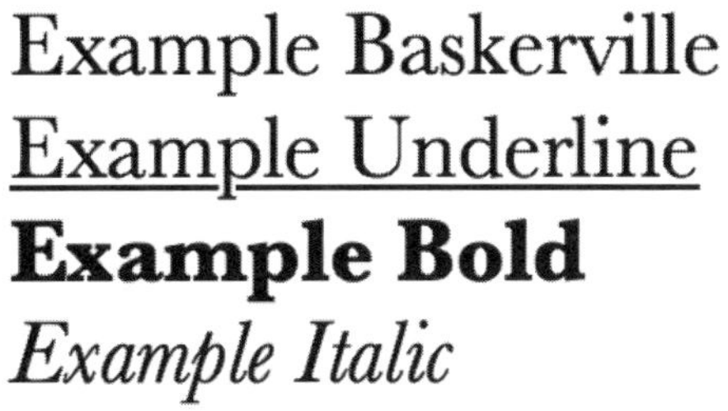

Example Baskerville
<u>Example Underline</u>
Example Bold
Example Italic

Figure 26-1 Variations and decorations of Baskerville

Understanding these subtleties will help you understand why `NSAttributedString` works the way it does. For instance, there is no "bold" attribute as you might expect. You bold text by changing its font.

Attributed Strings

The fundamental data type for rich text is `NSAttributedString`. An *attributed string* is a string that applies attributes to ranges of characters. The attributes can be any key-value pair, but for the purposes of rich text, they usually contain style information such as font, color, and indentation.

It's usually better to use `NSMutableAttributedString` so you can modify the attributes of various parts of the string. `NSAttributedString` requires that all of the strings have the same attributes.

In the following example, you create a basic rectangular layout to display some rich text in a `UITextView`. This project is available in the sample code named RichText. First, you add a `UITextView` in Interface Builder and select Attributed as its mode. Then you create an `NSAttributedString` and apply attributes to it. In this example, you create a string and apply a font attribute to all of it.

ViewController.m (RichText)

```
const CGFloat fontSize = 16.0;

// Create the base string.
// Note how you can define a string over multiple lines.
NSString *string =
@"Here is some simple text that includes bold and italics.\n"
@"\n"
@"We can even include some color.";

// Create the mutable attributed string
NSMutableAttributedString *attrString =
[[NSMutableAttributedString alloc] initWithString:string];

NSUInteger length = [string length];

// Set the base font
UIFont *baseFont = [UIFont systemFontOfSize:fontSize];
[attrString addAttribute:NSFontAttributeName value:baseFont
                range:NSMakeRange(0, length)];
```

As I discussed in the section "Understanding Bold, Italics, and Underline," to apply bold, you need to apply a different font. You can do this simply for the system font, because you can easily request the bold system font:

```
// Apply bold using the bold system font
// and seaching for the word "bold"
UIFont *boldFont = [UIFont boldSystemFontOfSize:fontSize];
[attrString addAttribute:NSFontAttributeName value:boldFont
                range:[string rangeOfString:@"bold"]];
```

But what if you're using a non-system font and need the bold or italic variation? Unfortunately, UIKit doesn't have a good way to determine variations of fonts, but Core Text does. After linking `CoreText.framework`, you can use the following function. It uses the name of the `UIFont` to find the correct `CTFont`. It then uses `CTFontCreateCopyWithSymbolicTraits` to add the requested trait (such as `kCTFontTraitItalic`). Finally, it uses the name of the resulting font to create the correct `UIFont`. Hopefully, this will become easier in future versions of iOS.

```
UIFont * GetVariationOfFontWithTrait(UIFont *baseFont,
                                     CTFontSymbolicTraits trait) {
  CGFloat fontSize = [baseFont pointSize];

  CFStringRef
  baseFontName = (__bridge CFStringRef)[baseFont fontName];
```

```
    CTFontRef baseCTFont = CTFontCreateWithName(baseFontName,
                                            fontSize, NULL);

    CTFontRef ctFont =
    CTFontCreateCopyWithSymbolicTraits(baseCTFont, 0, NULL,
                                    trait, trait);

    NSString *variantFontName =
    CFBridgingRelease(CTFontCopyName(ctFont,
                            kCTFontPostScriptNameKey));

    UIFont *variantFont = [UIFont fontWithName:variantFontName
                                    size:fontSize];
    CFRelease(ctFont);
    CFRelease(baseCTFont);

    return variantFont;
}
...
UIFont *italicFont = GetVariationOfFontWithTrait(baseFont,
                                        kCTFontTraitItalic);
[attrString addAttribute:NSFontAttributeName value:italicFont
                range:[string rangeOfString:@"italics"]];
```

Finally, you can add color by adding the appropriate attribute:

```
// Apply color
UIColor *color = [UIColor redColor];
[attrString addAttribute:NSForegroundColorAttributeName
                value:color
                range:[string rangeOfString:@"color"]];
```

Notice the use of `addAttribute:value:range:` rather than `setAttributes:value:range:`. This merges the new attributes with the existing attributes, which is more often what you want.

After you've set all the attributes, you can display this string by setting the `attributedText` property of `UITextView`:

```
self.textView.attributedText = attrString;
```

Most UIKit controls that have a `text` property now also have an `attributedText` property, making it generally easy to convert existing code.

It's important to note that `NSAttributedString` is not a subclass of `NSString`. It contains an `NSString`. Similarly, `NSMutableAttributedString` contains an `NSMutableString`. Although attributed strings have some of the more common string methods (like `length`), you often will need to fetch the underlying `string` using the string or `mutableString` methods.

Paragraph Styles

Some styles apply to paragraphs rather than characters. These include alignment, line break, and spacing. Paragraph attributes are bundled into an `NSParagraphStyle` object. You will almost always create an `NSMutableParagraphStyle` so that you can modify it. In this example, you modify the text alignment:

```
// Right justify the first paragraph
NSMutableParagraphStyle *
style = [[NSParagraphStyle defaultParagraphStyle] mutableCopy];
style.alignment = NSTextAlignmentRight;
[attrString addAttribute:NSParagraphStyleAttributeName
                   value:style
                   range:NSMakeRange(0, 1)];
```

Notice first that you create a `mutableCopy` of the `defaultParagraphStyle`. This is a very common pattern. You may also make a mutable copy of an existing paragraph style. Next, notice that the range for this style is just the first character. A "paragraph" starts at the beginning of the document or after a newline and continues to the next newline or the end of the document. The paragraph style of the first character applies to the entire paragraph. There is no way to change paragraph style within a paragraph.

Attributed Strings and HTML

Attributed strings are an incredibly convenient format for many tasks. By completely separating content from style information, it's easy to search and edit. The full power of Cocoa's strings handling routines is available. But attributed strings have one major problem: They are extremely cumbersome to build programmatically. In the RichText example, it took dozens of lines of code to format just a couple of sentences. On the Mac, it's possible to create an attributed string from HTML, but there's no built-in way to do this in iOS.

> The fact that Apple did not directly port the `NSAttributedString(AppKitAdditions)` category that handles HTML is actually a blessing. The Mac conversion routines are buggy, slow, generate very complicated HTML, and have bizarre side effects when you use them. When Apple finally provides a converter between HTML and `NSAttributedString`, we should all hope they don't use the Mac code.

Luckily, there is a very good solution: DTCoreText (formerly called NSAttributedString-Additions-for-HTML). DTCoreText can convert between simple HTML and `NSAttributedString`. It can also draw attributed strings in many ways that UIKit can't. For example, it can embed images. At this writing, DTCoreText is not compatible with the `NSAttributedString` attributes in iOS 6, but this is in development and hopefully will be in place shortly after iOS 6 becomes available. If your problem lends itself well to HTML, definitely consider this package. See the "Further Reading" section for more information.

Web Views for Rich Text

Although `UITextView` is highly useful for laying out simple rich text, and DTCoreText will allow you to display simple HTML, sometimes you need a really complicated layout that is difficult to solve with these tools. Regrettably, the best tool available in some of these cases is `UIWebView`. I say "regrettably" because `UIWebView` is filled with problems. It's slow, buggy, complicated to use, and difficult to debug. Even so, it's sometimes the best solution.

In this section, you will learn how to use `UIWebView` as a layout engine rather than as a web viewer.

Displaying and Accessing HTML in a Web View

The code to load HTML into a `UIWebView` is fairly straightforward:

```
NSString *html = @"This is <i>some</i> <b>rich</b> text";
[self.webView loadHTMLString:html baseURL:nil];
```

`webView` doesn't yet contain the HTML string. This is only a request to load the string at some point in the future. You need to wait for the delegate callback `webViewDidFinishLoad:`. At that point, you can read the data in the web view using JavaScript:

```
- (void)webViewDidFinishLoad:(UIWebView *)webView {
  NSString *
  body = [self.webView
          stringByEvaluatingJavaScriptFromString:
          @"document.body.innerHTML"];
  // Use body
}
```

The only way to access the data inside the web view is through JavaScript and `stringByEvaluatingJavaScriptFromString:`. I recommend isolating this JavaScript to a single object to provide a simpler interface. For instance, you could create a `MyWebViewController` object that owns the `UIWebView` and provides a `body` property to set and retrieve the contents.

Responding to User Interaction

It's fairly common in a rich text viewer to support some kind of user interaction. When the user taps a button or link, you may want to run some Objective-C code. You can achieve this by creating a special URL scheme. Rather than `http`, pick a custom scheme identifier. For example, you might have the following HTML:

```
<p>Click <a href='do-something:First'>here</a> to do something.</p>
```

In this example, the URL scheme is `do-something,` and the resource specifier is `First`. When this link is tapped, the entire URL is sent to the delegate, and you can act on it like this:

```
-  (BOOL)webView:(UIWebView *)webView
shouldStartLoadWithRequest:(NSURLRequest *)request
 navigationType:(UIWebViewNavigationType)navigationType {
   NSURL *URL = request.URL;
   if ([URL.scheme isEqualToString:@"do-something"]) {
     NSString *message = URL.resourceSpecifier;
     // Do something with message
     return NO;
   }
   return YES;
}
```

Returning YES from this delegate method allows UIWebView to load the request. You should return NO for your custom scheme. Otherwise, UIWebView will try to load it and pass an error to webView:didFailLoadWithError:.

URL schemes are case-insensitive. The result of request.URL.scheme will always be lowercase, even if you use mixed-case in the HTML. I recommend using a hyphen (–) to separate words in the scheme. You can also use a period (.) or plus (+). The rest of the URL is case-sensitive.

Drawing Web Views in Scroll and Table Views

UIWebView cannot be embedded in a UIScrollView or UITableView because the web view's event handling will interfere with the scroll view. This makes it effectively unusable for table view cells. Because web views have significant performance issues, they're not appropriate for table view cells in any case. You should generally draw table view cells using UIKit or Core Text.

Instead of directly embedding the web view, you can capture the web view as a UIImage and then draw that image in the scroll view or table view. To capture a web view as a UIImage, you need to call renderInContext: on its layer, as shown here:

```
UIGraphicsBeginImageContext(self.webView.bounds.size);
[webView.layer renderInContext:UIGraphicsGetCurrentContext()];
UIImage *image = UIGraphicsGetImageFromCurrentImageContext();
UIGraphicsEndImageContext();
```

You need to break web views larger than 1024 x 1024 into smaller pieces and render them individually. Web views are a useful addition to your toolbox, but I recommend avoiding them when possible. It's very difficult to get an iOS-native look and feel in WebKit. Communicating through JavaScript is cumbersome and error-prone. When things do go wrong, it can be very difficult to debug web view interactions. So generally use attributed strings or DTCoreText or gain full control and switch to Core Text.

Core Text

Core Text is the low-level text layout and font-handling engine in iOS. It's extremely fast and powerful. With it you can handle complex layouts like multicolumn text and even curved text.

Core Text is a C-based API that uses Core Foundation naming and memory management. If you're not familiar with Core Foundation patterns, see Chapter 27.

Core Text also uses attributed strings. In most cases, Core Text is quite forgiving about the kind of attributed string you pass. For instance, you can use the keys `NSForegroundColorAttributedName` or `kCTForegroundColorAttributeName`, and they will work the same. You can generally use a `UIColor` or a `CGColor` interchangeably, even though these are not toll-free bridged. Similarly, you can use `UIFont` or `CTFont` and get the same results. The SimpleLayout project in the example code for this chapter shows how to create a `CFMutableAttributedString`, but I won't go into detail about that here, since it's very similar to `NSMutableAttributedString`, and you can use the two interchangeably.

Simple Layout with CTFramesetter

After you have an attributed string, you generally lay out the text using `CTFramesetter`. A *framesetter* is responsible for creating frames of text. A `CTFrame` (*frame*) is an area enclosed by a `CGPath` containing one or more lines of text. Once you generate a frame, you draw it into a graphics context using `CTFrameDraw`. In the next example, you draw an attributed string into the current view using `drawRect:`.

First, you need to flip the view context. Core Text was originally designed on the Mac, and it performs all calculations in Mac coordinates. The origin is in the lower-left corner—lower-left origin (LLO)—and the *y* coordinates run from bottom to top as in a mathematical graph. `CTFramesetter` doesn't work properly unless you invert the coordinate space, as shown in the following code.

CoreTextLabel.m (SimpleLayout)

```
- (id)initWithFrame:(CGRect)frame {
  if ((self = [super initWithFrame:frame])) {
    CGAffineTransform
    transform = CGAffineTransformMakeScale(1, -1);
    CGAffineTransformTranslate(transform,
                               0, -self.bounds.size.height);
    self.transform = transform;
    self.backgroundColor = [UIColor whiteColor];
  }
  return self;
}
```

Before drawing the text, you need to set the text transform, or *matrix*. The text matrix is not part of the graphics state and is not always initialized the way you expect. It isn't included in the state saved by `CGContextSaveGState`. If you're going to draw text, always call `CGContextSetTextMatrix` in `drawRect:`.

```
- (void)drawRect:(CGRect)rect {
  CGContextRef context = UIGraphicsGetCurrentContext();
  CGContextSetTextMatrix(context, CGAffineTransformIdentity);

  // Create a path to fill. In this case, use the whole view
```

(continued)

```
CGPathRef path = CGPathCreateWithRect(self.bounds, NULL);

    CFAttributedStringRef
    attrString = (__bridge CFTypeRef)self.attributedString;

    // Create the framesetter using the attributed string
    CTFramesetterRef framesetter = CTFramesetterCreateWithAttributedString(a
ttrString);

    // Create a single frame using the entire string (CFRange(0,0))
    // that fits inside of path.
    CTFrameRef
    frame = CTFramesetterCreateFrame(framesetter,
                                     CFRangeMake(0, 0),
                                     path,
                                     NULL);

    // Draw the frame into the current context
    CTFrameDraw(frame, context);

    CFRelease(frame);
    CFRelease(framesetter);
    CGPathRelease(path);
}
```

There's no guarantee that all of the text will fit within the frame. `CTFramesetterCreateFrame` simply lays out text within the path until it runs out of space or runs out of text.

Creating Frames for Noncontiguous Paths

Since at least iOS 4.2, `CTFramesetterCreateFrame` has accepted nonrectangular and noncontiguous frames. The *Core Text Programming Guide* has not been updated since before the release of iPhoneOS 3.2 and is occasionally ambiguous on this point. Because `CTFramePathFillRule` was added in iOS 4.2, Core Text has explicitly supported complex paths that cross themselves, including paths with embedded holes.

`CTFramesetter` always typesets the text from top to bottom (or right to left for vertical layouts such as for Japanese). This works well for contiguous paths, but can be a problem for noncontiguous paths such as for multicolumn text. For example, you can define a series of columns this way.

ColumnView.m (Columns)

```
- (CGRect *)copyColumnRects {
  CGRect bounds = CGRectInset([self bounds], 20.0, 20.0);

  int column;
  CGRect* columnRects = (CGRect*)calloc(kColumnCount,
                                        sizeof(*columnRects));

  // Start by setting the first column to cover the entire view.
  columnRects[0] = bounds;
  // Divide the columns equally across the frame's width.
```

```
CGFloat columnWidth = CGRectGetWidth(bounds) / kColumnCount;
for (column = 0; column < kColumnCount - 1; column++) {
  CGRectDivide(columnRects[column], &columnRects[column],
         &columnRects[column + 1], columnWidth, CGRectMinXEdge);
}

// Inset all columns by a few pixels of margin.
for (column = 0; column < kColumnCount; column++) {
  columnRects[column] = CGRectInset(columnRects[column],
                     10.0, 10.0);

}
  return columnRects;
}
```

You have two choices for how to combine these rectangles. First, you can create a single path that contains all of them, like this:

```
CGRect *columnRects = [self copyColumnRects];

// Create a single path that contains all columns
CGMutablePathRef path = CGPathCreateMutable();
for (int column = 0; column < kColumnCount; column++) {
  CGPathAddRect(path, NULL, columnRects[column]);
}
free(columnRects);
```

This typesets the text as shown in Figure 26-2.

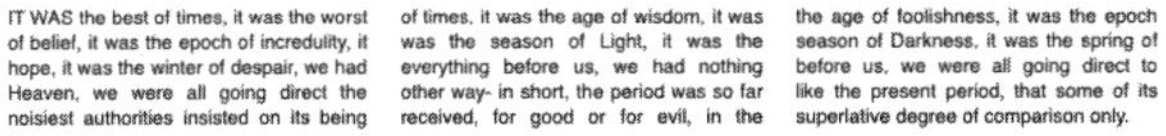

IT WAS the best of times, it was the worst of belief, it was the epoch of incredulity, it hope, it was the winter of despair, we had Heaven, we were all going direct the noisiest authorities insisted on its being   of times, it was the age of wisdom, it was was the season of Light, it was the everything before us, we had nothing other way- in short, the period was so far received, for good or for evil, in the   the age of foolishness, it was the epoch season of Darkness, it was the spring of before us, we were all going direct to like the present period, that some of its superlative degree of comparison only.

Figure 26-2 Column layout using a single path

Most of the time what you see in Figure 26-2 isn't what you want. Instead, you need to typeset the first column, then the second column, and finally the third. To do so, you need to create three paths and add them to a CFMutableArray called paths:

```
CGRect *columnRects = [self copyColumnRects];
// Create an array of layout paths, one for each column.
for (int column = 0; column < kColumnCount; column++) {
  CGPathRef path = CGPathCreateWithRect(columnRects[column], NULL);
  CFArrayAppendValue(paths, path);
  CGPathRelease(path);
}
free(columnRects);
```

You then iterate over this array, typesetting the text that hasn't been drawn yet:

```
CFIndex pathCount = CFArrayGetCount(paths);
CFIndex charIndex = 0;
for (CFIndex pathIndex = 0; pathIndex < pathCount; ++pathIndex) {
```

(continued)

```
    CGPathRef path = CFArrayGetValueAtIndex(paths, pathIndex);

    CTFrameRef
    frame = CTFramesetterCreateFrame(framesetter,
                                     CFRangeMake(charIndex, 0),
                                     path,
                                     NULL);
    CTFrameDraw(frame, context);
    CFRange frameRange = CTFrameGetVisibleStringRange(frame);
    charIndex += frameRange.length;
    CFRelease(frame);
  }
```

The call to `CTFrameGetVisibleStringRange` returns the range of characters within the attributed string that are included in this frame. That lets you know where to start the next frame. The zero-length range passed to `CTFramesetterCreateFrame` indicates that the framesetter should typeset as much of the attributed string as will fit.

Using these techniques, you can typeset text into any shape you can draw with `CGPath`, as long as the text fits into lines. You find out how to handle more complicated cases in the section "Drawing Text Along a Curve" later in this chapter.

Typesetters, Lines, Runs, and Glyphs

The framesetter is responsible for combining typeset lines into frames that can be drawn. The typesetter is responsible for choosing and positioning the glyphs in those lines. `CTFramesetter` automates this process, so you usually don't need to deal with the underlying typesetter (`CTTypesetter`). You will generally use the framesetter or move farther down the stack to lines, runs, and glyphs.

Starting at the bottom of the stack, a *glyph* (`CGGlyph`) is a shape that represents some piece of language information. This includes letters, numbers, and punctuation. It also includes whitespace, ligatures, and other marks. A *ligature* is when letters or other fundamental language units (*graphemes*) are combined to form a single glyph. The most common one in English is the *fi* ligature, formed when the letter *f* is followed by the letter *i*. In many fonts, these are combined into a single glyph to improve readability. The important thing is that a string may have a different number of glyphs than characters; the number of glyphs depends on the font and the layout of the characters.

A font can be thought of as a collection of glyphs, along with some metadata, such as the size and name. The `CGGlyph` type is implemented as an index into a `CGFont`. Don't confuse this with `CTFont` or `UIFont`. Each drawing system has its own font type. Core Text also has a `CTGlyphInfo` type for controlling how Unicode characters are mapped to glyphs. This is rarely used.

The typesetter is responsible for choosing the glyphs for a given attributed string and for collecting them into runs. A *run* (`CTRun`) is a series of glyphs that has the same attributes and direction (such as left-to-right or right-to-left). Attributes include font, color, shadow, and paragraph style. You cannot directly create `CTRun` objects, but you can draw them into a context with `CTRunDraw`. Each glyph is positioned in the run, taking into account individual glyph size and kerning. *Kerning* is small adjustments to the spacing between glyphs to make text more readable. For example, the letters *V* and *A* are often kerned very close together.

The typesetter combines runs into lines. A line (CTLine) is a series of runs oriented either horizontally or vertically (for languages such as Japanese). CTLine is the lowest-level typesetting object that you can directly create from an attributed string. This is convenient for drawing small blocks of rich text. You can directly draw a line into a context using CTLineDraw.

Generally in Core Text, you work with either a CTFramesetter for large blocks or a CTLine for small labels. From any level in the hierarchy, you can fetch the lower-level objects. For example, given a CTFramesetter, you create a CTFrame, and from that you can fetch its array of CTLine objects. Each line includes an array of CTRun objects, and within each run is a series of glyphs, along with positioning information and attributes. Behind the scenes is the CTTypesetter doing most of the work, but you seldom interact with it directly.

In the next section, "Drawing Text Along a Curve," you put all of these pieces together to perform a complex text layout.

Drawing Text Along a Curve

In this example, you use all the major parts of Core Text. Apple provides a somewhat simple example called CoreTextArcCocoa that demonstrates how to draw text along a semicircular arc. The Apple sample code is not very flexible, however, and is difficult to use for shapes other than a perfect circle centered in the view. It also forces the text to be evenly spaced along the curve. In this example, you learn how to draw text on any Bézier curve, and the techniques are applicable to drawing on any path. You also preserve Core Text's kerning and ligatures. The end result is shown in Figure 26-3. This example is available from the downloads for this chapter, in CurvyTextView.m in the CurvyText project.

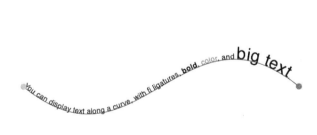

Figure 26-3 Output of CurvyTextView

Although CGPath can represent a Bézier curve and Core Graphics can draw it, there are no functions in iOS that allow you to calculate the points along the curve. You need these points, provided by Bezier(), and the slope along the curve, provided by BezierPrime():

CurvyTextView.m (CurvyText)

```
static double Bezier(double t, double P0, double P1, double P2,
                     double P3) {
  return
              (1-t)*(1-t)*(1-t)           *  P0
    + 3 *          (1-t)*(1-t)  *      t  *  P1
    + 3 *                (1-t)  *    t*t  *  P2
    +                              t*t*t  *  P3;
}


static double BezierPrime(double t, double P0, double P1,
                          double P2, double P3) {
  return
    -   3 * (1-t)*(1-t)  * P0
    + (3 * (1-t)*(1-t)  * P1)  -  (6 * t * (1-t)  * P1)
    - (3 *          t*t  * P2)  +  (6 * t * (1-t)  * P2)
    +   3 * t*t * P3;
}
```

P0 is the starting point, drawn in green by `CurvyTextView`. P1 and P2 are the control points, drawn in black. P3 is the endpoint, drawn in red. You call these functions twice, once for the x coordinate and once for y coordinate. To get a point and angle along the curve, you pass a number between 0 and 1 to `pointForOffset:` and `angleForOffset:`.

```
- (CGPoint)pointForOffset:(double)t {
  double x = Bezier(t, _P0.x, _P1.x, _P2.x, _P3.x);
  double y = Bezier(t, _P0.y, _P1.y, _P2.y, _P3.y);
  return CGPointMake(x, y);
}
- (double)angleForOffset:(double)t {
  double dx = BezierPrime(t, _P0.x, _P1.x, _P2.x, _P3.x);
  double dy = BezierPrime(t, _P0.y, _P1.y, _P2.y, _P3.y);
  return atan2(dy, dx);
}
```

These methods are called so many times that I've made an exception to the rule to always use accessors. This is the major hotspot of this program, and optimizations to speed it up or call it less frequently are worthwhile. There are many ways to improve the performance of this code, but this is fast enough for an example. See `robnapier.net/bezier` for more details on Bézier calculations.

With these two functions to define your path, you can now lay out the text. The following is a method to draw an attributed string into the current context along this path.

```
- (void)drawText {
  if ([self.attributedString length] == 0) { return; }

  // Initialize the text matrix (transform). This isn't reset
  // automatically, so it might be in any state.
```

```
CGContextRef context = UIGraphicsGetCurrentContext();
CGContextSetTextMatrix(context, CGAffineTransformIdentity);

// Create a typeset line object
CTLineRef line = CTLineCreateWithAttributedString(
                    (__bridge CFTypeRef)self.attributedString);

// The offset is where you are in the curve, from [0, 1]
double offset = 0.;

// Fetch the runs and process one at a time
CFArrayRef runs = CTLineGetGlyphRuns(line);
CFIndex runCount = CFArrayGetCount(runs);
for (CFIndex runIndex = 0; runIndex < runCount; ++runIndex) {
  CTRunRef run = CFArrayGetValueAtIndex(runs, runIndex);

  // Apply the attributes from the run to the current context
  [self prepareContext:context forRun:run];

  // Fetch the glyphs as a CGGlyph* array
  NSMutableData *glyphsData = [self glyphDataForRun:run];
  CGGlyph *glyphs = [glyphsData mutableBytes];

  // Fetch the advances as a CGSize* array. An advance is the
  // distance from one glyph to another.
  NSMutableData *advancesData = [self advanceDataForRun:run];
  CGSize *advances = [advancesData mutableBytes];

  // Loop through the glyphs and display them
  CFIndex glyphCount = CTRunGetGlyphCount(run);
  for (CFIndex glyphIndex = 0;
       glyphIndex < glyphCount && offset < 1.0;
       ++glyphIndex) {

    // You're going to modify the transform, so save the state
    CGContextSaveGState(context);

    // Calculate the location and angle. This could be any
    // function, but here you use a Bezier curve
    CGPoint glyphPoint = [self pointForOffset:offset];
    double angle = [self angleForOffset:offset];

    // Rotate the context
    CGContextRotateCTM(context, angle);

    // Translate the context after accounting for rotation
    CGPoint
    translatedPoint = CGPointApplyAffineTransform(glyphPoint,
                        CGAffineTransformMakeRotation(-angle));
    CGContextTranslateCTM(context,
                        translatedPoint.x, translatedPoint.y);

    // Draw the glyph
```

(continued)

```
CGContextShowGlyphsAtPoint(context, 0, 0,
                              &glyphs[glyphIndex], 1);

// Move along the curve in proportion to the advance.
offset = [self offsetAtDistance:advances[glyphIndex].width
                    fromPoint:glyphPoint offset:offset];
CGContextRestoreGState(context);
        }
    }
}
```

The translation at the end of `drawText` is particularly important. All the glyphs are drawn at the origin. You use a transform to move the glyph into the correct position. The transform includes a rotation and a translation, and the order matters. If you rotate the context halfway around, and then translate the context up one point, the net effect will be to translate *down* one point. To account for this, you need to apply the inverse transform to the translation using `CGPointApplyAffineTransform`. The inverse of rotating by `angle` radians is to rotate by `−angle` radians. You could also use `CGAffineTransformInvert` to get the same effect.

You need a few more methods. First, you need to apply the attributes to the context. In this example, you just set the font and color, but you could support any of the attributes listed in `CTStringAttributes.h`, or add your own. Note that this supports only a `CTFont`, not a `UIFont`. They're not interchangeable. If you want to support both the way `CTFramesetter` does, then you need to code for that. Similarly, this code supports only a `CGColor`, not a `UIColor`.

```
- (void)prepareContext:(CGContextRef)context forRun:(CTRunRef)run {
    CFDictionaryRef attributes = CTRunGetAttributes(run);

    // Set font
    CTFontRef runFont = CFDictionaryGetValue(attributes,
                                             kCTFontAttributeName);
    CGFontRef cgFont = CTFontCopyGraphicsFont(runFont, NULL);
    CGContextSetFont(context, cgFont);
    CGContextSetFontSize(context, CTFontGetSize(runFont));
    CFRelease(cgFont);

    // Set color
    CGColorRef color = (CGColorRef)CFDictionaryGetValue(attributes,
                                     kCTForegroundColorAttributeName);
    CGContextSetFillColorWithColor(context, color);
}
```

Fetching the glyph and advance information is easy. You just allocate a buffer and pass it to `CTRunGetGlyphs` or `CTRunGetAdvances`. The problem with these routines is that they copy all the data, which can be slow. There are faster versions (`CTRunGetGlyphsPtr` and `CTRunGetAdvancesPtr`) that return a pointer without making a copy. These versions can fail, however, if the data hasn't been calculated yet. `glyphDataForRun:`, shown in the following code, handles both cases and returns an `NSMutableData` that automatically handles memory management. `advanceDataForRun:` is nearly identical. You can find the source for it in `CurvyTextView.m`.

```
- (NSMutableData *)glyphDataForRun:(CTRunRef)run {
  NSMutableData *data;
  CFIndex glyphsCount = CTRunGetGlyphCount(run);
  const CGGlyph *glyphs = CTRunGetGlyphsPtr(run);
  size_t dataLength = glyphsCount * sizeof(*glyphs);
  if (glyphs) {
    data = [NSMutableData dataWithBytesNoCopy:(void*)glyphs
                              length:dataLength freeWhenDone:NO];
  }
  else {
    data = [NSMutableData dataWithLength:dataLength];
    CTRunGetGlyphs(run, CFRangeMake(0, 0), data.mutableBytes);
  }
  return data;
}
```

Finally, to maintain proper spacing, you need to find the point along the curve that is the same distance as the advance. This is not trivial for a Bézier curve. Offsets are not linear, and it's almost certain that the offset 0.25 will not be a quarter of the way along the curve. A simple solution is to repeatedly increment the offset and calculate a new point on the curve until the distance to that point is at least equal to your advance. The larger the increment you choose, the more characters tend to spread out. The smaller the increment you choose, the longer it takes to calculate. My experience is that values between 1/1000 (0.001) and 1/10,000 (0.0001) work well. Although 1/1000 has visible errors when compared to 1/10,000, the speed improvement is generally worth it. You could try to optimize this with a binary search, but that can fail if the loop wraps back on itself or crosses itself. Here is a simple implementation of the search algorithm:

```
- (double)offsetAtDistance:(double)aDistance
                 fromPoint:(CGPoint)aPoint
                    offset:(double)anOffset {
  const double kStep = 0.001; // 0.0001 - 0.001 work well
  double newDistance = 0;
  double newOffset = anOffset + kStep;
  while (newDistance <= aDistance && newOffset < 1.0) {
    newOffset += kStep;
    newDistance = Distance(aPoint,
                        [self pointForOffset:newOffset]);
  }
  return newOffset;
}
```

For more information on finding lengths on a curve, search the Web for "arc length parameterization."

With these tools, you can typeset rich text along any path you can calculate.

Summary

Apple provides a variety of powerful text layout tools, from `UILabel` to `UIWebView` to Core Text. In this chapter, you looked at the major options and how to choose among them. Most of all, you should have a good understanding of how to use Core Text to create beautiful text layout in even the most complex applications.

Further Reading

Apple Documentation

The following documents are available in the iOS Developer Library at `developer.apple.com` or through the Xcode Documentation and API Reference.

Core Text Programming Guide

Quartz 2D Programming Guide: "Text"

String Programming Guide: "Drawing Strings"

NSAttributedString UIKit Additions Reference

Text, Web, and Editing Programming Guide for iOS

WWDC Sessions

The following session videos are available at `developer.apple.com`.

WWDC 2011, "Session 511: Rich Text Editing in Safari on iOS"

WWDC 2012, "Session 222: Introduction to Attributed Strings for iOS"

WWDC 2012, "Session 226: Core Text and Fonts"

WWDC 2012, "Session 230: Advanced Attributed Strings for iOS"

Other Resources

Clegg, Jay. *Jay's Projects.* "Warping Text to a Bézier curves." Useful background on techniques for laying out text along a curve. The article is in C# and GDI+, but the math is useful on any platform.
`planetclegg.com/projects/WarpingTextToSplines.html`

Drobnik, Oliver. *DTCoreText.* Powerful framework for drawing attributed strings in iOS and converting between attributed strings and HTML.
`github.com/Cocoanetics/DTCoreText`

Kosmaczewski, Adrian. *CoreTextWrapper.* Wrapper to simplify multicolumn layout using Core Text.
`github.com/akosma/CoreTextWrapper`

## Chapter 27

Building a (Core) Foundation

As an iOS developer, you will spend most of your time using the UIKit and Foundation frameworks. UIKit provides user interface elements like UIView and UIButton. Foundation provides basic data structures like NSArray and NSDictionary. These can handle the vast majority of problems the average iOS application will encounter. But some things require lower-level frameworks. The names of these lower-level frameworks often start with the word "Core," Core Text, Core Graphics, and Core Video, for example. What they all have in common are C-based APIs based on Core Foundation.

Core Foundation provides a C API that is similar to the Objective-C Foundation framework. It provides a consistent object model with reference counting and containers, just like Foundation, and simplifies passing common data types to low-level frameworks. As you see later in this chapter, Core Foundation is tightly coupled with Foundation, making it easy to pass data between C and Objective-C.

In this chapter, you find out about the Core Foundation data types and naming conventions. You learn about Core Foundation allocators and how they provide greater flexibility than +alloc provides in Objective-C. This chapter covers Core Foundation string and binary data types extensively. You discover Core Foundation collection types, which are more flexible than their Foundation counterparts and include some not found in Objective-C. Finally, you learn how to move data easily between Core Foundation and Objective-C using toll-free bridging. When you're finished, you will have the tools you need to use the powerful Core frameworks, as well as more flexible data structures to improve your own projects.

All code samples in this chapter can be found in main.m and MYStringConversion.c in the online files for this chapter.

Core Foundation Types

Core Foundation is made up primarily of *opaque types*, which are simply C structs. Opaque types are similar to classes in that they provide encapsulation, and some inheritance and polymorphism. The similarity should not be overstated, however. Core Foundation is implemented in pure C, where there is no language support for inheritance or polymorphism, so sometimes the class metaphor can become strained. But for general usage, Core Foundation can be thought of as an object model with CFType as its root "class."

Like Objective-C, Core Foundation deals with pointers to instances. In Core Foundation, these pointers are given the suffix Ref. For example, a pointer to a CFType is a CFTypeRef, and a pointer to a string is a CFStringRef. Mutable versions of opaque types include the word Mutable, so a CFMutableStringRef is the mutable form of a CFStringRef. Generally, mutable types can be treated as though they were a subclass of the nonmutable type, just as in Foundation. For simplicity, and to match the Apple documentation, this chapter uses the term `CFString` to refer to the thing a `CFStringRef` points to, even though Core Foundation does not define the symbol `CFString`.

Because Core Foundation is implemented in C, and C has no language support for inheritance or polymorphism, how does Core Foundation give the illusion of an object hierarchy? First, `CFTypeRef` is just a `void*`. This provides a crude kind of polymorphism because it allows arbitrary types to be passed to certain functions, particularly `CFCopyDescription`, `CFEqual`, `CFHash`, `CFRelease`, and `CFRetain`.

Except for `CFTypeRef`, opaque types are structs. A mutable and immutable pair is usually of the form

```
typedef const struct __CFString * CFStringRef;
typedef struct __CFString * CFMutableStringRef;
```

This way, the compiler can enforce `const` correctness to provide a kind of inheritance. It should be clear that this isn't real inheritance. There is no good way to provide arbitrary subclasses of `CFString` that the compiler will type-check. For example, consider the following code:

```
CFStringRef errName = CFSTR("error");
CFErrorRef error = CFErrorCreate(NULL, errName, 0, NULL);
CFPropertyListRef propertyList = error;
```

A `CFError` is not a `CFPropertyList`, so line 3 should generate a warning. It doesn't because `CFPropertyListRef` is defined as `CFTypeRef`, which is required because it has several "subclasses," including `CFString`, `CFDate`, and `CFNumber`. Once something has several subclasses, it generally has to be treated as a `void*` (`CFTypeRef`) in Core Foundation. This isn't obvious from looking at the code, but luckily it doesn't come up that often. Most types are defined as a specific `struct` or `const struct`.

Naming and Memory Management

As in Cocoa, naming conventions are critical in Core Foundation. The most important rule is the Create Rule: If a function has the word `Create` or `Copy` in its name, you are an owner of the resulting object and must eventually release your ownership using `CFRelease`. Like Cocoa, objects are reference counted and can have multiple "owners." When the last owner calls `CFRelease`, the object is destroyed.

There is no equivalent of `NSAutoreleasePool` in Core Foundation, so functions with `Copy` in the name are much more common than in Cocoa. Some functions, however, return a reference to an internal data structure or to a constant object. These functions generally include the word `Get` in their name (the Get Rule). The caller is not an owner and does not need to release them.

> `Get` in Core Foundation is not the same as `get` in Cocoa. Core Foundation functions including `Get` return an opaque type or a C type. Cocoa methods that begin with `get` update a pointer passed by reference.

There is no automatic reference counting in Core Foundation. Memory management in Core Foundation is very similar to manual memory management in Cocoa:

- If you `Create` or `Copy` an object, you are an owner.

- If you do not `Create` or `Copy` an object, you are not an owner. If you want to prevent the object from being destroyed, you must become an owner by calling `CFRetain`.

- If you are an owner of an object, you must call `CFRelease` when you're done with it.

> `CFRelease` **is very similar to** `release` **in Objective-C, but there are important differences. The most critical is that you cannot call** `CFRelease(NULL)`. **This is somewhat unfortunate, and many specialized versions of** `CFRelease` **exist that do allow you to pass** `NULL` (`CGGradientRelease`, **for instance).** `CFRelease` **also behaves differently than** `retain` **in garbage-collected environments. This doesn't apply to iOS, so it isn't covered here, but you can read Apple's** *Memory Management Programming Guide for Core Foundation* **for more information.**

Some functions have both `Create` and `Copy` in their name. For example, `CFStringCreateCopy` creates a copy of another `CFString`. Why not just `CFStringCopy`? That's because `Create` tells you other things about the function than just the ownership rule. It indicates that the first parameter is a `CFAllocatorRef`, which lets you customize how the newly created object is allocated. In almost all cases, you pass `NULL` for this parameter, which specifies the default allocator: `kCFAllocatorDefault`. (I'll cover allocators in more depth in a moment.) Knowing that the function is a creator, the name also tells you that it makes a copy of the passed string.

Conversely, a function with `NoCopy` in its name does not make a copy. For example, `CFStringCreateWithBytesNoCopy` takes a pointer to a buffer and creates a string without copying the bytes. So who is now responsible for releasing the buffer? That brings us back to allocators.

Allocators

A `CFAllocatorRef` is a strategy for allocating and freeing memory. In almost all cases, you want the default allocator, `kCFAllocatorDefault`, which is the same as passing `NULL`. This allocates and frees memory in "the normal way" according to Core Foundation. This way is subject to change, and you shouldn't rely on any particular behavior. It is rare to need a specialized allocator, but in a few cases, it can be useful. Here are the standard allocators to give an idea of what they can do:

- `kCFAllocatorDefault`—The default allocator. It is equivalent to passing `NULL`.

- `kCFAllocatorSystemDefault`—The original default system allocator. This is available in case you have changed the default allocator using `CFAllocatorSetDefault`. This is very rarely necessary.

- `kCFAllocatorMalloc`—Calls `malloc`, `realloc`, and `free`. This is particularly useful as a deallocator for `CFData` and `CFString` if you created the memory with `malloc`.

- `kCFAllocatorMallocZone`—Creates and frees memory in the default `malloc` zone. This can be useful with garbage collection on the Mac, but is almost never useful in iOS.

- `kCFAllocatorNull`—Does nothing. Like `kCFAllocatorMalloc`, this can be useful with `CFData` or `CFString` as a deallocator if you don't want to free the memory.

■ kCFAllocatorUseContext—Only used by the CFAllocatorCreate function. When you create a CFAllocator, the system needs to allocate memory. Like all other Create methods, this requires an allocator. This special allocator tells CFAllocatorCreate to use the functions passed to it to allocate the CFAllocator.

See the later section "Backing Storage for Strings," for examples of how these can be used in a practical problem.

Introspection

Core Foundation allows a variety of type introspections, primarily for debugging purposes. The most fundamental one is the CFTypeID, which uniquely identifies the opaque type of the object, similar to Class in Objective-C. You can determine the type of a Core Foundation instance by calling CFGetTypeID. The returned value is opaque and subject to change between versions of iOS. You can compare the CFTypeID of two instances, but most often you compare the result of CFGetTypeID to the value from a function like CFArrayGetTypeID. All opaque types have a related GetTypeID function.

As in Cocoa, Core Foundation instances have a description for debugging purposes, returned by CFCopyDescription. This returns a CFString that you're responsible for releasing. CFCopyTypeIDDescription provides a similar string that describes a CFTypeID. Don't rely on the format or content of these because they're subject to change.

To write debugging output to the console, use CFShow. It will display the value of a CFString, or the description of other types. To display the description of a CFString, use CFShowStr. For example, given the following definitions

```
CFStringRef string = CFSTR("Hello");
CFArrayRef array = CFArrayCreate(NULL, (const void**)&string, 1,
                                 &kCFTypeArrayCallBacks);
```

here are the results for each kind of CFShow call:

```
CFShow(array);
<CFArray 0x6d47850 [0x1445b38]>{type = immutable, count = 1,
   values = (
      0 : <CFString 0x410c [0x1445b38]>{contents = "Hello"}
   )}

CFShow(string);
Hello

CFShowStr(string);
Length 5
IsEightBit 1
HasLengthByte 0
HasNullByte 1
InlineContents 0
Allocator SystemDefault
Mutable 0
Contents 0x3ba7
```

Strings and Data

CFString is a Unicode-based storage container that provides rich and efficient functionality for manipulating, searching, and converting international strings. Closely related are the CFCharacterSet and CFAttributedString classes. CFCharacterSet represents a set of characters for efficiently searching, including or excluding certain characters from a string. CFAttributedString combines a string with ranges of attributes. This is most commonly used to handle rich text, but can be used for a variety of metadata storage.

CFString is closely related to NSString, and they are generally interchangeable, as you'll see in "Toll-Free Bridging" section later in this chapter. This section focuses on the differences between CFString and NSString.

Constant Strings

In Cocoa, a literal NSString is indicated by an ampersand, as in @"string". In Core Foundation, a literal CFString is indicated by the macro CFSTR, as in CFSTR("string"). If you're using the Apple-provided gcc and the option -fconstant-cfstrings, this macro uses a special built-in compiler hook that creates constant CFString objects at compile time. Clang also has this built-in compiler hook. If you're using standard gcc, then an explicit CFStringMakeConstantString function is used to create these objects at runtime.

Because CFSTR has neither Create nor Copy in its name, you don't need to call CFRelease on the result. You may, however, call CFRetain normally if you like. If you do, you should balance it with CFRelease as usual. This allows you to treat constant strings in the same way as programmatically created strings.

Creating Strings

A common way to generate a CFString is from a C string. Here is an example:

```
const char *cstring = "Hello World!";
CFStringRef string = CFStringCreateWithCString(NULL, cstring,
  kCFStringEncodingUTF8);
CFShow(string);
CFRelease(string);
```

Although many developers are most familiar with NULL-terminated C strings, there are other ways to store strings, and understanding them can be useful in improving code efficiency. In network protocols, it can be very efficient to encode strings as a length value followed by a sequence of characters. If parsers are likely to need only a part of the packet, it's faster to use length bytes to skip over the parts you don't need than to read everything looking for NULL. If this length encoding is 1 byte long, then the buffer is a Pascal string and Core Foundation can use it directly, as shown in the following code:

```
// A common type of network buffer
struct NetworkBuffer {
  UInt8 length;
  UInt8 data[];
};

// Some data we pulled off of the network into the buffer
```

(continued)

```
static struct NetworkBuffer buffer = {
   4, {'T', 'e', 'x', 't'}};

CFStringRef string =
   CFStringCreateWithPascalString(NULL,
                                  (ConstStr255Param)&buffer,
                                   kCFStringEncodingUTF8);
CFShow(string);
CFRelease(string);
```

If you have length some other way, or if the length is not 1 byte long, you can use `CFStringCreateWithBytes` similarly:

```
CFStringRef string = CFStringCreateWithBytes(NULL,
                                             buffer.data,
                                             buffer.length,
                                      kCFStringEncodingUTF8,
                                             false);
```

The final `false` indicates this string does not have a *byte order mark* (BOM) at the beginning. The BOM indicates whether the string was generated on a big endian or little endian system. A BOM is not needed or recommended for UTF-8 encodings. This is one of many reasons to choose UTF-8 when possible.

> **Core Foundation constants begin with a** `k`**, unlike their Cocoa counterparts. For example, the Core Foundation counterpart to** `NSUTF8StringEncoding` **is** `kCFStringEncodingUTF8`.

Converting to C Strings

Although converting from C strings is very simple, converting back to C strings can be deceptively difficult. There are two ways to get a C string out of a `CFString`: Request the pointer to the internal C string representation or copy the bytes out into your own buffer.

Obviously, the easiest and fastest way to get the C string is to request the internal C string pointer:

```
const char *
cstring = CFStringGetCStringPtr(string, kCFStringEncodingUTF8);
```

This appears to be the best of all worlds. It's extremely fast, and you don't have to allocate or free memory. Unfortunately, it may not work, depending on how the string is currently encoded inside of the `CFString`. If there isn't an internal C string representation available, then this routine returns `NULL` and you have to use `CFStringGetCString` and pass your own buffer, although it isn't obvious how large a buffer you need. Here's an example of how to solve this problem:

```
char * MYCFStringCopyUTF8String(CFStringRef aString) {
   if (aString == NULL) {
      return NULL;
```

```
  }

  CFIndex length = CFStringGetLength(aString);
  CFIndex maxSize =
    CFStringGetMaximumSizeForEncoding(length,
                                        kCFStringEncodingUTF8);
  char *buffer = (char *)malloc(maxSize);
  if (CFStringGetCString(aString, buffer, maxSize,
                          kCFStringEncodingUTF8)) {
    return buffer;
  }
  return NULL;
}

...

CFStringRef string = CFSTR("Hello");
char * cstring = MYCFStringCopyUTF8String(string);
printf("%s\n", cstring);
free(cstring);
```

MYCFStringCopyUTF8String is not the fastest way to convert a CFString to a C string because it allocates a new buffer for every conversion, but it's easy to use and quick enough for many problems. If you're converting a lot of strings and want to improve speed and minimize memory churn, you might use a function like this one that supports reusing a common buffer:

```
#import <malloc/malloc.h> // For malloc_size()

const char * MYCFStringGetUTF8String(CFStringRef aString,
                                      char **buffer) {
  if (aString == NULL) {
    return NULL;
  }

  const char *cstr = CFStringGetCStringPtr(aString,
                                        kCFStringEncodingUTF8);
  if (cstr == NULL) {
    CFIndex length = CFStringGetLength(aString);
    CFIndex maxSize =
      CFStringGetMaximumSizeForEncoding(length,
                kCFStringEncodingUTF8) + 1; // +1 for NULL
    if (maxSize > malloc_size(buffer)) {
      *buffer = realloc(*buffer, maxSize);
    }
    if (CFStringGetCString(aString, *buffer, maxSize,
                            kCFStringEncodingUTF8)) {
      cstr = *buffer;
    }
  }
  return cstr;
}
```

The caller of `MYCFStringGetUTF8String` is responsible for passing a reusable buffer. The buffer may point to `NULL` or to preallocated memory. Keep in mind that the returned C string points into either the `CFString` or into `buffer`, so invalidating either of those can cause the returned C string to become invalid. In particular, passing the same buffer repeatedly to this function may invalidate old results. That's the trade-off for its speed. Here's how it would be used:

```
CFStringRef strings[3] = { CFSTR("One"), CFSTR("Two"),
                           CFSTR("Three") };
char * buffer = NULL;
const char * cstring = NULL;
for (unsigned i = 0; i < 3; ++i) {
  cstring = MYCFStringGetUTF8String(strings[i], &buffer);
        printf("%s\n", cstring);
}
free(buffer);
```

If you need conversion to be as fast as possible, and you know the maximum string length, then the following is even faster:

```
CFStringRef string = ...;
const CFIndex kBufferSize = 1024;
char buffer[kBufferSize];
CFStringEncoding encoding = kCFStringEncodingUTF8;
const char *cstring;
cstring = CFStringGetCStringPtr(string, encoding);
if (cstring == NULL) {
  if (CFStringGetCString(string, buffer, kBufferSize,
                         encoding)) {
    cstring = buffer;
  }
}
printf("%s\n", cstring);
```

Because this approach relies on a stack variable (`buffer`), it's difficult to wrap this into a simple function call, but it avoids any extra memory allocations.

Other String Operations

To developers familiar with `NSString`, most of `CFString` should be fairly obvious. You can find ranges of characters, append, trim and replace characters, compare, search, and sort as in Cocoa. `CFStringCreateWithFormat` provides functionality identical to `stringWithFormat:`. I won't explore all the functions here. You can find them all in the documentation for `CFString` and `CFMutableString`.

Backing Storage for Strings

Generally a `CFString` will allocate the required memory to store its characters. This memory is called the *backing storage*. If you have an existing buffer, it's sometimes more efficient or convenient to continue using it rather than copying all the bytes into a new `CFString`. You might do this because you have a buffer of bytes you want to convert into a string or because you want to continue to have access to the raw bytes while also using convenient string functions.

In the first case, where you already have a buffer, you generally use a function like
CFStringCreateWithBytesNoCopy.

```
const char *cstr = "Hello";
char *bytes = CFAllocatorAllocate(kCFAllocatorDefault,
                                  strlen(cstr) + 1, 0);
strcpy(bytes, cstr);
CFStringRef str =
  CFStringCreateWithCStringNoCopy(kCFAllocatorDefault,
                                  bytes,
                                  kCFStringEncodingUTF8,
                                  kCFAllocatorDefault);
CFShow(str);
CFRelease(str);
```

Because you passed the default allocator (kCFAllocatorDefault) as the destructor, the CFString
owns the buffer and will free it when it's done using the default allocator. This matches the earlier call to
CFAllocatorAllocate. If you were to allocate the buffer with malloc, the code would look like this:

```
const char *cstr = "Hello";
char *bytes = malloc(strlen(cstr) + 1);
strcpy(bytes, cstr);
CFStringRef str =
  CFStringCreateWithCStringNoCopy(NULL, bytes,
                                  kCFStringEncodingUTF8,
                                  kCFAllocatorMalloc);
CFShow(str);
CFRelease(str);
```

In both cases, the allocated buffer is freed when the string is destroyed. But what if you wanted to keep the
buffer for other uses? Consider the following code:

```
const char *cstr = "Hello";
char *bytes = malloc(strlen(cstr) + 1);
strcpy(bytes, cstr);
CFStringRef str =
  CFStringCreateWithCStringNoCopy(NULL, bytes,
                                  kCFStringEncodingUTF8,
                                  kCFAllocatorNull);
CFShow(str);
CFRelease(str);
printf("%s\n", bytes);
free(bytes);
```

You pass kCFAllocatorNull as the destructor. You still release the string because you created it with a
Create function. But now the buffer pointed to by bytes is still valid after the call to CFRelease. You are
responsible for calling free on bytes when you're done with the buffer.

There is no guarantee that the buffer you pass will be the actual buffer used. Core Foundation may call the
deallocator at any time and create its own internal buffer. Most critically, you must not modify the buffer after
creating the string. If you have a buffer that you want to access as a CFString while allowing changes to it,

then you need to use `CFStringCreateMutableWithExternalCharactersNoCopy`. This creates a mutable string that always uses the provided buffer as its backing store. If you change the buffer, you need to let the string know by calling `CFStringSetExternalCharactersNoCopy`. Using these functions bypasses many string optimizations, so use them with care.

CFData

`CFData` is the Core Foundation equivalent to `NSData`. It is much like `CFString` with similar creation functions, backing store management, and access functions. The primary difference is that `CFData` does not manage encodings like `CFString`. You can find the full list of functions in the `CFData` and `CFMutableData` references.

Collections

Core Foundation provides a rich set of object collection types. Most have Cocoa counterparts like `CFArray` and `NSArray`. There are a few specialized Core Foundation collections such as `CFTree` that have no Cocoa counterpart. Core Foundation collections provide greater flexibility in how they manage their contents. In this section, you learn about the Core Foundation collections that have Objective-C equivalents: `CFArray`, `CFDictionary`, `CFSet`, and `CFBag`. The other Core Foundation collections are seldom used, but I will introduce them so that you're aware of what's available if you need it.

Cocoa collections can hold only Objective-C objects and must retain them. Core Foundation collections can hold anything that can fit in the size of a pointer (32 bits for the ARM processor) and can perform any action when adding or removing items. The default behavior is very similar to the Cocoa equivalents, and Core Foundation collections generally retain and release instances when adding and removing. Core Foundation uses a structure of function pointers that defines how to treat items in the collection. Configuring these callbacks allows you to highly customize your collection. You can store nonobjects like integers, create weak collections that do not retain their objects, or modify how objects are compared for equality. The "Callbacks" section, later in this chapter, covers this topic. Each collection type has a default set of callbacks defined in the header. For example, the default callbacks for `CFArray` are `kCFTypeArrayCallBacks`. While introducing the major collections, I will focus on these default behaviors.

CFArray

`CFArray` corresponds to `NSArray` and holds an ordered list of items. Creating a `CFArray` takes an allocator, a series of values, and a set of callbacks, as shown in the following code.

```
CFStringRef strings[3] =
  { CFSTR("One"), CFSTR("Two"), CFSTR("Three") };
CFArrayRef array = CFArrayCreate(NULL, (void *)strings, 3,
                                 &kCFTypeArrayCallBacks);
CFShow(array);
CFRelease(array);
```

Creating a `CFMutableArray` takes an allocator, a size, and a set of callbacks. Unlike `NSMutableArray` capacity, which is only an initial size, the size passed to `CFMutableArray` is a fixed maximum. To allocate an array that can grow, pass a size of 0.

```
CFMutableArrayRef array = CFArrayCreateMutable(NULL, 0,
                              &kCFTypeArrayCallBacks);
```

CFDictionary

CFDictionary corresponds to NSDictionary and holds key-value pairs. Creating a CFDictionary takes an allocator, a series of keys, a series of values, a set of callbacks for the keys, and a set of callbacks for the values.

```
#define kCount 3
CFStringRef keys[kCount] =
   { CFSTR("One"), CFSTR("Two"), CFSTR("Three") };
CFStringRef values[kCount] =
   { CFSTR("Foo"), CFSTR("Bar"), CFSTR("Baz") };
CFDictionaryRef dict =
  CFDictionaryCreate(NULL,
                     (void *)keys,
                     (void *)values,
                     kCount,
                     &kCFTypeDictionaryKeyCallBacks,
                     &kCFTypeDictionaryValueCallBacks);
```

Creating a CFMutableDictionary is like creating a CFMutableArray, except there are separate callbacks for the keys and values. As with CFMutableArray, the size is fixed if given. For a dictionary that can grow, pass a size of 0.

CFSet, CFBag

CFSet corresponds to NSSet and is an unordered collection of unique objects. CFBag corresponds to NSCountedSet and allows duplicate objects. As with their Cocoa counterparts, uniqueness is defined by equality. The function that determines equality is one of the callbacks.

Like CFDictionary, CFSet and CFBag can hold NULL values by passing NULL as their callback structure pointer.

Other Collections

Core Foundation includes several collections that do not have a Cocoa counterpart:

- CFTree provides a convenient way to manage tree structures that might otherwise be stored less efficiently in a CFDictionary. There is a short example of CFTree in the section "Toll-Free Bridging," later in the chapter.
- CFBinaryHeap provides a binary-searchable container, similar to a sorted queue.
- CFBitVector provides a convenient way to store bit values.

Full information on CFTree is available in Apple's *Collections Programming Topics for Core Foundation*. See the Apple documentation on CFBinaryHeap and CFBitVector for more information on their usage. These are not often used and aren't heavily documented.

Callbacks

Core Foundation uses a structure of function pointers that define how to treat items in the collection. The structure includes the following members:

- `retain`—Called when an item is added to the collection. The default behavior is similar to `CFRetain` (you'll learn what "similar" means in a moment). If it's `NULL`, no action is performed.

- `release`—Called when an item is removed from the collection, and when the collection is destroyed. The default behavior is similar to `CFRelease`. If it is `NULL`, no action is performed.

- `copyDescription`—Called for each object in response to functions that want a human-readable description for the entire collection, such as `CFShow` or `CFCopyDescription`. The default value is `CFCopyDescription`. If this is `NULL`, the collection has some built-in logic to construct a simple description.

- `equal`—Called to compare a collection object with another object to determine if they're equal. The default value is `CFEqual`. If this is `NULL`, the collection will use strict equality (`==`) of the values. If the items are pointers to objects (as is the usual case), then this means that objects are only equal to themselves.

- `hash`—This only applies to hashing collections like dictionaries and sets. This function is used to determine the hash value of an object. A hash is a fast way to compare objects. Given an object, a hash function returns an integer such that if two objects are equal, then their hashes are equal. This allows the collection to quickly determine unequal objects with a simple integer comparison, saving the expensive call to `CFEqual` for objects that are possibly equal. The default value is `CFHash`. If this is `NULL`, the value (usually a pointer) is used as its own hash.

The default values for `retain` and `release` act like `CFRetain` and `CFRelease`, but are actually pointers to the private functions `__CFTypeCollectionRetain` and `__CFTypeCollectionRelease`. The `retain` and `release` function pointers include the collection's allocator in case you want to create a new object rather than retain an existing one. This is incompatible with `CFRetain` and `CFRelease`, which do not take an allocator. Usually this doesn't matter because, in most cases, you'll either leave `retain` and `release` as default or set them to `NULL`.

Each collection type has a default set of callbacks defined in the header. For example, the default callbacks for `CFArray` are `kCFTypeArrayCallBacks`. These can be used to easily modify default behavior. The following creates a nonretaining array, which could also hold nonobjects such as integers:

```
CFArrayCallBacks nrCallbacks = kCFTypeArrayCallBacks;
nrCallbacks.retain = NULL;
nrCallbacks.release = NULL;
CFMutableArrayRef nrArray = CFArrayCreateMutable(NULL, 0,
                          &nrCallbacks);
CFStringRef string =
  CFStringCreateWithCString(NULL, "Stuff",
                            kCFStringEncodingUTF8);
CFArrayAppendValue(nrArray, string);
CFRelease(nrArray);
CFRelease(string);
```

Another example of callback configuration is to allow NULL values or keys. Dictionaries, sets, and bags can hold NULL values or keys if the `retain` and `release` callbacks are NULL (this also makes them nonretaining). These types have CFTypeGetValueIfPresent functions to handle this case. For example, the function CFDictionaryGetValueIfPresent() allows you to determine whether the value was NULL versus missing, as shown in the following code:

```
CFDictionaryKeyCallBacks cb = kCFTypeDictionaryKeyCallBacks;
cb.retain = NULL;
cb.release = NULL;
CFMutableDictionaryRef dict =
  CFDictionaryCreateMutable(NULL, 0, &cb,
                            &kCFTypeDictionaryValueCallBacks);
CFDictionarySetValue(dict, NULL, CFSTR("Foo"));

const void *value;
Boolean fooPresent =
  CFDictionaryGetValueIfPresent(dict, NULL, &value);
CFRelease(dict);
```

Other collections, such as CFArray, cannot hold NULL values. As in Foundation, you must use a special placeholder NULL constant called kCFNull. It is an opaque type (CFNull), so it can be retained and released.

Core Foundation collections are much more flexible than their Cocoa equivalents. As you see in the next section, however, you can bring this flexibility almost transparently to Cocoa through the power of toll-free bridging.

Toll-Free Bridging

One of the cleverest aspects of Core Foundation is its capability to transparently exchange data with Foundation. For example, any function or method that accepts an NSArray also accepts a CFArray with only a bridge cast. A *bridge cast* is an instruction to the compiler about how to apply automatic reference counting.

In many cases, you only need to use the __bridge modifier, as shown in the following code:

```
NSArray *nsArray = [NSArray arrayWithObject:@"Foo"];
printf("%ld\n", CFArrayGetCount((__bridge CFArrayRef)nsArray));
```

This essentially tells the compiler to do nothing special. It should simply cast nsArray as a CFArrayRef and pass it to CFArrayGetCount. There is no change to the reference count of nsArray.

This works in reverse as well:

```
CFMutableArrayRef cfArray =
  CFArrayCreateMutable(NULL, 0, &kCFTypeArrayCallBacks);
CFArrayAppendValue(cfArray, CFSTR("Foo"));
NSLog(@"%ld", [(__bridge id)cfArray count]);
CFRelease(cfArray);
```

The __bridge cast works as long as there is no Core Foundation memory management involved. In the preceding examples, you aren't assigning the results to variables or returning them. Consider this case, however:

```
- (NSString *)firstName {
  CFStringRef cfString = CFStringCreate...;
  return (???)cfString;
}
```

How can you cast `cfString` correctly? Before ARC, you would have cast this to an `NSString` and called `autorelease`. With ARC, you can't call `autorelease`, and ARC doesn't know that `cfString` has an extra retain on it from `CFStringCreate....` You again use a bridge cast, this time in the form of a function as in this example:

```
return CFBridgingRelease(cfString);
```

This function transfers ownership from Core Foundation to ARC. In the process, it reduces the retain count by one to balance the `CFStringCreate....` You must use a bridge cast to achieve this. Calling `CFRelease` before returning the object would destroy the object.

When transferring an object from ARC to Core Foundation, you use `CFBridgingRetain`, which increases the retain count by one, as shown in the following code:

```
CFStringRef cfStr = CFBridgingRetain([nsString copy]);
nsString = nil; // Ownership now belongs to cfStr
...
CFRelease(cfStr);
```

The bridging functions can also be written in a typecast style as follows:

```
NSString *nsString = CFBridgingRelease(cfString);
NSString *nsString = (__bridge_transfer id)cfString;

CFStringRef cfString = CFBridgingRetain(nsString);
CFStringRef cfString = (__bridge_retained CFTypeRef)nsString;
```

> `CFTypeRef` **is a generic pointer to a Core Foundation object, and** `id` **is a generic pointer to an Objective-C object. You could also use explicit types here like** `CFStringRef` **and** `NSString*`**.**

The function form is shorter and, in my opinion, easier to understand. `CFBridgingRelease` and `CFBridgingRetain` should be used only when an object is being transferred between ARC and Core Foundation. They're not replacements for `CFRetain` or `CFRelease` or a way to "trick" the compiler into adding an extra `retain` or `release` on Objective-C objects. Once an object is transferred from Core Foundation to ARC, the Core Foundation variable should not be used again and should be set to `NULL`. Conversely, when converting from ARC to Core Foundation, you need to immediately set the ARC variable to `nil`. You have transferred ownership from one variable to another, so the old variable needs to be treated as invalid.

Not only is toll-free bridging very convenient for moving information between C and Objective-C, it also enables Cocoa developers to make use of certain Core Foundation functions that have no Objective-C equivalent. For example, `CFURLCreateStringByAddingPercentEscapes` allows much more powerful transformations than the equivalent `NSURL stringByAddingPercentEscapesUsingEncoding:`.

Even types that aren't explicitly toll-free bridged are still bridged to NSObject. This means that you can store Core Foundation objects (even ones with no Cocoa equivalent) in Cocoa collections, as shown in this example:

```
CFTreeContext ctx = {0, (void*)CFSTR("Info"), CFRetain,
                        CFRelease, CFCopyDescription};
CFTreeRef tree = CFTreeCreate(NULL, &ctx);
NSArray *array = [NSArray arrayWithObject:(__bridge id)tree];
CFRelease(tree);
NSLog(@"%@", array);
```

Toll-free bridging is implemented in a fairly straightforward way. Every Objective-C object structure begins with an ISA pointer to a Class:

```
typedef struct objc_class *Class;
typedef struct objc_object {
  Class isa;
} *id;
```

Core Foundation opaque types begin with a CFRuntimeBase, and the first element of that is also an ISA pointer:

```
typedef struct __CFRuntimeBase {
  uintptr_t _cfisa;
  uint8_t _cfinfo[4];
#if __LP64__
  uint32_t _rc;
#endif
} CFRuntimeBase;
```

_cfisa points to the toll-free bridged Cocoa class. These are subclasses of the equivalent Cocoa class; they forward Objective-C method calls to the equivalent Core Foundation function call. For instance, CFString is bridged to the private toll-free bridging class NSCFString.

If there is no explicit bridging class, then _cfisa points to __NSCFType, which is a subclass of NSObject, and forwards calls like retain and release.

To handle Objective-C classes passed to Core Foundation functions, all public toll-free functions look something like this:

```
CFIndex CFStringGetLength(CFStringRef str) {
  CF_OBJC_FUNCDISPATCH0(__kCFStringTypeID, CFIndex, str, "length");
  __CFAssertIsString(str);
  return __CFStrLength(str);
}
```

CF_OBJC_FUNCDISPATCH0 checks the _cfisa pointer. If it matches the Core Foundation bridging class for the given CFTypeID, then it passes the call along to the real Core Foundation function. Otherwise, it translates the call into an Objective-C message (length in this case, given as a C string).

Summary

Core Foundation bridges the gap between C and Objective-C code, providing powerful data structures for C and near-transparent data passing to and from low-level code. As Apple releases more low-level Core frameworks that require these types, Core Foundation is an increasingly important part of an iOS developer's toolkit.

Core Foundation data structures are generally more flexible than their Cocoa equivalents. They provide better control over how memory is managed through allocators and often include functions for more specialized problems like handling Pascal strings or very configurable URL percent substitutions. Core Foundation collections can be configured to be nonretaining and can even store nonobjects such as integers.

Although Objective-C is extremely powerful, you can still generally write code that is faster and more efficient in pure C, which is why the lowest-level APIs are all C APIs. For those parts of your programs that require the kind of performance you can only get from C, Core Foundation provides an excellent collection of abstract data types that you can easily exchange with the higher-level parts of your program. The vast majority of problems in iOS are best solved in Cocoa and Objective-C, but for those places that C is appropriate, Core Foundation is a powerful tool.

Further Reading

Apple Documentation

The following documents are available in the iOS Developer Library at `developer.apple.com` or through the Xcode Documentation and API Reference.

Collections Programming Topics for Core Foundation

Core Foundation Design Concepts

Data Formatting Guide for Core Foundation

Dates and Times Programming Guide for Core Foundation

Programming With ARC Release Notes: "Managing Toll-Free Bridging"

Memory Management Programming Guide for Core Foundation

Property List Programming Topics for Core Foundation

Strings Programming Guide for Core Foundation

Other Resources

Clang Documentation. "Automatic Reference Counting"
`clang.llvm.org/docs/AutomaticReferenceCounting.html`

ridiculous_fish, "Bridge." An entertaining introduction to toll-free bridging internals by one of the AppKit and Foundation team at Apple.
`ridiculousfish.com/blog/posts/bridge.html`

Chapter 28
Deep Objective-C

Much of Objective-C is very straightforward in practice. There is no multiple inheritance or operator overloading like in C++. All objects have the same memory-management rules, which rely on a simple set of naming conventions. With the addition of ARC, you don't even need to worry about memory management in most cases. The Cocoa framework is designed with readability in mind, so most things do exactly what they say they do.

Still, many parts of Objective-C appear mysterious until you dig into them, such as creating new methods and classes at runtime, introspection, and message passing. Most of the time, you don't need to understand how this works, but for some problems, it's very useful to harness the full power of Objective-C. For example, the flexibility of Core Data relies heavily on the dynamic nature of Objective-C.

The heart of this power is the *Objective-C runtime*, provided by `libobjc`. The Objective-C runtime is a collection of functions that provides the dynamic features of Objective-C. It includes such core functions as `objc_msgSend`, which is called every time you use the `[object message]` syntax. It also includes functions to allow you to inspect and modify the class hierarchy at runtime, including creating new classes and methods.

This chapter shows you how to use these features to achieve the same kind of flexibility, power, and speed as Core Data and other Apple frameworks. All code samples in this chapter can be found in the online files for Chapter 28.

Understanding Classes and Objects

The first thing to understand about Objective-C objects is that they're really C structs. Every Objective-C object has the same layout, as shown in Figure 28-1.

First there is a pointer to your class definition. Then each of your superclasses' *ivars* (instance variables) are laid out as struct properties, and then your class's ivars are laid out as struct properties. This structure is called `objc_object`, and a pointer to it is called `id`:

```
typedef struct objc_object {
    Class isa;
} *id;
```

The `Class` structure contains a metaclass pointer (more on that in a moment), a superclass pointer, and data about the class. The data of particular interest are the name, ivars, methods, properties, and protocols. Don't worry too much about the internal structure of `Class`. There are public functions to access all the information you need.

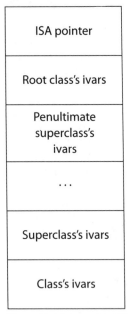

Figure 28-1 Layout of an Objective-C object

> The Objective-C runtime is open source, so you can see exactly how it's implemented. Go to the Apple Open Source site (`www.opensource.apple.com`), and look for the package `objc` in the Mac code. It isn't included in the iOS packages, but the Mac code is identical or very similar. These particular structures are defined in `objc.h` and `objc-runtime-new.h`. There are two definitions of many things in these files because of the switch from Objective-C 1.0 to Objective-C 2.0. Look for things marked "new" when there is a conflict.

`Class` is itself much like an object. You can send messages to a `Class` instance—for example, when you call `[Foo alloc]`—so there has to be a place to store the list of class methods. These are stored in the metaclass, which is the `isa` pointer for a `Class`. It's extremely rare to need to access metaclasses, so I won't dwell on them here; see the "Further Reading" section at the end of this chapter for links to more information. See the section "How Message Passing Really Works," later in this chapter, for more information on message passing.

The superclass pointer creates the hierarchy of classes, and the list of ivars, methods, properties, and protocols defines what the class can do. An important point here is that the methods, properties, and protocols are all stored in the writable section of the class definition. These can be changed at runtime, and that's exactly how categories are implemented (refer to Chapter 3 for more information about categories). Ivars are stored in the read-only section and cannot be modified (because that could impact existing instances). That's why categories cannot add ivars.

Notice in the definition of `objc_object`, shown at the beginning of this section, that the `isa` pointer is not `const`. That is not an oversight. The class of an object can be changed at runtime. The superclass pointer of

`Class` is also not `const`. The hierarchy can be modified. This is covered in more detail in the "ISA Swizzling" section later in this chapter.

Now that you've seen the data structures underlying Objective-C objects, you next look at the kinds of functions you can use to inspect and manipulate them. These functions are written in C, and they use naming conventions somewhat similar to Core Foundation (see Chapter 27). All the functions shown here are public and are documented in the *Objective-C Runtime Reference*. The following is the simplest example:

```
#import <objc/objc-runtime.h>
...
const char *name = class_getName([NSObject class]);
printf("%s\n", name);
```

Runtime methods begin with the name of the thing they act upon, which is almost always also their first parameter. Because this example includes `get` rather than `copy`, you don't own the memory that is returned to you and should not call `free`.

The next example prints a list of the selectors that `NSObject` responds to. The call to `class_copyMethodList` returns a copied buffer that you must dispose of with `free`.

PrintObjectMethods.m (Runtime)

```
void PrintObjectMethods() {
  unsigned int count = 0;
  Method *methods = class_copyMethodList([NSObject class],
                                         &count);
  for (unsigned int i = 0; i < count; ++i) {
    SEL sel = method_getName(methods[i]);
    const char *name = sel_getName(sel);
    printf("%s\n", name);
  }
  free(methods);
}
```

There is no reference counting (automatic or otherwise) in the runtime, so there is no equivalent to `retain` or `release`. If you fetch a value with a function that includes the word `copy`, you should call `free` on it. If you use a function that does not include the word `copy`, you must not call `free` on it.

Working with Methods and Properties

The Objective-C runtime defines several important types:

- `Class`—Defines an Objective-C class, as described in the previous section, "Understanding Classes and Objects."
- `Ivar`—Defines an instance variable of an object, including its type and name.
- `Protocol`—Defines a formal protocol.

- objc_property_t—Defines a property. Its unusual name is probably to avoid colliding with user types defined in Objective-C 1.0 before properties existed.

- Method—Defines an object method or a class method. This provides the name of the method (its *selector*), the number and types of parameters it takes and its return type (collectively its *signature*), and a function pointer to its code (its *implementation*).

- SEL—Defines a selector. A selector is a unique identifier for the name of a method.

- IMP—Defines a method implementation. It's just a pointer to a function that takes an object, a selector, and a variable list of other parameters (varargs), and returns an object:

```
typedef id (*IMP)(id, SEL, ...);
```

Now you use this knowledge to build your own simplistic *message dispatcher*. A message dispatcher maps selectors to function pointers and calls the referenced function. The heart of the Objective-C runtime is the message dispatcher objc_msgSend, which you learn much more about in the next section, "How Message Passing Really Works." The example myMsgSend is how objc_msgSend might be implemented if it needed to handle only the simplest cases.

The following code is written in C just to prove that the Objective-C runtime is really just C. I've added comments to demonstrate the equivalent Objective-C.

MyMsgSend.c (Runtime)

```
static const void *myMsgSend(id receiver, const char *name) {
  SEL selector = sel_registerName(name);
  IMP methodIMP =
  class_getMethodImplementation(object_getClass(receiver),
                                selector);
  return methodIMP(receiver, selector);
}

void RunMyMsgSend() {
  // NSObject *object = [[NSObject alloc] init];
  Class class = (Class)objc_getClass("NSObject");
  id object = class_createInstance(class, 0);
  myMsgSend(object, "init");

  // id description = [object description];
  id description = (id)myMsgSend(object, "description");

  // const char *cstr = [description UTF8String];
  const char *cstr = myMsgSend(description, "UTF8String");

  printf("%s\n", cstr);
}
```

With previous versions of the LLVM compiler, it was necessary to include a `__bridge` **cast on the** `id`. **With the compiler that ships with iOS 6 (LLVM 4.1), many of these extra** `__bridge` **casts are unnecessary.**

You can use this same technique in Objective-C using `methodForSelector:` to avoid the complex message dispatch of `objc_msgSend`. This only makes sense if you're going to call the same method thousands of times on an iPhone. On a Mac, you won't see much improvement unless you're calling the same method millions of times. Apple has highly optimized `objc_msgSend`. But for very simple methods called many times, you may be able to improve performance five to ten percent this way.

The following example demonstrates how to do this and shows the performance impact.

FastCall.m (Runtime)

```
const NSUInteger kTotalCount = 10000000;

typedef void (*voidIMP)(id, SEL, ...);

void FastCall() {
  NSMutableString *string = [NSMutableString string];
  NSTimeInterval totalTime = 0;
  NSDate *start = nil;
  NSUInteger count = 0;

  // With objc_msgSend
  start = [NSDate date];
  for (count = 0; count < kTotalCount; ++count) {
    [string setString:@"stuff"];
  }

  totalTime = -[start timeIntervalSinceNow];
  printf("w/ objc_msgSend = %f\n", totalTime);

  // Skip objc_msgSend.
  start = [NSDate date];
  SEL selector = @selector(setString:);
  voidIMP
  setStringMethod = (voidIMP)[string methodForSelector:selector];

  for (count = 0; count < kTotalCount; ++count) {
    setStringMethod(string, selector, @"stuff");
  }

  totalTime = -[start timeIntervalSinceNow];
  printf("w/o objc_msgSend  = %f\n", totalTime);
}
```

Be careful with this technique. If you do this incorrectly, it can actually be slower than using normal message dispatch. Since `IMP` returns an `id`, ARC will retain and later release the return value, even though this specific method returns nothing (see `http://openradar.appspot.com/10002493` for details). That overhead is more expensive than just using the normal messaging system. In some case the extra `retain` can cause a crash. That's why you have to add the extra `voidIMP` type. By declaring that the `setStringMethod` function pointer returns `void`, the compiler will skip the `retain`.

The important take-away is that you need to do testing on anything you do to improve performance. Don't assume that bypassing the message dispatcher is going to be faster. In most cases, you'll get much better and more reliable performance improvements by simply rewriting your code as a function rather than a method. And in the vast majority of cases, `objc_msgSend` is the least of your performance overhead.

How Message Passing Really Works

As demonstrated in the "Working with Methods and Properties" section earlier in this chapter, calling a method in Objective-C eventually translates into calling a method implementation function pointer and passing it an object pointer, a selector, and a set of function parameters. Like the example `myMsgSend`, every Objective-C message expression is converted into a call to `objc_msgSend` (or a closely related function; I'll get to that in "The Flavors of `objc_msgSend`" later in this chapter). However, `objc_msgSend` is much more powerful than `myMsgSend`. Here is how it works:

1. Check whether the receiver is `nil`. If so, then call the `nil`-handler. This is really obscure, undocumented, unsupported, and difficult to make useful. The default is to do nothing, and I won't go into it more here. See the "Further Reading" section for more information.

2. In a garbage-collected environment (which iOS doesn't support, but I include for completeness), check for one of the short-circuited selectors (`retain`, `release`, `autorelease`, `retainCount`), and if it matches, return `self`. Yes, that means `retainCount` returns `self` in a garbage-collected environment. You shouldn't have been calling it anyway.

3. Check the class's cache to see if it's already worked out this method implementation. If so, call it.

4. Compare the requested selector to the selectors defined in the class. If the selector is found, call its method implementation.

5. Compare the requested selector to the selectors defined in the superclass, and then its superclass, and so on. If the selector is found, call its method implementation.

6. Call `resolveInstanceMethod:` (or `resolveClassMethod:`). If it returns `YES`, start over. The object is promising that the selector will resolve this time, generally because it has called `class_addMethod`.

7. Call `forwardingTargetForSelector:`. If it returns non-`nil`, send the message to the returned object. Don't return `self` here. That would be an infinite loop.

8. Call `methodSignatureForSelector:`, and if it returns non-`nil`, create an `NSInvocation` and pass it to `forwardInvocation:`.

9. Call `doesNotRecognizeSelector:`. The default implementation of this just throws an exception.

Dynamic Implementations

The first interesting thing you can do with message dispatch is provide an implementation at runtime using `resolveInstanceMethod:` and `resolveClassMethod:`. This is usually how `@dynamic` synthesis is handled. When you declare a property to be `@dynamic`, you're promising the compiler that there will be an implementation available at runtime, even though the compiler can't find one now. This prevents it from automatically synthesizing an ivar.

Here's an example of how to use this to dynamically create getters and setters for properties stored in an `NSMutableDictionary`.

Person.h (Person)

```
@interface Person : NSObject
@property (copy) NSString *givenName;
@property (copy) NSString *surname;
@end
```

Person.m (Person)

```
@interface Person ()
@property (strong) NSMutableDictionary *properties;
@end

@implementation Person
@dynamic givenName, surname;

- (id)init {
  if ((self = [super init])) {
    _properties = [[NSMutableDictionary alloc] init];
  }
  return self;
}

static id propertyIMP(id self, SEL _cmd) {
  return [[self properties] valueForKey:
        NSStringFromSelector(_cmd)];
}

static void setPropertyIMP(id self, SEL _cmd, id aValue) {
  id value = [aValue copy];

  NSMutableString *key =
  [NSStringFromSelector(_cmd) mutableCopy];

  // Delete "set" and ":" and lowercase first letter
  [key deleteCharactersInRange:NSMakeRange(0, 3)];
  [key deleteCharactersInRange:
                    NSMakeRange([key length] - 1, 1)];
```

(continued)

```
    NSString *firstChar = [key substringToIndex:1];
    [key replaceCharactersInRange:NSMakeRange(0, 1)
                    withString:[firstChar lowercaseString]];

    [[self properties] setValue:value forKey:key];
}

+ (BOOL)resolveInstanceMethod:(SEL)aSEL {
    if ([NSStringFromSelector(aSEL) hasPrefix:@"set"]) {
      class_addMethod([self class], aSEL,
                      (IMP)setPropertyIMP, "v@:@");
    }
    else {
      class_addMethod([self class], aSEL,
                      (IMP)propertyIMP, "@@:");
    }
    return YES;
}
@end
```

main.m (Person)

```
int main(int argc, char *argv[]) {
    @autoreleasepool {
      Person *person = [[Person alloc] init];
      [person setGivenName:@"Bob"];
      [person setSurname:@"Jones"];

      NSLog(@"%@ %@", [person givenName], [person surname]);
    }
}
```

In this example, you use `propertyIMP` as the generic getter and `setPropertyIMP` as the generic setter.
Note how these functions make use of the selector to determine the name of the property. Also note that
`resolveInstanceMethod:` assumes that any unrecognized selector is a property setter or getter. In many
cases, this is okay. You still get compiler warnings if you pass unknown methods like this:

```
    [person addObject:@"Bob"];
```

But if you do it this way, you get a slightly surprising result:

```
    NSArray *persons = [NSArray arrayWithObject:person];
    id object = [persons objectAtIndex:0];
    [object addObject:@"Bob"];
```

You get no compiler warning because you can send any message to `id`. And you won't get a runtime error
either. You just retrieve the key `addObject:` (including the colon) from the `properties` dictionary and do
nothing with it. This kind of bug can be difficult to track down, and you may want to add additional checking
in `resolveInstanceMethod:` to guard against it. But the approach is extremely powerful. Although
dynamic getters and setters are the most common use of `resolveInstanceMethod:`, it can also be used to

dynamically load code in environments that allow dynamic loading. iOS doesn't allow this approach, but on the Mac you can use `resolveInstanceMethod:` to avoid loading entire libraries until the first time one of the library's classes is accessed. This can be useful for large but rarely used classes.

Fast Forwarding

The runtime gives you one more fast option before falling back to the standard forwarding system. You can implement `forwardingTargetForSelector:` and return another object to pass the message to. This is particularly useful for proxy objects or objects that add functionality to another object. The `CacheProxy` example demonstrates an object that caches the getters and setters for another object.

CacheProxy.h (Person)

```
@interface CacheProxy : NSProxy
- (id)initWithObject:(id)anObject
          properties:(NSArray *)properties;
@end

@interface CacheProxy ()
@property (readonly, strong) id object;
@property (readonly, strong)
                    NSMutableDictionary *valueForProperty;
@end
```

`CacheProxy` is a subclass of `NSProxy` rather than `NSObject`. `NSProxy` is a very thin root class designed for classes that forward most of their methods, particularly classes that forward their methods to objects hosted on another machine or on another thread. It's not a subclass of `NSObject`, but it does conform to the `<NSObject>` protocol. The `NSObject` class implements dozens of methods that might be very hard to proxy. For example, methods that require the local run loop, like `performSelector:withObject:afterDelay,` might not make sense for a proxied object. `NSProxy` avoids most of these methods.

To implement a subclass of `NSProxy`, you must override `methodSignatureForSelector:` and `forwardInvocation:`. These throw exceptions if they're called otherwise.

First, you need to create the getter and setter implementations, as in the Person example. In this case, if the value is not found in the local cache dictionary, you'll forward the request to the proxied object.

CacheProxy.m (Person)

```
@implementation CacheProxy

// setFoo: => foo
static NSString *propertyNameForSetter(SEL selector) {
  NSMutableString *name =
  [NSStringFromSelector(selector) mutableCopy];
  [name deleteCharactersInRange:NSMakeRange(0, 3)];
  [name deleteCharactersInRange:
                    NSMakeRange([name length] - 1, 1)];
```

(continued)

```
    NSString *firstChar = [name substringToIndex:1];
    [name replaceCharactersInRange:NSMakeRange(0, 1)
                        withString:[firstChar lowercaseString]];
    return name;
}

// foo => setFoo:
static SEL setterForPropertyName(NSString *property) {
  NSMutableString *name = [property mutableCopy];
  NSString *firstChar = [name substringToIndex:1];
  [name replaceCharactersInRange:NSMakeRange(0, 1)
                      withString:[firstChar uppercaseString]];
  [name insertString:@"set" atIndex:0];
  [name appendString:@":"];
  return NSSelectorFromString(name);
}

// Getter implementation
static id propertyIMP(id self, SEL _cmd) {
  NSString *propertyName = NSStringFromSelector(_cmd);
  id value = [[self valueForProperty] valueForKey:propertyName];
  if (value == [NSNull null]) {
    return nil;
  }

  if (value) {
    return value;
  }

  value = [[self object] valueForKey:propertyName];
  [[self valueForProperty] setValue:value
                             forKey:propertyName];
  return value;
}

// Setter implementation
static void setPropertyIMP(id self, SEL _cmd, id aValue) {
  id value = [aValue copy];
  NSString *propertyName = propertyNameForSetter(_cmd);
  [[self valueForProperty] setValue:(value != nil ? value :
                                    [NSNull null])
                            forKey:propertyName];
  [[self object] setValue:value forKey:propertyName];
}
```

Note the use of [NSNull null] to manage nil values. You cannot store nil in an NSDictionary. In the next block of code, you'll synthesize accessors for the properties requested. All other methods will be forwarded to the proxied object.

```
- (id)initWithObject:(id)anObject
          properties:(NSArray *)properties {
  _object = anObject;
  _valueForProperty = [[NSMutableDictionary alloc] init];
```

```
    for (NSString *property in properties) {
      // Synthesize a getter
      class_addMethod([self class],
                      NSSelectorFromString(property),
                      (IMP)propertyIMP,
                      "@@:");
      // Synthesize a setter
      class_addMethod([self class],
                      setterForPropertyName(property),
                      (IMP)setPropertyIMP,
                      "v@:@");
    }
    return self;
  }
```

The next block of code overrides methods that are implemented by NSProxy. Because NSProxy has default implementations for these methods, they won't be automatically forwarded by forwardingTarget ForSelector:.

```
  - (NSString *)description {
    return [NSString stringWithFormat:@"%@ (%@)",
            [super description], self.object];
  }

  - (BOOL)isEqual:(id)anObject {
    return [self.object isEqual:anObject];
  }

  - (NSUInteger)hash {
    return [self.object hash];
  }

  - (BOOL)respondsToSelector:(SEL)aSelector {
    return [self.object respondsToSelector:aSelector];
  }

  - (BOOL)isKindOfClass:(Class)aClass {
    return [self.object isKindOfClass:aClass];
  }
```

Finally, you'll implement the forwarding methods. Each of them simply passes unknown messages to the proxied object. See Chapter 4 for more details on message signatures and invocations.

Whenever an unknown selector is sent to CacheProxy, objc_msgSend will call forwardingTargetForSelector:. If it returns an object, then objc_msgSend will try to send the selector to that object. This is called *fast forwarding*. In this example, CacheProxy sends all unknown selectors to the proxied object. If the proxied object doesn't appear to respond to that selector, then objc_msgSend will fall back to normal forwarding by calling methodSignatureForSelector: and forwardInvocation:. This will be covered in the next section, "Normal Forwarding." CacheProxy forwards these requests to the proxied object as well. Here are the rest of the CacheProxy methods:

```
- (id)forwardingTargetForSelector:(SEL)selector {
  return self.object;
}

- (NSMethodSignature *)methodSignatureForSelector:(SEL)sel
{
  return [self.object methodSignatureForSelector:sel];
}

- (void)forwardInvocation:(NSInvocation *)anInvocation {
  [anInvocation setTarget:self.object];
  [anInvocation invoke];
}
@end
```

Normal Forwarding

After trying everything described in the previous sections, the runtime tries the slowest of the forwarding options: `forwardInvocation:`. This can be tens to hundreds of times slower than the mechanisms covered in the previous sections, but it's also the most flexible. You are passed an `NSInvocation`, which bundles the target, the selector, the method signature, and the arguments. You may then do whatever you want with it. The most common thing to do is to change the target and `invoke` it, as demonstrated in the `CacheProxy` example. `NSInvocation` and `NSMethodSignature` are explained in Chapter 4.

If you implement `forwardInvocation:`, you also must implement `methodSignatureForSelector:`. That's how the runtime determines the method signature for the `NSInvocation` it passes to you. Often this is implemented by asking the object you're forwarding to.

There is a special limitation of `forwardInvocation:`. It doesn't support *vararg methods*. These are methods such as `arrayWithObjects:` that take a variable number of arguments. There's no way for the runtime to automatically construct an `NSInvocation` for this kind of method because it has no way to know how many parameters will be passed. Although many vararg methods terminate their parameter list with a `nil`, that is not required or universal (`stringWithFormat:` does not), so determining the length of the parameter list is implementation-dependent. The other forwarding methods, such as Fast Forwarding, do support vararg methods.

Even though `forwardInvocation:` returns nothing itself, the runtime system will return the result of the `NSInvocation` to the original caller. It does so by calling `getReturnValue:` on the `NSInvocation` after `forwardInvocation:` returns. Generally, you call `invoke`, and the `NSInvocation` stores the return value of the called method, but that isn't required. You could call `setReturnValue:` yourself and return. This can be handy for caching expensive calls.

Forwarding Failure

Okay, so you've made it through the entire message resolution chain and haven't found a suitable method. What happens now? Technically, `forwardInvocation:` is the last link in the chain. If it does nothing, then nothing happens. You can use it to swallow certain methods if you want to. But the default implementation of `forwardInvocation:` does do something. It calls `doesNotRecognizeSelector:`. The default implementation of that method just raises an `NSInvalidArgumentException`, but you could override this behavior. That's not particularly useful because this method is required to raise `NSInvalidArgumentException` (either directly or by calling `super`), but it's legal.

You can also call `doesNotRecongizeSelector:` yourself in some situations. For example, if you do not want anyone to call your `init`, you could override it like this:

```
- (id)init {
    [self doesNotRecognizeSelector:_cmd];
}
```

This makes calling `init` a runtime error. Personally, I often do it this way instead:

```
- (id)init {
    NSAssert(NO, @"Use -initWithOptions:");
    return nil;
}
```

That way it crashes when I'm developing, but not in the field. Which form you prefer is somewhat a matter of taste.

You should, of course, call `doesNotRecognizeSelector:` in methods like `forwardInvocation:` when the method is unknown. Don't just return unless you specifically mean to swallow the error. That can lead to very challenging bugs.

The Flavors of objc_msgSend

In this chapter, I've referred generally to `objc_msgSend`, but there are several related functions: `objc_msgSend_fpret`, `objc_msgSend_stret`, `objc_msgSendSuper`, and `objc_msgSendSuper_stret`. The `SendSuper` form is obvious. It sends the message to the superclass. The `stret` forms handle most cases when you return a struct. This is for processor-specific reasons related to how arguments are passed and returned in registers versus on the stack. I won't go into all the details here, but if you're interested in this kind of low-level detail, then you should read *Hamster Emporium* (see "Further Reading"). Similarly, the `fpret` form handles the case when you return a floating-point value on an Intel processor. It isn't used on the ARM-based processors that iOS runs on, but it is used when you compile for the simulator. There is no `objc_msgSendSuper_fpret` because the floating-point return only matters when the object you're messaging is `nil` (on an Intel processor), and that's not possible when you message `super`.

The point of all this is not, obviously, to address the processor-specific intricacies of message passing. If you're interested in that, read *Hamster Emporium*. The point is that not all message passing is handled by `objc_msgSend`, and you cannot use `objc_msgSend` to handle any arbitrary method call. In particular, you cannot return a "large" struct with `objc_msgSend` on any processor, and you cannot safely return a floating point with `objc_msgSend` on Intel processors (such as when compiling for the simulator). This generally translates into: Be careful when you try to bypass the compiler by calling `objc_msgSend` by hand.

Method Swizzling

In Objective-C, *swizzling* refers to transparently swapping one thing for another. Generally, it means replacing methods at runtime. Using method swizzling, you can modify the behavior of objects that you do not have the code for, including system objects. In practice, swizzling is fairly straightforward, but it can be a little confusing to read. For this example, you add logging every time you add an observer to `NSNotificationCenter`.

Since iOS 4.0, Apple has rejected some applications from the AppStore for using this technique.

First you add a category on `NSObject` to simplify swizzling:

RNSwizzle.h (MethodSwizzle)

```
@interface NSObject (RNSwizzle)
+ (IMP)swizzleSelector:(SEL)origSelector
               withIMP:(IMP)newIMP;
@end
```

RNSwizzle.m (MethodSwizzle)

```
@implementation NSObject (RNSwizzle)
+ (IMP)swizzleSelector:(SEL)origSelector
               withIMP:(IMP)newIMP {
  Class class = [self class];
  Method origMethod = class_getInstanceMethod(class,
                                          origSelector);
  IMP origIMP = method_getImplementation(origMethod);

  if(!class_addMethod(self, origSelector, newIMP,
                    method_getTypeEncoding(origMethod))) {
    method_setImplementation(origMethod, newIMP);
  }

  return origIMP;
}
@end
```

Now, look at this in more detail. You pass a selector and a function pointer (`IMP`) to this method. What you want to do is to swap the current implementation of that method with the new implementation and return a pointer to the old implementation so you can call it later. You have to consider three cases: The class may implement this method directly, the method may be implemented by one of the superclass hierarchy, or the method may not be implemented at all. The call to `class_getInstanceMethod` returns an `IMP` if either the class or one of its superclasses implements the method; otherwise, it returns `NULL`.

If the method was not implemented at all, or if it's implemented by a superclass, then you need to add the method with `class_addMethod`. This is identical to overriding the method normally. If `class_addMethod` fails, you know the class directly implemented the method you're swizzling. You instead need to replace the old implementation with the new implementation using `method_setImplementation`.

When you're done, you return the original IMP, and it's your caller's problem to make use of it. You do that in a category on the target class, NSNotificationCenter, as shown in the following code.

NSNotificationCenter+RNSwizzle.h (MethodSwizzle)

```
@interface NSNotificationCenter (RNSwizzle)
+ (void)swizzleAddObserver;
@end
```

NSNotificationCenter+RNSwizzle.m (MethodSwizzle)

```
@implementation NSNotificationCenter (RNSwizzle)
typedef void (*voidIMP)(id, SEL, ...);
static voidIMP sOrigAddObserver = NULL;

static void MYAddObserver(id self, SEL _cmd, id observer,
                          SEL selector,
                          NSString *name,
                          id object) {
  NSLog(@"Adding observer: %@", observer);

  // Call the old implementation
  NSAssert(sOrigAddObserver,
           @"Original addObserver: method not found.");
  if (sOrigAddObserver) {
    sOrigAddObserver(self, _cmd, observer, selector, name,
                     object);
  }
}

+ (void)swizzleAddObserver {
  NSAssert(!sOrigAddObserver,
           @"Only call swizzleAddObserver once.");
  SEL sel = @selector(addObserver:selector:name:object:);
  sOrigAddObserver = (void *)[self swizzleSelector:sel
                                  withIMP:(IMP)MYAddObserver];
}
@end
```

You call swizzleSelector:withIMP:, passing a function pointer to your new implementation. Notice that this is a function, not a method, but as covered in "How Message Passing Really Works" earlier in this chapter, a method implementation is just a function that accepts an object pointer and a selector. Notice also the voidIMP type. See the section "Working with Methods and Properties" earlier in this chapter for how this interacts with ARC. Without that, ARC will try to retain the non-existant return value, causing a crash.

You then save off the original implementation in a static variable, sOrigAddObserver. In the new implementation, you add the functionality you want, and then call the original function directly.

Finally, you need to actually perform the swizzle somewhere near the beginning of your program:

```
[NSNotificationCenter swizzleAddObserver];
```

Some people suggest doing the swizzle in a +load method in the category. That makes it much more transparent, which is why I don't recommend it. Method swizzling can lead to very surprising behaviors. Using +load means that just linking the category implementation will cause it to be applied. I've personally encountered this when bringing old code into a new project. One of the debugging assistants from the old project had this kind of auto-load trick. It wasn't being compiled in the old project; it just happened to be in the sources directory. When I used "add folder" in Xcode, even though I didn't make any other changes to the project, the debug code started running. Suddenly the new project had massive debug files showing up on customer machines, and it was very difficult to figure out where they were coming from. So my experience is that using +load for this can be dangerous. On the other hand, it's very convenient and automatically ensures that it's only called once. Use your best judgment here.

Method swizzling is a very powerful technique and can lead to bugs that are very hard to track down. It allows you to modify the behaviors of Apple-provided frameworks, but that can make your code much more dependent on implementation details. It always makes the code more difficult to understand. I typically do not recommend it for production code except as a last resort, but it's extremely useful for debugging, performance profiling, and exploring Apple's frameworks.

> There are several other method swizzling techniques. The most common is to use `method_exchangeImplementations` to swap one implementation for another. That approach modifies the selector, which can sometimes break things. It also creates an awkward pseudo-recursive call in the source code that is very misleading to the reader. This is why I recommend using the function pointer approach detailed here. For more information on swizzling techniques, see the "Further Reading" section.

ISA Swizzling

As discussed in the "Understanding Classes and Objects" section earlier in this chapter, an object's ISA pointer defines its class. And, as discussed in "How Message Passing Really Works" (also earlier in this chapter), message dispatch is determined at runtime by consulting the list of methods defined on that class. So far, you've learned ways of modifying the list of methods, but it's also possible to modify an object's class (ISA swizzling). The next example demonstrates ISA swizzing to achieve the same NSNotificationCenter logging you did in the previous section, "Method Swizzling."

First, you create a normal subclass of NSNotificationCenter, which you'll use to replace the default NSNotificationCenter.

MYNotificationCenter.h (ISASwizzle)

```
@interface MYNotificationCenter : NSNotificationCenter
// You MUST NOT define any ivars or synthesized properties here.
@end

@implementation MYNotificationCenter
- (void)addObserver:(id)observer selector:(SEL)aSelector
             name:(NSString *)aName object:(id)anObject
```

```
{
  NSLog(@"Adding observer: %@", observer);
  [super addObserver:observer selector:aSelector name:aName
            object:anObject];
}
@end
```

There's nothing really special about this subclass. You could +alloc it normally and use it, but you want to replace the default NSNotificationCenter with your class.

Next, you create a category on NSObject to simplify changing the class:

NSObject+SetClass.h (ISASwizzle)

```
@interface NSObject (SetClass)
- (void)setClass:(Class)aClass;
@end
```

NSObject+SetClass.m (ISASwizzle)

```
@implementation NSObject (SetClass)
- (void)setClass:(Class)aClass {
  NSAssert(
    class_getInstanceSize([self class]) ==
      class_getInstanceSize(aClass),
    @"Classes must be the same size to swizzle.");
  object_setClass(self, aClass);
}
@end
```

Now, you can change the class of the default NSNotificationCenter:

```
id nc = [NSNotificationCenter defaultCenter];
[nc setClass:[MYNotificationCenter class]];
```

The most important thing to note here is that the size of MYNotificationCenter must be the same as the size of NSNotificationCenter. In other words, you can't declare any ivars or synthesized properties (synthesized properties are just ivars in disguise). Remember, the object you're swizzling has already been allocated. If you added ivars, then they would point to offsets beyond the end of that allocated memory. This has a pretty good chance of overwriting the isa pointer of some other object that just happens to be after this object in memory. In all likelihood, when you finally do crash, the other (innocent) object will appear to be the problem. This is an incredibly difficult bug to track down, which is why I take the trouble of building a category to wrap object_setClass. I believe it's worth it to include the NSAssert ensuring the two classes are the same size.

After you've performed the swizzle, the impacted object is identical to a normally created subclass. This means that it's very low-risk for classes that are designed to be subclassed. As discussed in Chapter 22, key-value observing (KVO) is implemented with ISA swizzling. This allows the system frameworks to inject notification code into your classes, just as you can inject code into the system frameworks.

Method Swizzling Versus ISA Swizzling

Both method and ISA swizzling are powerful techniques that can cause a lot of problems if used incorrectly. In my experience, ISA swizzling is a better technique and should be used when possible because it impacts only the specific objects you target, rather than all instances of the class. However, sometimes your goal is to impact every instance of the class, so method swizzling is the only option. The following list defines the differences between method swizzling and ISA swizzling:

- Method Swizzling

  - Impacts every instance of the class.

  - Highly transparent. All objects retain their class.

  - Requires unusual implementations of override methods.

- ISA Swizzling

  - Only impacts the targeted instance.

  - Objects change class (though this can be hidden by overriding `class`).

  - Override methods are written with standard subclass techniques.

Summary

The Objective-C runtime can be an incredibly powerful tool once you understand it. With it you can modify classes and instances at runtime, injecting new methods and even whole new classes. Used recklessly, these techniques can lead to incredibly difficult bugs, but used carefully and in isolation, the Objective-C runtime is an important part of advanced iOS development.

Further Reading

Apple Documentation

The following document is available in the iOS Developer Library at `developer.apple.com` or through the Xcode Documentation and API Reference.

Objective-C Runtime Programming Guide

Other Resources

Ash, Mike. *NSBlog*. A very insightful blog covering all kinds of low-level topics. `www.mikeash.com/pyblog`

- Friday Q&A 2009-03-20: Objective-C Messaging

- Friday Q&A 2010-01-29: Method Replacement for Fun and Profit. The method-swizzling approach in this chapter is a refinement of Mike Ash's approach.

bbum. *weblog-o-mat.* bbum is a prolific contributor to Stackoverflow, and his blog has some of my favorite low-level articles, particularly his four-part opcode-by-opcode analysis of `objc_msgSend`. `friday.com/bbum`

- "Objective-C: Logging Messages to Nil"

- "objc_msgSend() Tour"

CocoaDev, "MethodSwizzling." CocoaDev is an invaluable wiki of all-things-Cocoa. The MethodSwizzling page covers the major implementations out there. `www.cocoadev.com/index.pl?MethodSwizzling`

Parker, Greg. *Hamster Emporium.* Although there aren't a lot of posts here, this blog provides incredibly useful insights into the Objective-C runtime. `www.sealiesoftware.com/blog/archive`

- "[objc explain]: Classes and metaclasses"

- "[objc explain]: objc_msgSend_fpret"

- "[objc explain]: objc_msgSend_stret"

- "[objc explain]: So you crashed in objc_msgSend()"

- "[objc explain]: Non-fragile ivars"

Index